Praise for *Bring Back Our Girls*

DAILY TELEGRAPH BOOK OF THE YEAR

**OVERSEAS PRESS CLUB OF AMERICA'S
CORNELIUS RYAN AWARD**

"Everyone should read the testimonies of the Chibok girls who survived the capture. We need to help with efforts to liberate all of them and become more responsible for women and girls' protection in conflicts."   —Malala Yousafzai

"It's a really fascinating read, gripping as it's certainly been described, and the humanity of the girls that's brought to bear is really interesting, really, really interesting. . . . It is really a remarkable story. It is an amazing story."
                                        —Christiane Amanpour

"Phew. It is fascinating and gripping."
                                        —Archbishop Desmond Tutu

"Extraordinary. . . . A page-turner of a narrative; one that I found difficult to put down, even though I knew the eventual outcome before picking up the text. Their skill as writers builds suspense and conveys the sense of utter terror and despair the Chibok girls felt as they endured the unthinkable. At the same time, [the authors] never lose respect for their subjects, treating their stories with compassion and care."   —*Washington Post*

"This intimate and riveting account demonstrates the power of sustained international pressure in the name of human rights. Most important, it serves as a testament to the strength of the Chibok girls who resisted their captors and bravely asserted their humanity in the face of violent subjugation."

—Nadia Murad, recipient of the 2018 Nobel Peace Prize

"This account of the kidnapping of the Chibok schoolgirls and their courage and fortitude through the unspeakable brutality of their captivity is a nail-biter about survival told with Hitchcockian flair. Packed with their personal testimonies, along with fresh details of the hunt for them by a team of Swiss negotiators, Joe Parkinson and Drew Hinshaw have written a work of brilliant journalism."

—Lesley Stahl, correspondent, *60 Minutes*

"*Bring Back Our Girls* combines great reporting and first-rate storytelling, and authors Joe Parkinson and Drew Hinshaw deserve great credit for their effort. In their sensitive hands, the nightmarish ordeal of the Chibok girls has become a redemption tale, and a story about survival."

—Jon Lee Anderson, *The New Yorker*

"Extremely readable, written with empathy and grace . . . an incredibly well-written and well-researched book."

—Chika Unigwe, author of *On Black Sisters Street*

"Gripping."

—NPR

"Heart-stopping. . . . In a hostage crisis of this scale, there was never likely to be the kind of happy, virtue-signaling outcome that would have got universal likes on Twitter. What the influencers who endorsed #BringBackOurGirls might do, though, is encourage their followers to read this finely written, absorbing book."

—*Daily Telegraph* (London), Book of the Year

"It is a phenomenal book. . . . Powerful, gripping, and thoroughly well-researched." —Stephanie Busari, CNN

"A brilliant investigation. . . . Joe Parkinson and Drew Hinshaw, veteran Africa correspondents for the *Wall Street Journal*, have done an incredible job reconstructing life as captives of one of the world's most dangerous terrorist groups."
—*Sunday Times* (London), Book of the Week

"*Bring Back Our Girls* is a journalistic master class: a detailed and compelling story of the unsung heroes who won the release of Nigeria's schoolgirls after the social media circus had moved on."
—Tom Wright, *New York Times* bestselling coauthor of *Billion Dollar Whale*

"With this book, Parkinson and Hinshaw remind us why tenacious investigative journalism—and not reporting triggered by Twitter—is so essential to democracy. It's a most readable recounting of the immense passion and years of painstaking work that it took to return these kidnapped young girls to their mothers."
—Seymour Hersh, Pulitzer Prize–winning investigative reporter

"Dramatic and detailed . . . This is a brilliant work of investigative journalism that supplies all the missing puzzle pieces, uncovering for the first time intimate and crucial details about the girls' time in Sambisa Forest, and the brave men and women in Abuja and around the world who sacrificed so much to bring them back to their parents."
—Helon Habila, Commonwealth Writers Prize– and Caine Prize–winning novelist

"In light of the proliferation of hashtag activism by individuals and corporations following the Black Lives Matter protests in 2020, this exploration of the unintended impact of social media activism is both poignant and relevant."   —*Kirkus Reviews*

"A riveting chronicle of the 2014 kidnapping of a group of Nigerian schoolgirls by the terrorist group Boko Haram. . . . Written with compassion and insight, this deeply investigated account brings renewed attention to an ongoing tragedy."

—*Publishers Weekly*

"Told with factual rigor, crisp political insight, and deep compassion by Pulitzer-nominated American *Wall Street Journal* reporters Joe Parkinson and Drew Hinshaw. Their brilliant book interweaves raw, traumatic accounts of what happened to some of those individual hostages with the complex story of how the many, high-profile, high-tech international rescue efforts failed to recover them."   —*Daily Mail* (London), Book of the Week

"It is difficult to imagine a more thorough and significant piece of reportage for our troubled world than this new book."

—*The Scotsman*

"Remarkable and humanizing . . . this book puts these young women center-stage in a space that has made them victims."

—*Financial Times* (London), Book of the Week

"A must-read journalistic masterpiece . . . an outstandingly thorough and detailed account. . . . Buy this book *immediately*. It's the least you can do."   —*Strong Words*

Bring Back Our Girls

THE UNTOLD STORY OF THE GLOBAL SEARCH FOR NIGERIA'S MISSING SCHOOLGIRLS

Joe Parkinson and Drew Hinshaw

HARPER ● PERENNIAL

NEW YORK ● LONDON ● TORONTO ● SYDNEY ● NEW DELHI ● AUCKLAND

HARPER ⬤ PERENNIAL

FIRST HARPER PERENNIAL EDITION PUBLISHED 2022.

Designed by Nancy Singer

Map by Daniel Levitt

Library of Congress Cataloging-in-Publication Data has been applied for.

ISBN 978-0-06-293393-5 (pbk.)

22 23 24 25 26   LSC   10 9 8 7 6 5 4 3 2 1

It is a must for trials to come
But if we stand firm to the end
We will receive the crown
And be like the angels

—*Mama Agnes*

Inflexibility
Partiality
Arrogance
Haste
Impotence
Ignorance
False Promises

—*The Seven Deadly Sins of Mediation*

Contents

Authors' Note

The following is a work of reporting, conducted over six years and on four continents with hundreds of individuals involved in the search for the Chibok schoolgirls, including twenty of the young women themselves. It is also informed by a firsthand document, a diary, recorded at considerable personal risk during captivity and smuggled into freedom, which has been lightly edited for clarity.

List of Characters

**SENIORS AT CHIBOK GOVERNMENT
SECONDARY SCHOOL FOR GIRLS**

Naomi Adamu

Maryam Ali

Sadiya Ali

Saratu Ayuba

Hauwa Ishaya

Lydia John

Rebecca Mallum

Palmata Musa

Ladi Simon

Hannatu Stephen

PARENTS OF CHIBOK STUDENTS

Kolo Adamu—Naomi's mother

Rifkatu Ayuba—Saratu Ayuba's mother

Mary Ishaya—Hauwa's mother

THE MEDIATORS

Tijani Ferobe—a Quranic scholar and friend of Zannah's

Pascal Holliger—a Swiss diplomat from the Human Security
   Division

Fulan Nasrullah—a fundamentalist–turned–terrorism analyst

Ahmad Salkida—a freelance journalist and prolific tweeter

Tahir Umar—an elderly former disciple of Boko Haram leaders

Zannah Mustapha, "the Barrister"—a lawyer and orphanage
   director

NIGERIAN GOVERNMENT

Muhammadu Buhari—former general and president of Nigeria
   from 2015

Oby Ezekwesili—former education minister,
   #BringBackOurGirls protest leader

Goodluck Jonathan—president of Nigeria, 2010–2015

BOKO HARAM

Malam Abba, "Qaid"—a commander, and the Chibok girls'
   abductor

Malam Ahmed, "the Malam"—the Chibok girls' Islamic teacher
   and guardian

Ali Doctor—a Boko Haram physician

Aliyu, "Emirul Jesh"—Shekau's chief military commander and
   executioner

Abubakar Shekau, "the Imam"—leader of Boko Haram

Abu Walad, "Rijal"—a foot soldier and guard of the Chibok girls

Mohammad Yusuf—"The Martyr," slain founder of Boko Haram

Tasiu, aka Abu Zinnira—Boko Haram's spokesman

US GOVERNMENT

John Kerry—secretary of state

Barack Obama—POTUS

Michelle Obama—FLOTUS

John Podesta—senior counselor to President Obama

BACKGROUND FIGURES

Mama Agnes—a northern Nigerian gospel singer

Jack Dorsey, "@Jack"—founder and CEO of Twitter Inc.

Amina Rawaram, "The last Diva of the Sahara"—a Maiduguri-
   based singer popular in the 1960s

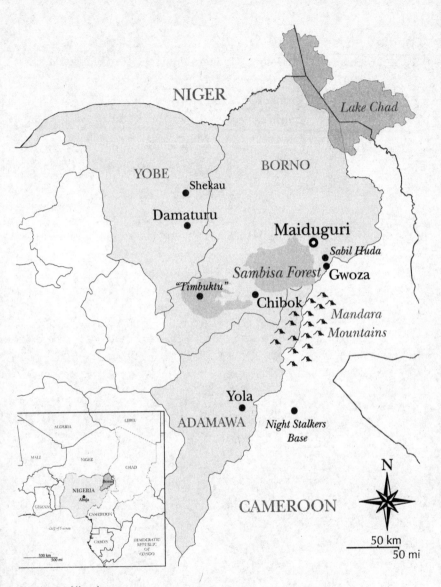

Northeast Nigeria

Prologue

Shortly before midnight on April 14, 2014, nearly 300 schoolgirls were taken from their dormitory on the outskirts of a remote Nigerian farming town called Chibok. Though it would become one of the world's most famous kidnappings, it began in obscurity, with nobody around to help.

The teenagers, as young as 15, had been dozing on bunk beds, studying notes, or reading the Bible by flashlight in the pastel-colored boarding house of the only girls' school for miles around. They were in the last weeks of their final year, mere hours of exam questions away from graduating as some of the few educated young women in an impoverished region where most girls never learned to read.

Then a group of militants barged in, bundled them onto trucks, and sped into the forest. The students had become captives of a little-known terrorist group called Boko Haram, which filled its ranks by seizing children. The girls' parents chased after them on motorbikes and on foot until the trail went cold. For weeks, hardly anybody seemed to notice. The schoolgirls looked set to be forgotten, new entries on a long list of stolen youth.

But this time, something mysterious would align inside the algorithms that power the attention economy, transforming this faraway crime into a global cause celebre. A small band of Nigerian activists on Twitter coined a hashtag calling for the hostages' immediate release. Through the unpredictable pinball mechanics of social media, it shot out from West Africa and into the celebrity-sphere, boosted by Hollywood and hip-hop royalty, before capturing the collective imagination. People from every corner of the map began tweeting the same clarion call: #BringBackOurGirls.

All at once, an ordinary group of teenagers, held against their will, became a test of social media's powers to influence events continents away, in a part of the world where most people had still never used the internet.

TV news anchors choked up retelling a story that seemed to connect the richest and poorest people on earth through the universal pain of parental loss. A class of teenagers had been studying toward a better life, chasing aspirations that made them relatable. Now these girls were trapped in a ghastly, faintly understood conflict far away, hostage to unambiguous evil.

More than that, they needed our help.

Many of the world's best-known names began rallying their fans to tweet on behalf of these faraway girls: Ellen de Generes, the Pope, Oprah, The Rock, Prince Harry, David Cameron, Harrison Ford, Mary J. Blige. All told two million people in May 2014 repeated the same demand. They included First Lady Michelle Obama who, after watching the morning news in the White House residential quarters, phoned her chief of staff, the two of them gathering next to a portrait of George Washington, where they prepared a tweet.

"Our prayers are with the missing Nigerian girls and their families," she wrote, a May 7 Twitter post that 179,000 people would share. "It's time to #BringBackOurGirls."

A day short of three years later, the skies above northeastern Nigeria were empty except for one solitary aircraft. A soft rain streaked the windows of a Russian helicopter juddering through gray clouds. Inside the cabin, a Nigerian lawyer lifted a list and a pen from the chest pocket of his crisply ironed ash-colored kaftan and studied the names through thick-framed Calvin Klein spectacles. Across from him sat a diplomat from Switzerland, nervously ticking through the final preparations for Phase Two. If everything went according to plan, their team could still make the rendezvous point by 4:00 p.m.

The helicopter headed southeast, rumbling over thorn forests and villages torched and abandoned during almost a decade of war. On the roads below, lookouts would be monitoring them, the passengers assumed, tracking their movements. One misstep could shatter the process, years in the making. The operation, hashed out over endless encrypted messages and meetings in safehouses, hung on a pair of delicate concessions. The first involved five militants released from jail now being driven to the front lines. Second was a black bag stuffed full of euros in high-denomination notes, the currency Boko Haram had demanded. Its contents were strictly secret.

Only a few senior officials in either of their governments knew the agreement the two men and their small team of mediators had painstakingly engineered. Along the way, they had lost friends and contacts to assassinations and imprisonment and had mourned when previous deals collapsed. Each assumed their phones were tapped, their routines followed. Both had sworn to observe a total information blackout at each stage of an operation that could be undone by a single errant tweet or a picture posted on Facebook. Not even the lawyer's family knew where he was.

They were the last in an army of would-be liberators, spies, and glory hunters that had descended on Nigeria to find a group of schoolgirl hostages that social media had transformed into a central prize in the global War on Terror. A few days of tweets had lit a fuse of unintended consequences that had burned for years, the forces of Silicon Valley disrupting a faraway conflict on Lake Chad. Satellites had spun in space, scanning thousands of square miles of tangled forests and desolate flatlands in vain. The air power and personnel of seven foreign militaries had converged around Chibok, buying information and filling the skies with the menacing hum of drones. Yet none of them had rescued a single girl. And somehow the fame that once started a race to free the young women had also prolonged their captivity.

The helicopter bumped to earth next to a military outpost

ringed by half-buried tires and sand-filled oil drums with a chain of white Toyota Land Cruisers parked on the cracked tarmac. The two men stooped as they stepped off the helicopter and went their separate ways.

The lawyer entered the first car of a convoy that rattled north over a dusty road, passing deserted farmland and the charred mud-brick walls of villages whose few remaining residents were too old to leave. Outside the window lay fallow fields, overlain with discarded tools and an upside-down rust-coated wheelbarrow.

The area was notorious for land mines and roadside bombs. Each driver steered carefully into the tread marks left by the car in front, their bumpers mounted with fluttering Red Cross flags. The lawyer told himself there was nothing to fear: "The prayers of the orphans will protect you."

His car halted, and its flashing brake lights signaled the convoy behind to stop. Fighters in fatigues, their heads wrapped in turbans, were gathered on the other side of the dirt road, standing alert beside waist-high grass. In the branches of acacia trees, and crouching behind bushes, the lawyer could see other figures, training their rifles. He held his list, the silence broken by the warning chime of a Toyota door left ajar.

In the distance he could see a snaking line of silhouettes, dozens of women wrapped in dark, floor-length hooded shrouds billowing in the breeze. They were stepping through long grasses flanked by armed men. The figures looked exhausted, each trudging awkwardly toward him. Two of them were walking on crutches, and one was missing her left leg below the knee. Another had her arm draped in a sling. One carried a baby boy on her back.

These were the students that millions had tweeted about, then forgotten, but none of them had any idea about the social media campaign, and they lacked the faintest notion that anybody except their parents had been advocating for their release. These schoolgirls, almost all Christians, had come of age in captivity. To keep their friendships and faith, they had whispered prayers together at night, or into cups of water, and memorized Bible passages in secret. At risk of beatings and torture, they had softly sung gospel songs, fortifying each other with a hymn from Chibok: "We, the children of Israel, will not bow."

The women, eighty-two in all, walked onto the road and halted opposite the lawyer, huddling into two lines, staring ahead with their eyes fixed. Some linked arms, others squeezed hands, their baggy clothes concealing the few possessions they'd managed to accumulate, strips of colored fabric and small twigs for pinning their hair.

One of the women was trailing behind, dressed in a gray shroud and walking with a slight hunch. Tied around her thigh, hidden from view, was something the men with guns had never found, an article of defiance. It was a secret diary, filling three notebooks, a firsthand record of the women's ordeal.

Her name was Naomi Adamu. It was her 1,118th morning in captivity.

PART I

Kidnapped

1

"Gather Outside"

The class of 2014 at the Chibok Government Secondary School for Girls were only four weeks from finishing senior year when almost three hundred of them were seized by armed men and packed onto pickup trucks that disappeared into the night. It was a Monday, and the students had spent the afternoon finishing a three-hour civics exam and the evening relaxing on campus, studying in their dorms, or gathering in small circles in the prayer room. Some had been singing cathartic *sabon rai*, or "reborn soul" hymns or practicing a local dance called "Let's Shake Our Waist." The following day they were to sit for a math final, which their teacher, Mr. Bulama, had warned would be so hard it would make them cry.

The Chibok seniors, each wearing blue-and-white-checkered uniforms sewn with the motto "Education for the Service of Man," were only a few tests away from becoming some of the first in their families to graduate. Nigeria, over their parents' lifetime, had become one of the most literate societies in the developing world, but in their corner of it, just 4 percent of girls finished high school. To complete a basic education, the students each had needed to master two second languages, Hausa and English, in addition to their mother tongues, all on top of their chores at home: cleaning, babysitting, or cooking over firewood stoves. For the better part of a decade they had shared and fought over bunk beds in the over-crowded dormitories of a school with no electricity, trying to pass exams that were challenging for even the most high-achieving students in West Africa. But the real struggle of their lives was just about to begin.

At twenty-four, Naomi Adamu was one of the oldest, a student who had prayed and fasted more times than she could count to get through high school. Petite, with a teardrop-shaped face and slender arched eyebrows she liked to outline with an angled brush, she was known to her younger classmates as Maman Mu, Our Mother. Yet when the incident began, she would find no guidance to offer them, and it would be more than a year before she discovered herself able to be the leader she wished to be.

The air that night was stiff and muggy, cut by a breeze that rippled over the flat brushland outside the school gate and through the window, next to the thin foam mattress Naomi shared with her cousin Saratu Ayuba. Well over one hundred students were lying on rows of rusting spring iron bunk beds in the Gana Hostel, one of the school's four single-story dorms. Above them, a dust-coated ceiling fan hovered idle. The teachers had all gone home for the night, and it was pitch dark, and quiet, until Naomi's eyes opened shortly before midnight to the sound of distant popping.

Half asleep, her dorm mates turned to each other, confused as the crackling noise faded then returned. It was sporadic but growing louder and presumably closer, echoing off the cement walls. From adjacent bunks Naomi could hear the rustling of bedsheets, then anxious whispers. Her classmates were talking about Boko Haram. On campus, there wasn't a single girl who hadn't heard of what this group of fundamentalist militants had done at boys' schools in distant towns. But the teenagers had been told by their teachers that a girls' campus would be safe, protected by the military.

"Should we run?"

"No. They told us to stay."

For a moment the girls debated.

Then came a thunderous roar, followed by the crash of deafening explosions, sending the hostel into pandemonium. Naomi threw herself flat on the cement floor, her cheeks next to flip-flops, open textbook pages, and scattered study notes. In one corner, a small group of students began running through a prayer, their heads bowed as classmates scampered between bunks or hid beneath beds. Some yanked the wire frames toward the wall to barricade themselves in. In the distance, the light trails of rockets cut across the night. Chibok was under attack.

"We should run!"

"No! We should not run! We should pray!"

The school's security guard, a frail man in his seventies whom the girls called Ba, or Dad, hobbled into the dorm and asked for help.

"What should we do?"

The militants might spare the young women, but any man found would be slaughtered, he explained. "They won't have mercy on me, let me go and hide," he said, and disappeared into the night.

Girls began bracing themselves to run, stuffing clothes into bags, flashlights weaving skittishly in trembling hands. Students with phones were furiously dialing family or friends. But the cell service was weak, and for those who could connect, the clatter of gunshots and bursting grenades made it almost impossible to hear. Mary Dauda stared into a phone that had dropped a call to her brother, sobbing at her failure to reach him and mumbling, "God have mercy." Margret Yama managed to briefly speak to her brother, Samuel. But she was so frightened she could barely talk, and as he begged her to flee, the line dropped. Naomi's phone was ten years old, a battered but trusty gray Nokia 3200 handset held together with a rubber band, and she frantically dialed her contacts until she managed to reach Yakuba Dawa, a neighbor. "Stay where you are," he shouted immediately, before the call cut out.

Separated from Saratu in the commotion, Naomi had only her

ears to guide her. From beyond the school gates, she could hear
the sound of motorbike engines approaching, then coming to a halt.
Outside, two men's voices were chatting over the clanging sound
of metal on metal, trailed by the jangle of a chain. The rusted
hinges squealed as the gate eased open.

"There is nobody here," a male voice said.

"No," said the other. "There are girls here."

The school campus was an island unto itself, a yellow-painted
congregation of one-story buildings, perched on the vacant scrubland on the far outskirts of town. Its low cement walls were the
last structure on a road that meandered into a flat horizon, and
there was nobody around to help as a convoy of Toyota Hilux
trucks and motorcycles halted at the school's green iron gate. An
army of men began exploring the campus, their boots and sandals
sinking into the sand of a courtyard lit by a full moon. One of the
last was their leader, a stocky man with jutted front teeth wearing a red cap, who strode toward the dormitory and hollered to the
teenagers inside. He hadn't expected them to be there.

"Gather outside," he shouted.

"Don't worry!" said another man through the window. "We are
soldiers."

The students began to slowly move across their rooms, swim-
ming through darkness toward the voices in the doorway. There
was no alternative but to take these men at their word. Rhoda
Emmanuel, a pastor's daughter, grabbed a blue Bible and tucked
it inside her clothes. Maryamu Bulama carried her own copy of
the Bible. Naomi furiously stuffed a bag with as many essentials
as she could think to collect: a red shawl, her school uniform, *her*
blue Bible, 3,600 naira (about $23) in pocket cash, and her Nokia.
"I'm very sure it's Boko Haram," she said as she followed the crowd
filing into the courtyard. Naomi found her cousin Saratu in the
scrum and in a whisper revealed the fear running through her
mind. "I'm not baptized yet."

2

The Day of the Test

Until the third week of April 2014, hardly anyone had ever heard of Chibok, a grain-and-vegetable-farming community of pink tin-roofed homes and whitewashed churches that stands alone under the hard desertlike skies of Nigeria's northeast. Built at the foot of a solitary hill studded with jagged gray boulders, its patchwork of single-story cement buildings blends into a parched landscape of dusky yellow and brown grasses. It is home to around seventy thousand people, with the feel of a small town. The nearest village lies a half day's walk over the sole dirt road available to cars, or along footpaths that skirt sand-clogged streambeds where creeks used to flow. Those trails thread through peanut, bean, or maize plots that fade away into the distance. After that, there is only tough soil for miles, punctured by a lonely set of four telecom masts providing patchy phone service for one of the dwindling communities on Earth still not reliably connected to the Internet.

At dawn, Chibok's families step out together, or pedal single-speed bicycles, to reach small farms and tend their crops before the sun provokes daily highs that often top 100 degrees. Neighbors sell each other the crops they don't eat themselves in the stalls of the town's open-air market, where mechanics clang metal tools against the broken parts of motor scooters. The rhythm of commerce is slow. The grocery-stall shopkeeper, among the few residents wealthy enough to own a refrigerator, is nicknamed Dangote, after Nigeria's richest billionaire. For entertainment, teenage girls dreaming of adventures beyond town often meet at a portrait studio where they pose in front of aspirational backdrops of gilded mansions, manicured gardens, or the skyline of Dubai.

For Chibok's residents, April is traditionally a month of celebration. It is the time of year when the air is no longer spiked with the coarse, eye-aggravating sand brought by the harmattan, a harsh wind that blows south from the Sahara during the long dry season. It is also the month, along with December, when the town holds weddings. Local custom holds that each groom must give his bride a bicycle, for her personal mobility, and later, grant each of his daughters a parcel of land, for financial independence. May brings *damina*, the rainy season, when the streets, unnamed and unpaved, transform from dust to clay-like sludge. An apocryphal story holds that the word Chibok comes from the sound of a sandal slapping in mud: *chi-bok*.

The town is so detached from its neighbors that it speaks its own language, Kibaku. Just one-tenth of 1 percent of the country understands it. An hour's drive beyond the boundaries of what is called the Chibok Local Government Area, it is rare to find anyone who can muster even the basic greetings of a community that might have remained in its pleasant obscurity were it not for an agonizing series of unexpected events that began one night while its parents were sleeping.

Mimigai? "How are you?" *Yikalang.* "All is well."

SECLUDED AND RARELY VISITED EVEN BY ITS ELECTED OFFICIALS, THIS small town might not have mattered to the world if its story didn't trace the fate of a country whose success or failure will shape the next century. Nigeria is Africa's most populous nation, home to 206 million people, half of them not yet adults, a citizenry that will double and outnumber Americans by 2050. Some five hundred languages are spoken on its soil.

And yet a picture is often drawn that Nigeria is *two* separate countries tucked inside the same borders, each complicating the fortunes of its opposite half. The south, verdant and tropical, is largely Christian and relatively more educated, with prosperity trickling into port cities along an oil rig–studded shore. The British Empire, which conquered Nigeria in the nineteenth century, considered the southerners more rebellious and pushed them to adopt Christianity, the English language, and the colonial school system.

The north, hotter, drier, and sitting on the southern fringe of the Sahara, was different. There, the imperial government coerced Muslim emirs and sultans into carrying out what Nigerian scholars would later call subcolonialism: a colonialism of the colonized. So long as northern elites obeyed the British, they were empowered to spread *their* religion—Islam—and *their* language, Hausa. By the time the empire fell, in 1960, the flag of an independent Nigeria would rise above a country that had been converted en masse to the world's two largest faiths, a vast experiment to test whether any nation so conceived could long endure.

Chibok, however, became a Christian town in the Muslim north, one of the small communities complicating that broad and simplistic divide. It had been founded in the 1700s by refugees fleeing bands of Muslim slave raiders. The settlers built houses beneath a single rocky hill, a safe distance removed from the slow and violent decline of the Borno Empire, a trans-Saharan kingdom fading into sand. As locals would boast, not even the nineteenth-century revolutionary Usman dan Fodio had subjugated Chibok when his army converted much of Nigeria's north to Islam. The slave traders, filling ships headed to the New World, never made it as far upland as their town. Local children learned that their community had been the last in the country to submit to the British Empire.

For centuries, the residents had kept their ancestral religions, praying for full harvests and sacrificing livestock on sanctified rocks. But by the time Nigeria declared independence, Chibok had made itself an exception to the country's religious geography.

It was a devoutly Christian town in a region of Muslim villages and cities. The catalyst was a group of American missionaries who came to build a school.

IT IS SAID THAT A VISITOR CAN TELL WHAT A COMMUNITY VALUES MOST BY its most prominent building, and in Chibok, that structure was a sprawling boarding school visible from miles away, its pink tin roofs bleached by the blazing sun. Boxed in by a low wall with a single metal gate, the Chibok Government Secondary School was the first to educate girls in all of the surrounding Borno State, a point of local pride. Hundreds of young women came from miles around to study at CGSS, wearing the blue-and-white checkered uniform, each sewn with an emblem containing the three symbols the town chose to define itself: a book, a pencil, and a chili pepper, a fruit hardy enough to survive Chibok's nine months of dry, sandswept weather.

CGSS had once been an arid, weedy grass patch until a family of missionaries arrived in a Ford pickup on an April evening in 1941. The Petres of Hagerstown, Maryland, brought a cook, a gardener, and a local man they called a wash boy, along with a gas lantern they carried door-to-door to introduce themselves and an entirely new worldview. Over years, Ira Petre made two efforts that endeared him to the town: he learned its history, and he raised his children to speak fluent Kibaku. The town's parents eventually allowed him to teach the alphabet and the Bible to a small group of about thirty farm boys who studied each morning on a row of mudbrick benches. That temporary outdoor classroom, washed away each rainy season, would gradually evolve into a boarding school for girls.

The faith continued to spread through song after a local early convert and chorister, Mai Sule, organized Chibok's children into a choir that met each night under the stars. They sang hymns translated into Kibaku and set to a pattering beat, while recently converted parents listened with pride. After the Petres returned to Maryland, a Christian community grew and flourished. Residents built cavernous whitewashed churches decorated with brightly colored banners proclaiming evangelical messages. They held competitive Bible quizzes. Muslim and Christian neighbors would get together to trade holy books, and at an Islamic wedding it was not uncommon to hear Christian hymns.

All in all, the town, though small and barely connected to the rest of the country, was an expression of one vision for modern Nigeria—coexistence of Christians and Muslims progressing together in small steps while battling the odds to raise and educate their children. On April 13, the last day of Chibok's peaceful seclusion, the community held an interfaith wedding, the reception under a pavilion pitched over a dusty courtyard, the tables decorated with polyester flowers. The owner of a local welding shop, a Muslim, married a Christian woman whose parents were prominent members of the local church. Flower girls in matching glossy dresses sat on plastic chairs while the tall and handsome groom, Wamdo Dauwa, gave his bride her customary bicycle. But there was another vision for Nigeria lurking in the towns and villages beyond Chibok, and on the next day, it began driving toward the school.

CHIBOK, APRIL 14, 2014, NOON

Naomi Adamu penciled her name into the long blank box on the cover page of her final exam, draping the decorative apostrophe she liked to add between the *a* and the *o*: Na'omi. It was just about noon on a Monday in the final stretch of senior year for the young woman, who mouthed a prayer as she opened the thin, rustling pages.

"Instructions to Candidates," began the test, a three-hour exam that would assess students' knowledge of Nigeria's government.

"Answer all the questions. Each question is followed by four options lettered A to D."

All around Naomi sat hundreds of students dressed in blue-and-white checkered uniforms, squeezed into rows of wooden desks. From the open windows, she could hear the voices of classmates whose finals weren't scheduled until the following day chattering, while students in the nearby science lab were conducting a practical exam. The midday heat was approaching its 105-degree high. Idle ceiling fans hung motionless over the desks, which faced a blackboard that a teacher, probably Mr. Bulama, had ordered a student to wipe clean. In a few weeks, the seniors would gather in this room for the last time and sing farewell to one another with the end-of-year anthem, "Departure Is Not Death My Sister." But today was another day of exams.

Naomi was sure that God had finally delivered her into the last chapter of a long struggle. At twenty-four, she was an awkward eight years older than the youngest seniors, and had spent several of her teenage years at home, nursing a lower back pain that the town's clinic had neither the equipment nor expertise to treat. Friends her age had long ago graduated without her. To slog through the curriculum, she had relied on classmates much younger and had compensated with a protective instinct, issuing instructions and encouragement to her juniors, and sometimes chastising them with a suck of the teeth.

The nickname they called her, Maman Mu, Our Mother, also meant the Preacher's Wife. It was a subtle tease and a poke at her self-righteous streak. Her favorite singer was a local gospel star, Mama Agnes, whose lyrics called for Christians to keep their faith. Naomi's extended family included Christians, Muslims, and converts between the two, but by senior year her mind was made up. She told classmates she planned to marry a pastor.

"I'm married to Jesus Christ," went the chorus of a bubbly, syncopated Mama Agnes song that she knew by heart. "And nobody can separate us."

Students around her began to work their way through the multiple-choice questions. Her cousin Saratu Ayuba expected to pass with flying colors, as did the class valedictorian, Lydia John. Maryam Ali had studied for months, waking before dawn to shine a flashlight on her textbooks. Her sister Sadiya just wanted to be done with exams and plan her wedding. This was the class of 2014, schoolmates who had spent years shuffling together between classrooms and the mess hall, the prayer room, and dorms, and whose fates were now set to diverge based on the answers they shaded in front of them. They were on the verge of a new chapter. Although they couldn't have known it, the first question on the test was an omen.

1. Government protects the lives and property of the citizens of a state through the:
   a. courts and the police
   b. legislature and prisons
   c. ministers and the police
   d. customs and the police

EN ROUTE TO CHIBOK, 5:00 P.M.

Malam Abba's men had packed enough gasoline to start several fires but had forgotten to bring enough water to drink. Dozens of his Boko Haram fighters were squeezed onto the back of pickup trucks that same Monday afternoon, trundling in convoy under a baking sun, a jumble of wiry bodies in mismatched uniforms and ragged T-shirts, holding rifles and rocket-propelled grenade launchers. The pickups carved a path along sandy tracks snaking through a stark and desolate landscape interrupted by an occasional tuft of grass or a lonesome baobab. Flanking the trucks was a fleet of motorbikes whose tires struggled to grip, some men riding two or three to a saddle, the cheap engines scraping and sputtering. Their journey had started in a forest camp less than a hundred miles north of their target, driving slowly through the day, then late into the night, bumping across rutted back roads to avoid military outposts or soldiers on patrol.

It was the fifth year that Malam Abba had been waging jihad against the Nigerian state, a government he considered more wicked than all the others in the world. Just one month after escaping from the country's most notorious jail, the Giwa Barracks, he was free again, leading another operation from the seat of a motorbike that trailed a safe distance behind the convoy. Short and stocky with front teeth that jutted over his lower lip and two long facial marks sculpted under each eye like tear tracks, he was in his late twenties and already a commander, or qaid, and demanded to be treated as such. While his foot soldiers wore tattered secondhand clothes or stolen army fatigues, Malam Abba would go into battle wearing a bulletproof vest over a fresh kaftan and a red cap he called No Sulhu, or No Reconciliation. He was born Abba Mustapha, but he insisted that his three wives and many subordinates address him as a *malam*, a Nigerian honorific meaning "wise teacher." In three years and fourteen days, he would be dead from an airstrike, his body shattered by a bomb intended for his leader.

"Allah says the best martyrs are those who fight on the front," he would tell his men, grizzled veterans and fresh-faced boys as young as thirteen, before an operation. "They are the ones who do not turn back until they are killed."

These men were waging a war to purge their society of the sins they believed colonialism had brought, especially *boko*, a Hausa word they used to mean all non-Quranic learning, especially Western-inspired humanistic secular education, which was all *haram*, forbidden. On some nights they directly confronted the military, driving their trucks in circles around outposts and firing rockets toward soldiers trapped inside. They rode motorbikes into villages suspected of housing so much as a single informant, firing Kalashnikovs in every possible direction. Lately, Boko Haram had begun kidnapping boys as child soldiers, and girls as chattel, packing them onto trucks to be driven into the forest for *reeducation*. On operations there were always casualties, and before battle, Malam Abba would gird his fighters to lose comrades and to kill. Years in an overcrowded Nigerian prison, where inmates died of cholera and lay on top of each other for lack of space, had redoubled his conviction that his leader, the Imam, was right: unbelievers were collateral damage in a righteous war that needed to be won.

This evening, his unit was driving to a school, but on a mission to steal, not kidnap. They were a looting crew, ordered to find the supplies the insurgency needed to keep its members fed and equipped. Their main target was a rare piece of equipment they had long struggled to find: a brickmaking machine, to construct new houses. Their ranks were swelling, with young men volunteering en masse or being forced to enlist at gunpoint. The new recruits were overwhelming Boko Haram's forest encampments, where commanders lived in grass huts and had only the branches of tamarind trees to offer as shelter to their foot soldiers.

Information had reached Malam Abba that there was a brickmaker stored at a girls' school on the edge of a small, mostly Christian town that he had never been to, called Chibok. Boko Haram had a sympathizer on the campus, his comrades would say. His unit knew little about the school and doubted there would be anybody still there after dark. They had a flatbed truck waiting in a nearby village to take home the sacks of food they hoped to find. On the edge of town, there was a small army outpost, a cinder-block garage housing weapons to seize. Malam Abba had prepared for resistance from the detachment of a dozen or so soldiers stationed there, but if those men refused to run, Boko Haram would kill them.

Whatever the case, they would be there by midnight.

THE CHIBOK GIRLS' SECONDARY SCHOOL WAS MEANT TO BE AN OUTPOST through which Western education spread through Nigeria's conservative northeast. But it was now almost the last of Borno State's 1,947 schools left open. For months, Boko Haram's men had been attacking campuses, torching at least 50 schools and murdering teachers they accused of polluting children with godless knowledge. Educators were shutting their gates. The voices of broadcasters from plastic radios scattered around Chibok's courtyards had recently begun to report a new horror, terrible enough that residents tried to push the thought of it out of their mind. The militants were kidnapping children. Most of those seized were boys, hundreds of them taken by force. But scores of young women were reported missing as well.

As rumors swelled, Chibok's parents had debated whether to keep their children in school, and teachers questioned whether to close the campus entirely. After some deliberation, the school's administration had decided the classrooms should remain open, but only long enough for the students to take their final exams. Faculty had reassured the parents and students that the town was too isolated and too far south for the militants to access. Plus, it was a girls' school.

Keeping the school open was a calculated risk. If the students missed their exams, they would have to wait an entire year to take them again, and some might drop out. The exam season would be over on May 15, and with prayer, maybe the worst outcomes could be avoided. The girls were braving just forty-five days of tests. And to many parents, the school simply seemed like the safest place for their daughters to be.

CHIBOK, 7:00 P.M.

As the sunset cast shadows through the lead-lined windows of Gana Hostel, one of the seniors picked up a plastic water bucket, flipped it over, and started to beat the twitching rhythm of "A Girgiza Kwankwaso," a Hausa dance that translated as "Let's Shake Our Waist." The day's tests had ended, and dusk offered a time students could gather in their bustling dorms to relax. Naomi and Saratu rose slowly and began rolling their hips and popping their shoulders to the drumbeat. The two tilted their heads toward each other's feet and slowly slid into the same pattern. It was a release and a way to suppress the uneasy feeling sweeping a school they knew in their hearts had become unsafe.

Saratu was Naomi's dearest friend, a first cousin she loved more than all her siblings. They had been close since early childhood, when they would play *katezi*, a game of trust. Saratu, the younger, would close her eyes, and fall backward. Naomi was usually the one catching and would let her cousin's tall and slender body get just close enough to the ground so that each time Saratu fell she thought she might collide. Then Naomi would grab her and they would both laugh, and repeat the game. Saratu was quiet and studious, and by senior year, she had begun talking about going to college to become a doctor. "Maybe I could be a nurse," Naomi would suggest, and then the two could work together.

Naomi, Saratu, and most of the girls taking classes at CGSS lived far enough away that they would spend school nights in one of its four dorms—"Houses," as in the British school tradition. Naomi and Saratu were in Gana Hostel, a jumble of seventy-two rusting bunk beds, whose iron frames would squeak as girls clambered up and down. There were far more students than beds, which had to be booked in advance.

The dorms were meant to be a safe space for the girls, especially those whose homes were a long walk away. But recently they had become a cauldron of rumor and ghoulish stories about violence that seemed to be edging closer. There had been false alarms. On more than one occasion, girls had run from their dorms in a panicked stampede after a student reported hearing unfamiliar voices at the gate. The collective anxiety was beginning to affect Naomi. She had been having nightmares, and days earlier had dreamed of a group of faceless men chasing her as she tried to climb a wall but fell on the ground. In just a few more days, she would be able to go home, she told herself.

As the darkness fell over Chibok, Naomi, Saratu, and a friend, Hauwa Ishaya, slipped off to the school's prayer room. The young women sat cross-legged in a circle and took turns:

"I pray for success on my final exams."

"I pray for my family."

"I pray for our safety."

"We put our fate in the hands of God," said Naomi.

It was getting late and the friends tiptoed back to their bunks, where some of their classmates had already begun to fall asleep while others were shining their phones onto textbooks for a few final hours of reading. Naomi curled up in a corner and dialed her mom to say good night, but the reception was poor and the call couldn't connect.

Her phone was a battered old Nokia that could only send texts and calls. Scrolling through its contacts, she debated whether to dial Andrew, a boy from the neighborhood whom she liked. She did. "Why are you staying there?" Andrew asked. "Isn't it dangerous?"

"They don't attack girls," Naomi said, echoing the faculty's arguments as if they were her own. "Only boys." Saying it out loud somehow made it better and made the teachers' reassurances more real.

Naomi wanted to call her mom. The two had fought the day before, and the guilt of upsetting her mom ate at her, but the network kept dropping her calls. Saratu was already climbing into the bottom bunk they shared near the window, a prime location that meant they caught the evening breeze. Students were drifting off, squeezed two to a mattress, some awake but lying silently in the heat. Naomi and Saratu lay there, arms around each other, as their eyelids closed. It was almost midnight when her eyes snapped open.

CHIBOK, 11:55 P.M.

"Gather outside."

"Don't worry! . . . We are soldiers."

The young women filtered into the courtyard and found themselves face-to-face with dozens of young men who they could immediately see were not soldiers. They wore unkempt beards, flip-flops, and tattered uniforms. They stood or paced behind their leader, a stocky man Naomi could see was wearing a red cap.

Malam Abba fired a gunshot into the air and issued a demand to the throng of schoolgirls gathering in front of him, "Where is your brickmaking machine?" The students froze and stared back at him, or at the ground, silent. Malam Abba raised his voice, scanning the crowd for a collaborator. "Find it and we will leave you in your hostel," he said.

"We don't know where it is-o!" a student shouted. Another began to lift her arm, but the girl next to her slapped it down. A militant rushed over and pointed his rifle at the first student's head. "Please don't kill me," she cried.

"Anyone who decides to run, we will shoot her!" another man shouted.

One girl spoke up, suggesting they should check the storage depots behind the dormitories. Malam Abba ordered his men to follow her there, as more fanned out across the campus. On the other side of the courtyard, past the crooked guava tree, his fighters were raiding the school cafeteria, hauling away sacks of grain and rice and containers of cooking oil. He could see it wasn't going to be enough to fill their trucks. "What is in your hostel?" he shouted.

"Our shopping. Clothes, mats," a student replied. Malam Abba scoffed, unsure whether to believe them. Then several of his men appeared from behind the storage depot with the brickmaker, a tall and cumbersome contraption with a long metal handle to press cinder blocks. They loaded it onto a truck, while others shuttled back and forth from the pantry, gathering the food. The looting was taking hours, as the students stared on. Then, one by one, the men returned to their trucks to retrieve their gasoline, each grabbing a yellow jerrican and walking toward a different building. The stream of fuel gargled as it fell from the mouths of the jugs onto the concrete walls. They were going to leave this campus in ash.

Naomi watched, standing in a row of other schoolmates, choking back tears as flames began to climb the walls of her school. The fire ripped through the principal's office, the teachers' quarters, the library, the lab, the classrooms, the exam hall, her dorm and the others, then finally the staff office, which was next to where she and the rest of the students were standing. Smoke rushed into her face, filling her eyes with an acrid sting.

The voice of a militant broke her trance. "Shift a little! . . . Don't you know that the heat is increasing?" The students inched away from the fire, unsure where to go. The men corralled them under a tree where a gunman trained a rifle toward them. "You should say your final prayers," he said.

He fired the weapon in the air: "Have you finished praying? Lie down."

The young women began to scramble to hide under one another, each trying to burrow under a squirming stack of bodies. One girl held a sandal against her chest, in tears pleading: "Don't shoot."

But now the men were arguing over what they should do with the students.

"Put them into different places and burn them," one said.

"Let them lead us to their parents' homes."

"No," said another. "Let's just take them."

Malam Abba hadn't come to kidnap these girls. But he could see that his men, after traveling over the course of a day to reach Chibok, were disappointed with the night's haul. One of his militants was complaining, frustrated with the long distance and meager returns: "I came with my vehicle empty; I cannot go back with my vehicle empty."

Malam Abba considered what the Imam would want. This revered place of learning that had stood on the edge of Chibok for more than seventy years was now bathed in orange flame, coughing up smoke. He turned to the crowd and made his judgment. He ordered the hostages to climb into pickup trucks or to walk at gunpoint until they reached a flatbed Mercedes truck.

"Shekau will know what to do with them."

3

The Perfume Seller

Abubakar Shekau was supposed to be dead. On two separate occasions, Nigeria's military had issued statements confirming that they had killed him. But in March 2014, a month before the Chibok kidnapping, the bearded and bellowing leader of Boko Haram reappeared to threaten his countrymen from the one place where the army could never reach him, YouTube. Sitting in front of a golden curtain, wearing a bulletproof vest over a starched white robe and gripping a Kalashnikov in his right hand, Africa's most wanted man had thirty-seven minutes of angry invective to perform for an online audience.

"In the name of the almighty Allah, killing is my mission now!" he screamed out of the website's white-paneled video player. "Let's kill them, my brothers, let's kill them all!"

Days earlier, Shekau's men had fought their way into the Giwa Barracks, freeing hundreds of men and young boys, followers and new recruits, from Nigeria's most infamous jail, an overflowing cellblock so crowded that some of the prisoners running free struggled to restraighten their legs and had to be carried. His video statement was a declaration of victory and a vow to bring more bloodshed to Nigeria. Cackling and screeching death threats, Shekau worked himself into a caricature of rage, pushing his face so close to the camera that the lens struggled to focus. "Just pick up your knives, break into their homes, and kill!"

For five years, Nigeria had been fighting an insurgency whose central riddle was this enigmatic warlord. Abubakar Shekau had grown up a child beggar and perfume seller wandering the same streets of Nigeria's northeast that he, Boko Haram's self-declared Imam, was now tormenting. On his orders, teenagers strapped with suicide bombs were detonating themselves in city markets where he had panhandled as a boy. Thousands of militants who had sworn him loyalty were rampaging in Nigeria's northern states, overwhelming army outposts, dynamiting bank vaults, and breaking hundreds of followers out of prisons. For his cause, children as young as nine were dousing their schools in gasoline and burning them down. His hardened face, frozen in a near-wink and a curling grin, looked out onto cities throughout Nigeria from wanted posters plastered on lampposts and government buildings. With murderous ingenuity, Shekau had transformed Boko Haram from an obscure radical sect into a militant army whose war with the government had left thousands dead.

And yet most Nigerians knew very little about him. By 2014, the government was lavishing a quarter of its entire budget to security agencies tasked with capturing or killing Shekau, and still they couldn't agree on the most basic details of his biography. Intelligence officials argued over whether he was alive, or dead and replaced by a body double. The US government had offered a $7 million reward for his capture, the largest bounty for any terrorist on the continent, but couldn't say if their target was born in 1965, 1969, or 1975. Commentators in Nigeria's capital shared rumors that Shekau was living comfortably with his wives across Nigeria's northeastern border, in Chad, at the behest of that country's president, or in Libya, or that he was hiding in plain sight, in a mansion in Nigeria's northeast. Another urban legend held that he'd escaped from a mental asylum before climbing his way to the top of Boko Haram's militant army.

In reality, the fugitive was operating from a hideout somewhere in a remote corner of a forest not too far from the mountainous border with Cameroon, never using a phone and never allowing his visitors to bring one.

On YouTube anyone could watch this mass murderer speak his mind and threaten millions of Nigerians however he pleased. His hourlong video sermons, which would rack up thousands of views, appeared to be the stream-of-conscious ramblings of a madman taking glee in the preceding month's bombs, jailbreaks, and slaughter, and promising more to come. Jabbing his finger at the camera, he shouted and screeched while gnawing on his *miswak* chewing stick, the toothbrush used in the time of the Prophet. He threw vitriol at historical figures from the distant past, and to presidents and prime ministers thousands of miles away, from the pharaohs of Egypt to Margaret Thatcher. "It was said that I was killed, but here I am," he laughed in one video, after the army declared him dead for the second time.

"Obama must be fuming with rage! Queen Elizabeth must be fuming with rage!"

"I am Shekau, the problem child!"

Nigeria's intelligence analysts couldn't decide whether he was insane or a strategic genius, or both. Emissaries from the government and the northern Muslim elites had tried to negotiate with him but come back in disbelief. Shekau had demanded that all of Nigeria, Christians and Muslims alike, implement Sharia law as a precondition for talks. One group of community leaders had gone to a distant village near the border with Niger to find a woman who claimed to be his mother and begged her to talk her son out of militancy. She rebuffed them, declaring there was no use trying to reason with her son. "Even his own mother said he was insane," a security official reported back.

Shekau seemed to relish the attention. "You can see I'm a radical," he said in one of his videos, flashing a bright and enormous grin. "You should kill me!"

THE MAN WHOSE ARMY WOULD ONE DAY KIDNAP 276 STUDENTS FROM
Chibok had something in common with them: his life was defined
by a decision to send him away to school. It was 1980 when Abubakar Shekau's parents sent him to study the Quran, the young boy
arriving in the outskirts of Maiduguri, the regional capital, in the
backseat of a car with only an uncle as a companion. He was seven
years old. Home had been a thatched-roof cottage in a village near
the border with Niger, six miles down a dirt track from the nearest
main route. Its streets were paved with sand, stamped with the
hoofprints and tire tracks of cattle-drawn carts. His mother, Falmata, sold peanuts from a roadside tabletop. Few children went to
school or learned how to read. They dashed around, chasing each
other across barren grazing land or between picket fences made
of sticks bound with wire, until they became old enough to farm.
But Falmata could see that her son was brilliant.

Her husband, Mohammad, was an imam, versed in scripture,
who would read their child Quranic passages to test his memory.
The new parents had been amazed at how much Shekau could
recite back, and agreed their son was talented. Yet he was also
troubled. Outside the home, he was argumentative and extremely
aggressive, a loner who seemed to be living in his own mind. Most
of all, he was utterly intractable. Years later, his neighbors would
all say the same thing about the young Abubakar Shekau: there
was no use trying to reason with such a quarrelsome child. *Head-
strong* was the term his mother would use to describe him.

Her son was born into a national experiment that would leave
Nigeria further divided in two. Two decades into its independence
from British rule, Nigeria declared public elementary school free.
The country's military government had hoped that even the poorest families from every corner of the country would enroll their
boys and girls, forging a new generation of Nigerians who would
come of age with a common identity. In some regions, the number
of students in school went up tenfold in two years, children of
different creeds all learning the Roman alphabet and beginning

each morning singing the newly adopted exultant national anthem, "Arise, O Compatriots." But Shekau's parents, like many of
their neighbors, preferred their child to absorb the tradition and
humility of the Al-Majiri schools, where young boys learned the
Quran. Some of the best of these schools were located in the place
he would come to call home.

The people of Maiduguri liked to say their town was both "the
land of a thousand kings," but also "hotter than hellfire," and in
1980, the largest city in Nigeria's northeast was sizzling with
the energy of a fast-growing metropolis. Perched near the borders of three countries—Chad, Niger, and Cameroon—its economy turned on the fortunes of smugglers who loaded trucks with
kerosene tanks, colorful wax print fabric, car parts, and dried fish
from Lake Chad. Maiduguri was once a British garrison post and
had been home to just eighty-eight thousand people in 1960, the
year Nigeria gained its independence. By the time Shekau arrived two decades later, its population had more than tripled, and
the sleepy open-air market town had become a sprawl of cement
shops, tin-roofed homes, and whitewashed minarets under a Saharan sun so relentless that locals joked they never used a kettle
because water fell boiling from the tap.

The clamor of commerce reverberated through Maiduguri's
sand-heaped alleys, and its commercial boulevard, Babban Layi—
Broadway—buzzed with talk and shouts, thwacks of butcher
knives, squeaking wheelbarrow tires, and calls to prayer from
jousting muezzins. At night, drummers stood under the stars, performing rhythms that drifted into the jail cell of the imprisoned Nigerian band leader and Black Power–inspired dissident Fela Kuti.
Boomboxes played cassette tapes of Diana Ross, Stevie Wonder, or
the hauntingly beautiful praise songs of Amina Rawaram, "the last
Diva of the Sahara," but also broadcast fiery sermons from radical preachers, local and foreign. Subversive ideas were starting to
find an audience in the city nicknamed "the home of peace."

The year 1980 was the first of a new century in the Islamic

calendar and appeared to herald a revolutionary era. In Iran, a turbaned preacher named Ayatollah Khomeini had toppled 2,500 years of monarchy and was implementing the Quran into a constitution for a new Islamic Republic. The Soviet troops crossing into Afghanistan had provoked a young Osama bin Laden to join a resistance movement that would ultimately use the language of jihad to wage war against a superpower. These events felt far away yet somehow close to a pious, landlocked city across the Sahara.

That same year, as the young Shekau was starting his Al-Majiri studies, a fundamentalist preacher calling himself "The One Who Damns" was touring Maiduguri. His sermons castigated the corruption he saw in modern Nigeria and called for a break with modern consumerism and a return to Islamic purity. Thousands of mostly male supporters cheered him through the streets. Nigeria was in tumult, lurching between military dictatorships, and the state had limited patience for even the most eccentric resistance movements. After his supporters began to riot, the preacher was shot by the police, who reportedly kept his ashes displayed in a bottle on a station shelf. But his radical ideas would prove harder to contain.

THERE ARE SEVENTY-SEVEN THOUSAND WORDS IN THE QURAN, AND AS A child, Shekau, a gifted student, memorized them quickly, outdoing the other children, who would miss a line and feel a teacher's whip across their thumbs. After class, he spent afternoons knocking on doors and enunciating Quranic verses for alms or rice. He earned spare change sweeping the streets of the city's Government Reserved Area, head bowed in silence as he brushed dirt from the sidewalks where civil servants lived.

Shekau was bright and a polyglot, learning to speak Arabic, Hausa, Fulani, and English. But he was also bursting with anger, quick to pick fights over the Holy Book, a loner who rocked back and forth mumbling scripture to himself, who made few friends and whose parents never visited. As the teenager neared graduation, his teacher deemed him his most troublesome student. In 1991, when he was eighteen, Shekau was expelled. He would have to make his own way in the streets of Maiduguri.

Nigeria sits on trillions of dollars' worth of oil and natural gas, such a large treasure that when the country's first oil boom began in 1973, its military ruler General Yakubu Gowon declared, "Nigeria's problem is not money, but how to spend it." And yet by the early 1990s, the price of oil had collapsed, the windfall from years of plenty had been frittered away, and the country had toppled into its final humiliating decade of military dictatorship. The central bank had been so thoroughly looted that the currency, the naira, once stronger than the dollar, was now worth a penny on the black market. Prices for food were doubling every three months, and shelves were bare. Blackouts lasted for weeks, government workers went unpaid, and the middle class fled through airports where armed robbers swarmed planes on the runway, relieving departing passengers of their suitcases in the baggage hold. Nigeria—once the world's thirty-third richest country per capita, not far behind the industrialized nations—had two decades later become the thirteenth poorest.

By day, the expelled Shekau would walk downtown, through a city growing both larger and poorer, trekking down the checkered-painted curbs of Bank Road, passing men lined up in the hopes of securing a few hours of manual labor. Boys from the countryside prodded cattle through the traffic, trying to replicate a pastoral lifestyle in the choke of urban smog. Diesel engines growled at downtown's main commercial plaza, Monday Market, where shopkeepers nicknamed Shekau "Kauwa Dima," or Perfume Seller, the strange man who sold Arabian fragrances to women but refused to look them in the eyes. The Prophet Mohammad himself had encouraged believers to wear musk to Friday prayers, the young salesman, doused in the earthy of scent of Oud, would tell his customers. Years later, he would dispatch several bombs into Monday Market.

Dressed only in white to signify moral purity, Shekau would harangue shoppers, starting arguments over the Quran, a book he sometimes balanced on the handlebars of his motor scooter while driving. He would often stop by a nearby technical college where graduates of Quranic schools received the secular education they needed to land jobs. But there were no jobs to be had, so what good was their learning, and Shekau would barge into their library to taunt them. Lifting a Quran from the shelves, he would loudly recite passages, and repeat the judgment of a radical new group he was helping to form: Western education—*Boko*—was forbidden—*Haram*.

AT NIGHT, SHEKAU WOULD TAKE HIS FUNDAMENTALIST MESSAGE TO Maiduguri's poorest corner, the Railway Quarters, a shantytown built along abandoned train tracks, trapping the unwanted in a chockablock grid of tin roofs knifing into each other. The housing stock was a daily deluge of indignities against an underclass of villagers made to feel like outcasts for migrating into Maiduguri to support their families. Roofs leaked and bedrooms flooded. Rains left behind puddles where mosquitoes bred and spread malaria. Maiduguri's population had doubled in the two decades since Shekau had moved to the city, and for the next round of fresh arrivals, Railway Quarters was the only place to live. It was a downtrodden community primed for a revolutionary message.

In its unlit streets, organized gangs ran prostitution rackets while police, busy extorting bribes from motorists, looked the other way. By the early 2000s, Nigeria had become a democracy, but the voting was rigged, and election winners, from top to bottom, continued to plunder, setting an example followed by police officers up and down the chain of command. Typhoid and measles festered, and at one point a cholera epidemic that coursed through the area killed more than a thousand people.

As the neighborhood collapsed, Shekau began to follow a baby-faced cleric promising an Islamic rebirth. Mohammad Yusuf was a bright and charismatic village-born theology student who had dropped out of high school to study under Quranic teachers in neighboring Chad and Niger. He returned to Nigeria to preach, diagnosing a society so broken that the only way to redeem it was through a moral reeducation of its youth. Public schools were the root of the country's ills, he argued, an incubator where generations of thieving elites had been groomed to prey on their own people. Against a backdrop of catastrophic corruption and daily humiliation, the message struck a nerve.

When institutions fail, charismatic power often fills the void. By 2006, Maiduguri's poor and rich alike had begun to visit Yusuf in the tumbledown complex of buildings that was the headquarters of his growing movement, Boko Haram. Members of the political elite parked expensive SUVs next to the destitute cupping begging bowls. Inside, hundreds would squeeze onto plastic mats in dimly lit concrete rooms to listen, swatting the small insects that came out at dusk. Many were fresh arrivals to the city, villagers who converted from ancestral religions to Islam. Others were privileged students at the University of Maiduguri, radicalized by guilt and disgust with the gambling, cocaine, and liquor at their school's bacchanalian "King of the Campus" parties. Most spoke Kanuri, the language of a people whose empire had stretched across the Sahara in medieval times. The congregation would sit enraptured by Yusuf, who preached cross-legged with a Quran on his lap, rattling off his worldview through a cheap sound system that quivered with distortion as he condemned Nigeria's love for the West. The knowledge those Western schools taught was all trickery, cover for a hidden curriculum that advocated worship of a predatory government. The Earth was flat, Yusuf insisted. Evaporation was a lie, for Allah caused rain. It was blasphemy to sing the national anthem, a praise song to a secular state.

Shekau would sit next to him, his face contorted in concentration. At times he would chime in, as Yusuf riled up large crowds in Maiduguri and other cities of the northeast with stories of soldiers raping women in Iraq or cartoonists mocking the Prophet in Europe: "Why would you love those people? Because they're good at football?" The audience would chant in agreement. "We cannot back down!"

Then, after the rallies had ended and supporters headed home, Shekau would ask Yusuf the same question: "When are we going to war?"

IN THE SWELTERING RAINY-SEASON WEATHER OF JUNE 2009, A MONTH when the global financial crisis was fueling double-digit inflation and spiking food prices, rifle-clutching police stopped a convoy of Boko Haram members on their way to a funeral. The fundamentalists weren't wearing helmets, disobeying a new road-safety regulation. For years, the tensions between the state and the sect had been building. Boko Haram's most radical members had begun to stockpile weapons and had at one point established a sandbagged commune outside Maiduguri that the local press dubbed "Afghanistan." Another group had tried to arrange a flight to Afghanistan to fight alongside the Taliban, only to be arrested before they could leave Nigeria. In makeshift home laboratories, Yusuf's followers had been concocting firebombs, experiments that seemed to dare the authorities to respond. Nigeria's police and military had teamed up to prepare a crackdown codenamed Operation Flush II, "to ensure the security and lives of the people in the hinterlands." But when police stopped the funeral convoy, an argument erupted. Shouts escalated into a gunfight and when the shooting stopped, seventeen men were injured.

A cycle of retaliatory violence and revenge was set in motion.

Days later, hundreds of militants swarmed police stations, firing Kalashnikovs and throwing crude gasoline bombs at patrol vehicles, then breaking open a prison cell. Officers watched comrades die and retaliated in blind anger, shooting people on sight and raking a mosque with gunfire. Troops fired mortar shells into the clinic and living quarters at Boko Haram's headquarters.

The insurgents melted into Maiduguri's labyrinthine side alleys, engaging army and police patrols in hours-long gun battles, while sucking dates to ward off their thirst. When the five days of fighting were over, the local Red Cross alone removed nearly eight hundred bodies baking under Maiduguri's blazing sun. On the last day of July, police captured Mohammad Yusuf, then paraded him shirtless before a crowd of jeering spectators, filming his interrogation on an early-model camera phone. Yusuf was composed, at ease as he batted away the officers' questions: "Why should you say *boko* is *haram*?"

"Of course it is," he replied. "The reasons are so many." Minutes later he was shot in the chest and thrown in the dirt, the officers unaware that their grainy video would be seen by millions on a website, YouTube, that had just caught on in Nigeria. The preacher had become a martyr. But before they executed him, they had demanded to know who was his second-in-command.

"Abubakar Shekau."

Shot in the thigh during the Maiduguri street battles, Shekau recuperated in a safehouse and eventually fled into the countryside. Yusuf's most die-hard supporters and an army of new recruits seeking safety and revenge followed him, melting into camps and villages dotted over the vast stretch of sparsely populated scrubland and rock-strewn hills that spilled over Nigeria's eastern border with Cameroon. Shekau became Nigeria's most wanted man, moving between hideouts in the Sambisa Forest, a sprawling thorn forest of arching tamarind and baobab trees branching over neck-high reeds that stretched hundreds of square miles. Parts of Sambisa had once been a game reserve during colonial times, but now it was the perfect base for those who could bear its deprivations. From here, fighters on motorbikes could sneak out, launch attacks in towns anywhere in the northeast, then retreat back down tracks too thin for tanks and armored vehicles, and where plant life camouflaged them from the air.

The war that followed exploded with such ferocity that nearly every day brought new reports of unfathomable cruelty, as each side, neither able to defeat the other, pursued revenge on the civilians it suspected of sympathizing with the enemy. In 2011, Boko Haram sent car bombs into churches during Christmas services and suicide bombers into crowded markets during morning traffic. Shekau's followers threw grenades into beer gardens at happy hour and brutalized civilians for acts of collaboration as slight as selling a car to an off-duty policeman. By 2012, militants began to systematically attack schools, the institution they most blamed for their country's sinfulness. Gunmen ran through campuses, executing teachers and setting classrooms on fire as their leader popped up on YouTube to promise more violence. "Teachers who teach Western education? We will kill them! We will kill them in front of their students."

In reply, soldiers drawn from every other corner of the country stormed through cities and villages hunting Shekau and his followers, but most were unable to understand Kanuri and relied on civilian vigilantes who often lashed out in rage and paranoia. Soldiers barged into mosques and homes and rounded up thousands of men, women, and children, accusing them of loyalty to Boko Haram. They threw them into the city's most crowded jail, the Giwa Barracks, a place that soldiers called the die house, where at least 4,700 inmates would perish. Prisoners fought over plastic bags of water, tossed by guards into dank ten-meter cells, the concrete floor crammed with bodies sleeping head-to-toe. Desperate family members from neighborhoods throughout the city came to the jail to beg or bribe for the release of their sons, nephews, and husbands, but many had disappeared. In theory, Nigeria's legal system owed its citizens a trial, but in practice Boko Haram suspects often simply languished in detention. By the end of 2012, as Nigeria neared its third year of war, state prosecutors had charged fewer than six people for terrorism.

At the end of that year, police burst through the door of a Boko Haram safehouse and arrested one of Shekau's wives. The Imam, furious, appeared on YouTube with a bone-chilling threat. If the government was going to take his women, he would take theirs and force them into slavery. Through 2013, reports of kidnappings and forced marriages conducted at gunpoint or the blade of a knife came trickling out from the northeast. Girls as young as twelve began to disappear, from markets or family farms in the countryside.

"Since you are now holding our women, just wait and see what will happen to your own women," Shekau said in a video. "Just wait and see what will happen to your own wives according to Sharia law. Just wait and see if it is sweet and convenient for you."

4

Kolo and Naomi

Jammed into the middle of a truck bed close to the front of a large convoy, Naomi could see the faces of her captors only when the headlight beams caught their features. There were dozens of gunmen perched on the front and sides of each vehicle, holding rifles. A swarm of motorcycles rasped alongside the trucks as if they were a governor's motorcade. It was still dark and oddly quiet except for the rattle of engines and the squeaks of brakes. The trucks were bumping along a rutted dirt road and knifing through branches that sliced the moonlight into strips.

The men wore musty, mismatched uniforms—military camouflage shirts, police-unit pants—some with turbans covering their faces, others with nothing more than ordinary clothes and a gun. Many of them appeared to be boys, barely teenagers, wiry and sporting wispy attempts at facial hair. The acrid smell of body odor hung over the truck. *There are so many, you couldn't spit without hitting one*, Naomi thought.

It was the early hours of Tuesday morning, and the young men were jubilant. They had taken no casualties and were returning with sacks of pasta, rice, and beans stolen from the high school pantry. Bunched in with the students taken from Chibok was another young woman the men had found cowering at a military outpost on their way into town. One of the men shouted that they were taking their hostages to "our soldiers' barracks."

Some of the fighters were chanting, others jabbering in the adrenaline rush. "We thought Chibok was a big town with big defenses," crowed one. "But it's a village, full of simple farmers." Inside the truck bed, many of the young women were quietly sobbing, their shoulders quivering. Their bodies had wedged into one another, a jumble of elbows digging into ribs. Naomi was thinking of her family. Her mom had high blood pressure, and Naomi worried that she would blame herself. She wished she could sneak a call home with the phone hidden in her skirt, but she knew it wasn't safe. One of the men had asked the students to submit their cell phones, promising that anyone found with a device would be beaten. "We know girls like you are holding Nokia," he said.

Other students had obeyed, surrendering their devices through a chain of classmates' hands. Naomi hid hers. She could feel its plastic and metal stashed under her wrapper skirt pressing into her abdomen. One of the young men was leering at the girls, talking loud enough for his comrades to hear over the rumble of engines. "I've been looking for a wife!" he said. "We should all pick one."

Naomi was squeezed into the middle of the truck bed next to her cousin Saratu, trying not to make eye contact with their captors. Some of the men were starting to fall asleep. And as the vehicles plunged ahead into the emptiness of the inky predawn, she could see classmates near the edges taking action. Some were tearing strips off their school uniform and throwing them off the truck onto the road. Others were throwing brightly colored hair ties. It was a bread-crumb trail. Naomi watched as a handful of girls seated on the edge of the truck bed panels peered over the rims and braced themselves for a daring escape.

One by one, they jumped.

AT THE FIRST HINT OF DAWN THAT MORNING, KOLO ADAMU RUSHED TO her daughter's school and found ash floating over the campus.

She had spent the night lying low under a tree just beyond town, listening to the clatter of gunfire, while pounding her phone's call button in a desperate bid to reach her daughter Naomi, and telling herself to keep hope. Her first steps through the squeaking green gate of the Chibok Government Secondary School for Girls confirmed her worst fears.

Fire had gutted the pastel green exam hall, and the blackened remnants of its corrugated iron roof now dangled from beams exposed to the sunlight. Empty desks sat in four rows facing a blackboard that hung undamaged on the front wall. Outside in the courtyard, bullet cases lay in the dirt. Dozens of parents had arrived and were stepping through the dormitories, dazed as they rifled through burnt textbooks and charred skeletons of iron-spring bunk beds. More families were hurrying onto the campus, some still in their sleeping attire, asking the others what happened. Some paced in circles desperately trying to reach their children on cell phones, as they had been doing since midnight. The cafeteria kitchen appeared to have been ransacked; canned food and cooking oil had been looted from its windowless storage shed.

Kolo stepped onto a small concrete stoop entering the Gana Hostel and weaved around the rusting green door now falling off its hinges, its paint blistered. Inside, girls' shoes, books, and other belongings lay tossed across the cement floor caked in a crunch of ash and dirt. In the debris, she could make out a metal lunchbox and realized it belonged to her daughter.

Two days earlier, she had watched Naomi load the canister with spaghetti, then pack her bags for boarding school. On their last day together, they'd had an argument. Kolo's second-born was the stubbornest of her six children, she would say, and the one she worried about most. The two were close in age, only sixteen years apart, almost sisters. Kolo called her Talatu, meaning both "born on a Tuesday" and "full of grace." After settling their disagreement, Naomi had braided her mom's hair on the veranda outside their home. "Just a few more days and you can graduate," Kolo had told her.

Gathering in the courtyard outside the dormitories, a small group of mothers anxiously debated what they could do, while fathers, uncles, and older brothers held their own conference in an opposite corner. The men said they would go to the military outpost to petition the soldiers, who had fled during the night but had now returned, to beg them to mount a rescue mission. But the women wanted to act immediately. There were tire tracks in the sand, leading from the gate along a road that headed north into the open plains. One mother proposed to take that trail on foot, and follow the treadmarks until they found their daughters.

The others agreed and set off in a silent march. Beyond the school walls lay a stretch of emptiness that rolled toward a horizon shimmering in heat haze, a desolate landscape of thornbushes, ankle-high shrubs, and knotty baobab trees. Somewhere out there were their children.

Kolo was near the front, her feet pounding the ground, walking with such force that the strap on one of her sandals snapped off. She continued, barefoot. The women were looking for any kind of clue as to their daughters' whereabouts, a footprint or maybe a dropped possession. Kolo stared into the thin soil, trying to decipher the twisting tire tracks. Her sister-in-law and one of her closest friends, Rifkatu Ayuba, was moving alongside her, trembling. Rifkatu's daughter, Saratu, was also missing. Another mother, Mary Ishaya, whose child Hauwa was gone as well, was telling Rifkatu to be calm, and had begun repeating aloud, "Let's hope for the best. Let's hope for the best."

It was another day when the temperature would top 100 degrees, and the women pushed against the heat for hours, the sun directly on their shoulders with no shade along the route. Some of them began to feel nauseated, their tongues swelling with thirst. Several were weeping uncontrollably as they walked. There was nothing to see except one burned, bullet-riddled vehicle on the side of the road and an occasional herdsman tending his cattle. Around two p.m. Kolo, Rifkatu, and the other mothers accepted that they could not go on. They had to turn back.

"I was like a living corpse," Kolo would say of her grief. "I did not know if I would survive it."

FROM THE DAY NAOMI WAS BORN, KOLO'S GREATEST FEAR WAS TO LOSE her. Nearly a quarter of children die before their fifth birthday in Nigeria's northeast, the worst rate in the country, and her daughter had almost slipped away in the womb. In the minutes after Naomi was born underweight, teenage Kolo had prayed frantically while a nurse resuscitated the body on the table of Chibok's poorly stocked clinic. Sometimes her infant daughter looked so frail that Kolo would look at her and just cry.

Kolo had scanned the Bible for names and chosen Naomi because a pastor told her it meant "gentle child." But her childhood was tough, troubled by endless health problems the town had no means to treat. When Naomi turned two, Kolo rushed her back to the clinic with a bout of whooping cough. At twelve, after falling on the soccer field, Naomi suffered a kidney problem that sent pains shooting across her back. None of the nurses could identify or explain it. The pain was so severe that at one point, Naomi, balled up at home on a plastic mat, threatened to cut the organ out herself.

The Adamus lived on the southern side of Chibok in a sun-dried mudbrick home that opened onto a shaded courtyard, where six children kept their shoes in an unruly pile. The family would return from their garden, where they harvested beans, switch on the radio, and sit together on the porch, listening to the British Broadcasting Corporation's Hausa service. The house had no running water and relied on a gas lamp for light after dark. In the mornings, Naomi fed the chickens, ordering them to eat as they scuttled and pecked around the yard. She helped her mother with chores and bossed her younger siblings into sweeping sand from

the house while she fetched water from the well and ground durum wheat into *semo* porridge. She carried her younger siblings
on her back while she made spicy *kuka* soups from baobab leaves
and simmered dried okra in heavy pots over a fragrant wood-fired
stove. A glossy picture on the wall showed the family posing awkwardly in their Sunday attire, ahead of a service at the Church of
the Brethren.

Sunday worship was a six-hour extravaganza that stretched
into the afternoon, a rapturous flock of hundreds of parishioners
jostling for blue and green plastic chairs or standing in the back
with their palms facing up to the pointed zinc roof. Women in
brightly colored *atampa* dresses sat with children on their laps,
sometimes walking up to slip folded naira bills into the collection
box. The most active church women were called prayer warriors,
frontline soldiers in the community's spiritual battle against
evil; they wore color-coordinated dresses with sequins glued on.
Men in collared shirts, some in suit jackets and bow ties, filled
the third and smallest aisle, crunched up against the cream
painted wall. Next to the altar, women kept time on tambourine-
like *kachau-kachau* drums wrapped in plastic beads. A dance line
filled the front row, their shoulders dipping in unison as the whole
congregation swang while belting out piercing high-octave hymns
like "Thank God," a church standard. The pastor, Enoch Mark,
whose daughter Monica was in Naomi's class, would close his eyes
as he asked the congregants to settle into their seats. During the
sermon, some pulled out ballpoint pens to take notes on the Bible
verses they should reflect on in the upcoming week.

When it came to religion, Kolo's life had been a series of ad-
aptations, quiet concessions she made to keep the family peace.
In three generations, her family had moved through three faiths.
Her grandparents had practiced an ancestral religion, her father
had been a Muslim, and her husband was an Evangelical. Kolo
had switched back and forth between Islam and Christianity, try-

ing to keep her husband happy without disappointing her father. Nigeria's religious divide had been turning more bitter as the economy stalled. Eventually, she chose Christianity, and one Sunday in 1998, as eight-year-old Naomi looked on, Kolo was baptized at Chibok's Church of the Brethren.

Kolo had attended only a few years of school, leaving before her twelfth birthday, and she dedicated herself to elevating her children into the life that had eluded her. Naomi was a project for her ambitious mother, who dreamed up big plans, like medical school. But Naomi was stubborn, "like an ox," said Kolo, who found it impossible to change her daughter's mind on even the smallest issues. Kolo believed that Naomi's grandmother, who died when Naomi was in the womb, lived in spirit through the granddaughter she never met, so Kolo could never bear to discipline her with the rod. Naomi, she learned, was willing to escalate a fight to win it.

The mother and her daughter were close, but as adolescence intruded on their relationship, they fought often. As Naomi's health problems worsened each year, the pain drained her of patience, though not her resolve. When she finally felt well enough to return to class, teachers put her in a lower grade at a junior high school. It was one humiliation too much. Naomi walked to the market and bought a canister of rat poison, a powdered pesticide called Atomic Bomb that could be mixed with water to form a lethal cocktail. Newspapers were full of stories of teenage girls who'd opted out of arranged marriages by drinking it. Naomi threatened to swallow it if she wasn't enrolled in the school of her choice: the Chibok Government Secondary School for Girls. Kolo was terrified enough to call the police, who led her daughter to a cell on suicide watch. There the young woman tossed rocks out the window in protest until two days passed by, and the officers reached the same conclusion. Naomi was in severe trouble and something else had to be done. The next morning, Kolo dialed the principal with a request to admit her eldest daughter.

If not for that decision, Kolo thought when she had returned from her effort to trace the tire track, *my daughter never would have been kidnapped.*

NAOMI HAD LOST TRACK OF TIME, WATCHING THE TRUCK ROLL OVER AN unfamiliar landscape of thickening forest as her first day as a hostage wore on, when she heard the crash of metal against wood. One of the trucks had plowed into a tree trunk, throwing several schoolgirls to the ground and drawing yelps of pain. The militants scrambled down, screaming insults at the sleep-deprived driver, then nearly came to blows over how to repair the wheezing engine. Three students were crying out that they had broken limbs. One had vomited all over her skirt, and a friend had to lend her a wrapper. Another complained of a headache and her friends asked a militant if there were any painkillers.

"You're just thinking too much," he replied. "You should all stop thinking."

The men passed around packets of biscuits and cans of Malta Guinness, a thick, syrupy corn drink they had looted from somewhere before they reached the school. Naomi stared at the ground, unwilling to accept the stolen food. After what felt like hours, the engine returned to life, and the men loaded the girls back on. At one point, the convoy passed a pair of young boys waving from the roadside and chanting "Allahu Akbar!"

Then the vehicles came to a halt in a village where Naomi saw an elderly man stepping out toward them. He was shouting at the militants in a language she couldn't understand, gesturing toward the girls then angrily confronting their captors. A group of men, more than a dozen, hopped down from their trucks and surrounded the villager, arguing back and hurling insults. They walked closer, in a tightening circle, as the man's weathered face began to register fear, then panic. Naomi winced as she watched them jab assault rifles into his side, his knees buckling in pain. He began shouting, "I'm a Muslim! I'm a Muslim." She heard one voice say, "Petrol!" which was affirmed by a chorus of cheers. "Al-lahu Abkar!"

One of the militants returned to the truck, grabbed a yellow canister, then walked back to the condemned man and excitedly began to douse him. Another fighter followed, emptying his canister as the old man pleaded through gulps for air, his arms raised as if he could somehow push the liquid away. He stood there, forlorn and drenched in fuel. The fighters stepped back and pointed their rifles. It was a choreography that looked well rehearsed. One of them lit a match and set the old man on fire. Naomi stared, unable to turn away as he screamed then fell onto the ground grabbing his face and rolling until his charred body went limp. It happened so quickly. His killers were celebrating.

The teenagers sat in the trucks, mute with shock.

The militants climbed victorious into their pickups, which revved back to life and drove on. Some of the schoolgirls began to sob, their whimpers reverberating around the truck bed. Naomi heard one of her classmates whisper, "God will punish them."

"We should leave everything to God," another said. Naomi was exhausted and thirsty, and her mind was racing. She had heard news reports about how these men forced women into marriage. The girls around her were wearing dazed expressions. Naomi turned to one of them and confided what she thought might happen next: "They are going to force us to convert and then marry them."

5

The Caretaker of an Orphan

APRIL 15, 2014, ONE DAY INTO CAPTIVITY

Zannah Mustapha liked to say that when he wasn't in Maiduguri, he wasn't himself. In all his fifty-five years, the tall, bespectacled law professor had never so much as rented a bedroom outside the city he was now flying home to. For many of his countrymen, the crowded and sunbaked conurbation outside the airplane window had become known for the curse it brought Nigeria, as it was the birthplace of Boko Haram. But to Zannah, it was still the hometown he called Yerwa, Blessed Land, a center of Islamic scholarship that had once held a reputation for hospitality, like its feathery neem trees that offered welcome shade from the sun.

While the Nigerian state had been trying to kill Abubakar Shekau, Zannah Mustapha had been one of the few people trying to talk to the Boko Haram leader. Gentle-mannered with an affable, gap-toothed smile and an aristocratic bearing, Zannah had come to see it as his destiny to end the war tearing his city apart. He had all but closed the Maiduguri law firm he'd run for decades, first to start a school housing orphans of the violence, now to pursue another role most people didn't know existed, a conflict mediator. To his closest friends, he had divulged a privately held dream. First, he would moderate peace talks with Boko Haram. Then he would win the Nobel Peace Prize.

For the moment, the professor faced the more immediate challenge of finding his luggage in an airport that looked like a battleground, not an arrivals hall. Soldiers in bulletproof vests and helmets rested their rifles on sandbags positioned around the dilapidated cement building that called itself Maiduguri International Airport. Others hung off camouflage-colored flatbed trucks that sped along a single runway lacerated by damage the maintenance staff called "alligator cracks." Locals claimed that the landing strip had been lengthened in the 1990s, when Nigeria was under a corrupt military dictatorship, so that the thieves in power could charter jets to smuggle gold bullion directly to Switzerland.

But by 2014, the terminal had suffered so much damage from terrorist bombings and rocket attacks that the only way for civilians to fly in was aboard an aid-agency propeller plane or a politician's private jet. The airport had become more dilapidated as the war became worse, a portrait of the daily drumbeat of attacks and reprisals playing out across the northeast. Shortly before take-off, the lawyer's phone had received a message from a contact in Maiduguri, forwarding vague rumors of a large abduction overnight of dozens of schoolgirls.

Something about a kidnapping in the town of Chibok struck Zannah as different. Boko Haram had never attacked a girls' school before. As he left the airport and squeezed into his Toyota Corolla, parked outside, a thought circled in his mind. *Those could be* my *daughters.*

Born the year before independence, his father a judge in the colonial-era courts, Zannah moved easily in Nigeria's highest circles. He served as an attorney to some of its most prominent figures and was a founding member of the People's Democratic Party, the political machine that controlled the country. He wore

finely stitched kaftans and delicately embroidered kufi caps and
walked the same way as he talked; slowly, with the unhurried
pace of a man of privilege. Former Nigerian president Olusegun
Obasanjo called him "my son." Zannah was a stalwart of a Maiduguri elite that had prospered during the hardships of military
rule and Nigeria's chaotic shift into democracy. But he had made
two choices that distinguished him from his well-heeled neighbors along the sequestered palm-lined sidewalks of the city's
Government Reserved Area. One was representing the family of
Boko Haram members who had been killed, arguing in court that
children could not be denied their inheritance on account of their
father's crimes. The other was opening an orphanage in a tumbledown Maiduguri mansion that he'd filled with wooden desks
then rollered with paint the same color green as Nigeria's flag.
More than five hundred children now lived and learned behind
its metal gates, in classrooms painted with apples, balls, cats, and
other alphabetical aids.

The orphans there called him Barrister, a courtroom lawyer
in Nigeria's British English. Their spiritual energy kept him safe,
he believed, because the Prophet Mohammad had once wrapped
his two fingers together and said, "Myself and the caretaker of an
orphan will be in Paradise like this."

In a city polarized by war, Zannah had mastered the art of
floating above the fray. He had a habit of humming and mumbling "that's good" to whatever his neighbors said or garbling non-
answers to sidestep sensitive questions. He practiced a cheerful,
vacant look that gave orphanage visitors the impression he wasn't
always listening. His students came from *both* sides of the conflict, a delicate and risky balancing act that no other school in the
city had attempted. Some had lost their parents to Boko Haram
while others had lost parents who *were* Boko Haram. He'd also let
a few widows of slain members live on the premises. To the rare
Westerner who visited, he called his school Future Prowess. But
to the Boko Haram sympathizers on campus, he called it Islami-

yah school. The Barrister would say his greatest accomplishment was convincing these women that math and science were Islamic subjects, along with English, a necessary language for students to communicate with other Muslims around the world. The rhetorical juggling act had helped him hone the art of suspending judgment, at least in everyday conversation. He felt he had a talent for reasoning with fundamentalists, and he was eager to advance his skills to another level.

As it happened, Zannah had stumbled into an opportunity a few months earlier, when a peculiar foreigner visited his orphanage. The guest was a Swiss diplomat with a pouf of red hair, who'd been sent by his government to understand a war moving toward its fifth year and to assess whether the faraway Alpine nation could play a role in ending it. Zannah wasn't sure what to make of this man from Switzerland, but the two had begun to chat regularly about whether they could find an opening that could move Nigeria toward peace. When the Barrister heard the news from the girls' school, he tapped out a message.

Had he seen the news of a kidnapping in Chibok?

BEHIND THE IRON GATES OF A BARELY MARKED OUTDOOR GRILL IN ABUJA, a hangout where Nigeria's intelligence officers and military men talked shop over *suya* grilled kebabs of spiced meat, the regulars had begun to notice a newcomer who some assumed to be a spy. On first sight, the Swiss diplomat looked unremarkable. He was somewhere in his thirties, of average height and build, and though he wasn't an intelligence officer, he didn't dress like a senior member of Switzerland's button-down diplomatic service, either. He wore unfussy-looking box-cut suits and ties that bunched over an unfastened top button. His name was Pascal Holliger, a suggestion that he came from somewhere in his country's French-speaking west. But the Nigerians who sat for a beer with him all immediately noticed one thing: Switzerland's newest envoy spoke exactly like them. Mingling among tables of bureaucrats, holding hands with other men he'd just met, as Nigerians often do, this gregarious foreigner had cultivated a pitch-perfect sing-song Nigerian accent. His counterparts would smile as he cycled through earnest, slanglike greetings in pidgin English, throwing an *o* at the end of his sentences for emphasis: "How far?" (What's up?) "How you dey?" (How are you?) "I dey greet you-o!" (Hello!). On the phone, some first-time callers to his office on the second floor of the Swiss embassy assumed they were speaking to his assistant. Other expats found him off-putting, and an American diplomat who squeezed him into a busy meeting schedule walked away thinking he was some kind of do-gooder, the sort of gentle soul Nigeria would break. But many Nigerians who met Pascal were touched by his enthusiasm for their country, and some began to joke that he should have a local passport. Behind his back, and sometimes to his face, they had started calling him "the White Nigerian."

Pascal had spent his first few months glad-handing new contacts at the capital's nightspots, but the substance of his work was considerably more sensitive. He had been deployed into Nigeria by a division of the Swiss foreign ministry that most of his countrymen didn't even know it existed. His employers back in Bern ran a global program that quietly inserted peacemakers into the world's most intractable conflicts. Switzerland, small, landlocked and neutral, needed a way to maintain its influence abroad. Each year, the department reviewed situations in which it could play a role, using Swiss expertise to help warring parties settle their grievances over secret backchannels of communication. Nigeria, Pascal knew, was what his superiors had determined a tier-one target. He was the first officer to be deployed to Abuja, and he would need to keep his mission relentlessly discreet.

The Barrister and the Diplomat made an odd partnership, but the two men began trading texts, sizing up how they might work together. One of the earliest messages Zannah sent was a question. Had he seen the news about the kidnapped schoolgirls?

It didn't seem like a big moment. Atrocities and abductions were being reported in the north almost daily, and to Pascal, this was just one more tragedy. He went back to the important work of the day, advising Zannah to do the same. "Let's stay in our lane."

6

@Oby

Two thumbs furiously hammered out a tweet, the thirtieth in a matter of hours, this time in response to a BBC news alert: "I hope this is NOT TRUE? The @BBCAfrica reports that some 200 school girls were abducted LAST NIGHT in Borno State? Any Tweeps in Borno?"

Oby Ezekwesili, known to her 128,000 followers as @ObyEzeks, was glued to her phone as usual, drowning Nigerian Twitter in a thunderstorm of 140-character statements, replies, and retweets. She had been walking across the lobby of her air-conditioned office on a sticky Tuesday afternoon in Nigeria's capital, when the news alert had pinged her phone. Stopping in her tracks, she fired off a single-word tweet at the global news agency to question the accuracy of the report: "Verified?" A few steps later, she felt the need to sit down. "It's BBC-o!" she thought. It's probably true. A handful of her followers replied to her, confirming her worst fears. Sensing she might faint, Oby walked out of the office and into the parking lot in a haze of shock and rising alarm.

A tall and imposing public servant with close-cropped hair and a low-pitched voice that boomed with righteous indignation, Oby Ezekwesili had spent the first chapter of her career rising to the top of Nigeria's testosterone-charged political system, and the second chapter railing against it. Before her fiftieth birthday, she had built up one of the country's most impressive CVs, including stints as a cabinet minister, a vice president at the World Bank, and an adviser to a clutch of African leaders. Married to a pastor, Oby liked to pepper her speeches with Bible verses and lessons on morality. Her reputation as a no-nonsense stickler for the rules earned her the nickname Madame Due Process. She'd left government in disgust after a 2007 election so fraught with ballot-box theft and concocted results that the European Union called it "the worst they had ever seen anywhere in the world."

The Ezekwesili family lived in Abuja, the planned capital whose long, sedate boulevards sought to filter out the clamor and poverty so common in Nigeria's other cities. The "Centre of Unity" was a government town, consciously modeled to be West Africa's answer to Brasília or Washington, DC, and located almost exactly in the middle of the country to discourage a civil war or political unrest. A mile of checkpoints manned by elite troops stretched along the tree-lined road to the Aso Rock Presidential Villa, a secluded arabesque palace on the city's eastern edge, situated under a granite outcrop. SUPPORT THE PRESIDENT, read pro-government billboards all over town, plastered with soothing reassurances, like REAL NIGERIANS ARE NOT TERRORISTS and THIS TOO SHALL PASS.

It was on Twitter where Oby encountered a more scornful view of her country and its governing elite. There, through the looking glass, she could connect with compatriots spilling their angriest innermost voices, followers who didn't "support the president" and weren't sure Nigeria's problems would soon pass. By 2009, Oby Ezekwesili, public servant, had become @ObyEzeks, a Nigerian Twitter mainstay. The replies to her musings could be livid, sometimes personal. More often, they were in furious agreement, especially when she pointed out her government's failings. When replies tested the boundaries of decorum, she'd tweet back with a smiley face :) or a quote from the New Testament. Her followers reverently called her Ma or Ma Oby, and her Twitter bio summed up her philosophy: "Validated by God and my Dad who taught me to never dignify nonsense." Her children rolled their eyes at the addiction: "She is literally always on Twitter," Chine, one of her three sons, said.

In the afternoon after that first BBC report appeared on her feed, Oby began to piece together fragmented details of a high-school kidnapping in a small town most Nigerians had never heard of—Chibok. She had immediately headed home and sat with her phone until late in the night, compulsively scrolling and tweeting, testing Twitter's crowdsourcing power by asking her followers if anybody had some intelligence on the missing girls from a community too remote for steady Internet.

"What great joy it would be if tomorrow brings us GOOD NEWS that our girls abducted in Borno are found&safely released to their families," she posted before falling asleep.

The first few hours of a kidnapping are statistically determinative. If law enforcement doesn't find the missing within seventy-two hours, the odds of swift and safe rescue plummet. But on the next day, to Oby's disbelief, there was still not even an official confirmation of the kidnapping, and the number of reported victims kept changing, from two hundred down to eighty-five, then shooting up to three hundred. Time was of the essence. How could hundreds of schoolgirls have just disappeared without so much as an exact number of the missing? The government needed to take action, she urged, firing off dozens more tweets about what she saw as a national emergency. Where was the military? Why were the details so sketchy? She laced her posts with hashtags to see if one would catch fire. #CitizensSolutionToEndTerrorism called for fresh ideas to tackle Boko Haram. #Whereisempathy? waved an accusatory finger at a government seemingly unmoved. #KeepOn pleaded with the military to keep up the search.

But the tweetstorm didn't seem to be working. Only a smattering of people engaged, and many didn't even seem shocked.

Hours ticked by, then another day. Other hashtags lit up Nigerian Twitter, which seemed to collectively shrug its shoulders. The nation had become inured to atrocity.

By Thursday night, Oby, exhausted and deflated, had retired to her bedroom alone. She couldn't shake the nightmarish thoughts of where the unnamed schoolgirls might be and what was happening to them. As a former education minister, she felt she bore some personal responsibility for their safety. In silence, she sat there, refreshing her Twitter feed over and over in the hope that new posts might finally bring some positive news for the surely heartbroken parents. She had always wanted a daughter, and maybe, her family would later wonder, that desire helped explain the cause she was about to grab on to.

As she sat in her room, her son knocked at the door. Chine had just come back to Nigeria after completing a sociology major at Boston University. He found his mother, one of the country's most decorated public servants, crying alone. "These people's children are missing and nobody is talking about it," she said to him, her voice cracking with emotion. "Or doing anything?"

THAT SAME EVENING, NIGERIA'S SECURITY CHIEFS FILED INTO A WOOD-paneled meeting hall inside the Aso Rock Presidential Villa to deliver a classified briefing for their commander in chief on the whereabouts of the missing girls. President Goodluck Jonathan had been at a campaign rally on the afternoon the news broke, launching his bid for reelection, before attending a birthday party that evening. The information reaching his ear from advisers had been scattered and at times contradictory. Strangely, Washington seemed intensely interested. The State Department had called to say Secretary John Kerry wanted to speak to President Jonathan, after his own deputy for African affairs had asked, in a meeting, "What if these were *our* daughters?" Beleaguered by the news of constant terrorist attacks, Jonathan had instructed his military chiefs to provide an update with more precise intelligence on what had happened in Chibok on Monday night. Now these men had arrived, almost exactly seventy-two hours after the abduction.

A fifty-six-year-old zoologist with a jovial demeanor and a soft, pillowy handshake, Goodluck Ebele Jonathan was the son of a canoe maker who had become Nigeria's president almost by accident. He'd grown up in a poor creekside village, studying diligently by candlelight and eventually earning a doctorate with a 215-page thesis contrasting six different species of amphibious crustaceans. Almost immediately after becoming a professor, he ventured into politics. Nigeria's political godfathers who found him nonthreatening helped rocket him into the vice presidency.

When President Umaru Yar'Adua died in office in 2010, Jonathan became president, Nigeria's first from the South-South, a swamplike coastal region that produced almost all of the country's oil but reaped few of its benefits, with potholed roads and foul tap water. For most of his life, military dictators had governed his country, most of them from the Muslim north. The soft-spoken professor, dressed daily in his trademark black fedora, hoped to lead Nigeria in a gentler way. "I have come to preach love, not hate," he had said in one of his first speeches as president.

Then, just as he'd taken power, the Boko Haram insurgency exploded, killing thousands in a part of the country where he had never lived, whose languages he didn't speak. These fundamentalists who killed in the name of God were a mystery to him. They were like ghosts, everywhere and nowhere, and no source of information seemed to satisfy his suspicions. He was skeptical that an army of poor, unemployed youth could pull off such spectacular attacks without powerful sponsors. An opposition politician, or a group of them, must be funding these crimes. The president and his advisers had begun to coin a dark conspiracy theory: that the carnage was a political plot to deny him his re-election.

The kidnapping in Chibok didn't fit easily into the president's

crowded agenda. April 2014 was supposed to be a crowning moment for Jonathan's transformative presidency. After years of
high oil prices and prosperity in the big cities of the country's
southwest, Nigeria recently had dethroned South Africa as the
continent's largest economy. Within weeks, Nigeria would welcome
prime ministers, presidents, and captains of industry as it hosted
the World Economic Forum, an investors conference, in Abuja.
Jonathan's security services had been tasked with ensuring Boko
Haram would not ruin the party. Helicopters were circling over
the center of the capital as troops set up watch posts made from
sandbags in front of luxury hotels.

Now his security chiefs were sitting across the table to brief
him on a mysterious kidnapping in a remote town somewhere in
the northeast. One by one they assured him that whatever had
happened in Chibok was a distraction from a successful presidency. The chief of the defense staff said the details of the abduction remained murky and something about it didn't add up. How
could anybody fit so many girls into trucks? Why was the school
still open? Hours earlier, a military division had released a statement claiming to have freed all but eight of the missing students,
then retracted it hours later, blaming the school's principal for the
unfortunate mix-up. There was only one conclusion to be drawn
from this fog of misinformation: The kidnapping was a hoax, "an
effort to discredit the administration," one official in the room
chimed in. Several others agreed. Jonathan thanked his security
chiefs. His office replied to Secretary Kerry with an assertion that
Nigeria's military had the situation firmly under control.

OBY EZEKWESILI HAD BEEN MONITORING THE GOVERNMENT'S RESPONSE
to the kidnapping, her disbelief curdling into a fury. "Anyone of us
that belong to the Elite Class in Nigeria, should be ASHAMED
that this is what has become of beloved country & people. I AM,"
she tweeted from the @ObyEzeks handle. Each post seethed with
capitalized condemnations. "SHAME!"

She had watched aghast as the president's Thursday night security meeting terminated with no update on the schoolgirls' whereabouts. By Friday, she'd concluded that the president not only didn't have a plan to rescue the girls, but that his administration didn't seem to care. If anybody was going to do something, she would have to do it herself. "Our entire Political Class is behaving not just SELFISHLY BUT IRRESPONSIBLY & TOYING with the FATE of NIGERIA & NIGERIANS. This MUST STOP!!!"

On April 23, nine days after the abduction, Oby left town for a book festival in the southern city of Port Harcourt, where she would inadvertently light the fuse for a digital campaign that would ricochet around the world. The topic of her session was "Bring Back the Books," and the atmosphere was celebratory. Port Harcourt had been chosen by the United Nations as the 2014 World Book Capital, and hundreds of delegates had squeezed onto gold-rimmed chairs in a cavernous marquee next to the city's Hotel Presidential. Baby blue–and–white fabric was draped artfully onto a red carpet in front of a large stage bracketed by national flags of the countries represented at the fair.

But when Oby took the microphone, she was in no mood to celebrate. She shunned the wooden lectern and walked to the front of the stage to address the crowd directly. "We would like you to please rise and make a collective demand," she instructed the hall. "Join us in declaring that Port Harcourt, as the book capital of the world, makes a collective demand for the rescue of our schoolgirls." The crowd listened silently, sharing quizzical glances about their role. "We call together: bring back all our daughters!" she said. The room politely applauded, and she walked off the stage.

Oby Ezekwesili never said the words "bring back our girls," but when she left the stage and checked her phone, she saw a tweet from a man who had been in the audience. Ibrahim Abdullahi, a lawyer with a penchant for books and bow ties, had traveled from Abuja to represent an oil company in a waste-management case and was staying at the hotel where the book fair was being held.

When he returned to the hotel, still wearing his white wig and black gown, he had listened to Oby's speech and been inspired to tweet his support. He had two young daughters in school, and the news of the kidnapping had made him hug them closer at night. At 12:49 p.m. he thumbed out a tweet on his BlackBerry to a few dozen followers that would unwittingly create the first ripple of a tsunami: "Yes #BringBackOurDaughters #BringBackOurGirls declared by @ObyEzeks and all people at Port Harcourt World Book Capital 2014."

Oby saw the post, retweeted by just twenty-three people. "As soon as I saw it, I knew," she said. "I started to tweet madly." A request went out from @ObyEzeks to her hundreds of thousands of followers. "Use the hashtag #BringBackOurGirls to keep the momentum UNTIL they are RESCUED."

A gratifying plink of tweets and retweets began to light up her phone. This hashtag was a slogan that relayed a human tragedy. The four words articulated a broader critique of government ineptitude and an urgent call for action, transcending sectarian divides and political squabbles. Nigerians began to retweet it in the thousands, sending it from Abuja to Lagos and dozens of other Nigerian cities. A digital momentum was building.

But the online campaign needed an analog dimension, a real-world focus, to wrestle the government into action. Oby called the Kibaku Area Development Association in Abuja, a tiny group that in ordinary times helped Chibok residents who moved to the capital to get settled. Its leaders had grim news. Back home, some of the girls' mothers, stricken with grief, had been rushed to the hospital. One father had been struck dumb and was refusing to eat. Over the patchy phone line, Oby asked for their trust: "Join me," she said.

THE FIRST DAY OF PROTEST BEGAN IN ABUJA WITH ONLY A FEW DOZEN people under a darkening sky that threatened rain. It had been exactly two weeks after the students had been abducted, and the small crowd included a few of the missing students' parents, bussed in from Chibok, their faces etched with sorrow and drawn from lack of sleep. They were standing on an overgrown highway median in the capital, next to Abuja's stained and derelict Unity Fountain, built to symbolize Nigerians' desire to be a nation. There was no water in the fountain.

Oby had brought them onto a strategic site. Across the street was the nexus of political and financial power in Nigeria, the Abuja Hilton hotel, whose $1,000-a-night tenth floor was frequented by the country's richest men. Journalists with expense accounts liked to network there, she knew, and it was the *New York Times* correspondent's hotel of choice. Plus, in a few days, the Hilton would host the World Economic Forum, the president standing in the lobby to greet major investors and the foreign media who were already flying in. But first, Oby knew, these guests would have to pass this highway median, sitting in snail-like traffic. They would be forced to witness the protest from behind the tinted windows of their climate-controlled sedans. And they would have to ask themselves a question: Is a country where two hundred girls can go missing a place where you would put your money?

Almost all of the protesters were women, dressed in matching red to express their common grief. As the sky unleashed a ferocious downpour, their clothes were instantly soaked. The protesters waved placards scrawled in all caps and passed around a microphone that amplified their pleas through a tinny speaker. "May God curse every one of those who failed to free our girls!" screamed the pastor of Chibok's Church of the Brethren, Enoch Mark, whose daughter Monica was among the missing. "If this happened anywhere else in the world, the country would be brought to a standstill," said Pogu Bitrus, the chairman of the Chibok residents' association, his tears running into the rain.

Oby stood stone-faced at the head of demonstration, unsure where all this was going. She cradled a yellow placard with four pieces of A4 paper affixed, each emblazoned with a single word.

Bring. Back. Our. Girls. As pools of water formed in the grass, a small group of journalists arrived and started filming. The cinematic scene unfolding on the humble highway median had the elements of great literature, a David and Goliath standoff of bereaved parents against an evil terror group. The raw emotion of parents who had lost their children contrasted with the easy privilege of the luxury hotel opposite. Cameras flashed and reporters jotted down the names of the weeping mothers, recording the ingredients of a story with a profound and universal power: a mother's loss, a struggle for justice, and hope.

Some of the protesters began wailing. Others, overcome, needed family members to prop them up. One of the fathers fell at Oby's feet, kneeling in a muddy puddle that was swelling from the downpour. "Ma," he said, "tell me you will never stop until all the girls are back." The next day, an article appeared in the *New York Times*: "Hundreds of protesters marched in the streets of the Nigerian capital on Wednesday, demanding that the government do more to find scores of schoolgirls abducted by armed militants more than two weeks ago."

7

"Timbuktu"

A tall, muscular figure in military fatigues and a black turban stood before the Chibok girls and held up two books, a Quran in one hand and in the other, a Bible. This man looked different from the more bedraggled fighters who had taken them from their school. He had a lighter complexion; his beard was neatly trimmed, and the top and bottom halves of his uniform matched and appeared freshly laundered. Several fighters flanked him on each side, carrying rifles and standing beside yellow jerricans. Across his waist, he had tied a belt stuffed with bullets, and a small gun hung strapped to his thigh. He scanned the crowd, then recited the opening words of the Quran, his long fingers wrapped around both holy books, as if to weigh them against each other.

Naomi was watching him, wedged into a throng of more than two hundred classmates seated cross-legged or on their knees atop a patch of dirt. It was midmorning and the sun was burning. At first, she couldn't follow the Arabic phrases he was reciting, but as he switched into the language of Hausa, she understood. Jesus was a prophet, the man said, but not the son of God, and to say so was a blasphemy so enormous that it made the heavens shake. He had once been a Christian himself, he explained, but had converted to Islam when he accepted that the Quran was the final revelation. The state of Nigeria was an abomination, he said, and Boko Haram were the only ones fighting for "the true Islam," the only ones who were prepared to die to honor Allah.

"Take a look at yourselves," he asked, his eyes narrowing in anger. "Don't you know going to school is forbidden?"

The girls stayed silent as the man paused. He had not introduced himself, but they had been told they would be addressed by Aliyu, Shekau's Emirul Jesh, or military chief, an extremist so fanatical he made even the other guards nervous. On YouTube, Aliyu had once appeared unmasked, executing prisoners with an axe. He had fought with al Qaeda in the Sahara. Now he was lecturing the Chibok girls, his voice booming with righteous indignation, a few feet from their faces. Their schools had been teaching them science, Aliyu complained, and raising them to sing Nigeria's national anthem, which was an offense to Allah. In the classrooms, they learned "geography," a subject of lies and heresy. Those who believed in it would suffer in hell. Democracy was sinfulness, and the only correct laws came from the Almighty. "Government of the Allah for the Allah by the Allah," he said.

"Islam is the true religion!" He was shouting now, jutting his finger toward the teenagers as if they had personally offended him. "All of you are infidels!"

The girls said nothing. Most of them were staring at the ground. Naomi could hear someone weeping softly in the crowd, but she couldn't tell who.

Aliyu lowered his voice and addressed the young women directly. "I have come here to offer you a choice," he said.

"You can convert to the true Islam, join us, and please God. Or you will be executed."

The ultimatum hung heavy over the crowd of hungry, exhausted, disoriented young women. Some panicked. Naomi saw several bodies in front of her collapse and began convulsing in a violent fit. She looked at the yellow jerricans and wondered if the men would pour gas on them as they had on their schoolhouse and the old villager. Voices cried out, shouting over one another.

Aliyu was screaming now to make himself heard over the hostages, working himself into a rage.

"If you don't convert, you are an infidel! We will starve you, then remove your head from your body," he said. "I swear by Allah that any one of you who refused to convert to Islam, I will have her beheaded."

One by one, students begged him: Yes! Yes, they would convert if it would spare their lives. Naomi said nothing. Then she turned to her cousin, in a whisper. "He is playing a game with us," she said. "We must hold our ground."

THE CHIBOK GIRLS HAD BEEN DRIVEN ALONG DIRT TRACKS FOR MORE than two days, stopping to sleep in clumps on the hard ground, until they arrived at the patch of remote forest that was to be their first open-air prison. They had reached the camp in the middle of the night, as a hail of gunfire and celebratory cries of "Allahu Akbar" rang out through the darkness. Naomi and several of her classmates had clambered down from the truck and thrown themselves on the ground in terror. "We were lying down to avoid the bullets," she said. "They were coming from all sides."

Dawn light revealed the full scale of the camp.

The kidnappers had corralled their victims under a vast tamarind tree whose branches arched out like the skeleton of a blown-out umbrella, wreathed with bright green leaves and black fruit pods. Fenced around the base of the tree was a ring of thorn bramble, curled and contorted like razor wire. More than two hundred Chibok seniors, many still in their blue-and-white school uniforms, were crowded onto the ground, hard and uncomfortable, with rocks and weedy tufts of grass alive with small bugs. All around them stretched an unfamiliar landscape of yellows, greens, and browns, a tangled maze of roots, leaves, and thorny branches. There were no roads in sight, and they could only see a short distance before their view was blocked by rubbery date palms and fields of dry reeds that rose above waist height. In the distance, dozens of figures were stepping out of grass and mudbrick cabins built under tree branches. Black-and-white flags, imprinted with Arabic letters, fluttered from trees or were affixed to tarpaulin tents. Hilux trucks and armored vehicles with the markings of the Nigerian army were parked nearby. Under a tree trunk, Naomi spotted a large gun with a long barrel mounted on wheels. Young men sat nearby, watching them.

The hostages had no clue where they were.

This was the Sambisa Forest, the vast sprawl of wilderness to the north of Chibok that stretched over hundreds of square miles that the militants had made their own. No longer the wildlife reserve it had been during colonial times, the forest was at turns desolate and abundant, where desertlike brush gave way to sharp thorn-studded reeds that grew higher than six feet. It had been years since anybody had tried to map its varied terrain.

The only villages were tiny settlements, some that spoke languages understood by just a few thousand people. Beyond that, abandoned game-hunting lodges sat atop a landscape that mutated from sparse thorn forest to lush thickets and rolling green hills that flowed across the border with Cameroon then climbed into the Mandara Mountains, a 120-mile range honeycombed with caves and crested with huge shards of rock that jutted skyward like flat hands.

Boko Haram saw this topography as the perfect base for its brand of warfare. Gunmen riding motorcycles and pickups could launch deadly raids on military outposts, then retreat into a topography only they could navigate. Most important, the broad-branched tamarind trees kept their members, their weapons, and their hostages hidden from the Nigerian warplanes and helicopters that would sometimes cut through the sky above.

By 2014, the militants had erected camps throughout the Sambisa, making a home for thousands of members and their families.

The camps housed makeshift mosques and Quranic schools where preachers instructed the faithful how to follow the "true Islam." Tin-roofed shacks built under the shade of trees doubled as clinics, stocked with stolen medicine and staffed by doctors and nurses kidnapped or recruited into the cause. These were where wounded fighters would be treated and new children would be born into the movement. Diesel electric generators powered MacBook laptops where men uploaded and edited footage of their latest attacks from Sony Camcorders. Markets sold produce, clothes, and toiletries, looted by fighters or smuggled into the forest by villagers eager to trade. The camps were a state unto themselves, and many of them were mobile. They could easily be packed up and moved if the army tried to mount an incursion.

The Chibok girls had been brought to one of Boko Haram's most fortified bases, a training ground they called Timbuktu. It was less than a hundred miles from their homes. But like the fabled Sahara trading town, it would be difficult for any outsider to find.

FOR THE FIRST FEW DAYS, THE INSURGENCY SEEMED UNSURE OF WHAT to do with its captives. Their guards, bored-looking young men and teenage boys, toyed with Kalashnikovs or sparred with one another, and spoke no more than a few words to the prisoners. Some were lanky, and looked to the students like Chadians, men from the desert. Several tied their silky hair up in a turban. They conversed in different languages, Hausa, Kanuri, Arabic, and one spoke Igbo, the language of Nigeria's southeast, but they had nothing to say to the girls and avoided eye contact. Monitoring these new captives seemed to be tedious for boys who'd joined to fight on the front line. Some sat under the trees, peering into laptop screens or listening to the Voice of America radio news' Hausa service, while cleaning their teeth with their *miswaks*.

The only explanation for their kidnapping came from the

stocky commander in black fatigues who'd led their abduction. On one of their first mornings he approached them.

"Since many of our Boko Haram members are in government prisons, it will also be good if the government can feel the same pain," said Malam Abba, the man they had last seen in the courtyard of their school barking orders to his men. "Don't worry. . . . If Shekau doesn't want you, you will all be released."

Many of the young women sobbed for hours each day. Some called for their parents. The space around the tamarind was full of schoolgirls lying face down, heads tucked in the crook of their elbow, trying not to look at their new setting. Some fiddled with pebbles they found in the dust and tossed them. The hostages tried to keep their backs turned to the men. When they weren't looking, Naomi would tap her side to check for the cell phone hidden in her clothes, a reminder of home.

One afternoon, when she was lying under the enormous tree and pretending to sleep, she removed the Nokia from her wrapper and turned it on. Checking to make sure none of the men were watching, she began scrolling through the names of friends and neighbors back in Chibok, hovering over each name and picturing their faces. One by one, she memorized the numbers of neighbors and relatives, mouthing them to herself, until she was sure she could dial them if she managed to reach a working phone. When she'd finished, she opened the Games folder and played Snake, tapping the sticky number pad of her forbidden Nokia to send a serpent racing after apples, trying to avoid its own tail.

Even though there was no reception, she typed out a text to her mother, as if she were speaking to her: "Be patient," the message said. "Pray that God should touch the heart of Boko Haram terrorists so we can be set free."

8

#BringBackOurGirls

THE FRENCH ANTILLES
APRIL 30, 2014, SIXTEEN DAYS INTO CAPTIVITY

Russell Simmons was finishing his morning yoga routine on a yacht floating in the turquoise Caribbean waters off St. Barths island, peacefully unaware that he was about to provoke a tidal wave. The man who founded the boom-bap hit machine Def Jam Records in a cramped Manhattan dorm room and made it one of the world's top record labels with a roster that included Public Enemy and the Beastie Boys was now an evangelist for what he called "operating from a quiet place," practicing transcendental meditation or yoga for hours daily. His latest book, *Success Through Stillness*, published the previous month, advocated silence as the place "from which all creativity springs." But Simmons's obsession with mindfulness had not stopped him developing an addiction to what had become the most cacophonous corner of the Internet. His @UncleRush Twitter account, combined with the Facebook and Instagram pages it cross-posted onto, had more than ten million followers, and he updated them constantly from TweetDeck, an application that let him promote his media and fashion brands while scanning four columns of streaming content at once. His posts were retweeted, liked, or shared thousands of times. The mogul, now nearly sixty, had become an influencer: a second act that would falter years later after a dozen women accused him of sexual assault.

Simmons's tweets hardly ever mentioned foreign affairs, usually opting for self-help quotes, vegan recipes, and aphorisms for the aspiring entrepreneur. And the early hours of April 30 were no different, as he sent a flurry of posts on subjects from mindfulness to gun violence and the writings of Maharishi Mahesh Yogi, a white-bearded yoga master who had once been guru to the Beatles.

Shortly after breakfast, Simmons read an article on Global-Grind, a black entertainment website he owned and often used his social media accounts to promote. It was about a kidnapping and protests in Nigeria. *Good intentions lead to good outcomes*, he thought, as @UncleRush popped out an eighty-one-character tweet: "234 Nigerian girls have gone missing and no one is talking about it . . . Please RT! #BringBackOurGirls."

The number he cited was inaccurate, but the hashtag was correct. He didn't know it, but his tweet, sent at 9:11 a.m. East Coast time, 2:11 p.m. in Nigeria, marked the beginning of a celebrity-driven cycle that would catapult #BringBackOurGirls from a distant campaign about a forgotten war to the very pinnacle of American cultural and political power. Within hours, the hashtag would explode, mushrooming out through social media, captivating millions, and ultimately raising intricate questions over why some causes catch fire and others don't. Simmons had no idea his tweet held any special significance. A few minutes later, he was tweeting again to plug Mission: G-Rok, his new video game pitting an eleven-year-old rapper against an evil boss named Smogorex.

But in the background, the attention economy had begun to react. @UncleRush's followers included a potent mix of Twitter's superfamous and almost famous users: talk-show hosts and recording artists, pop-culture critics and activists with a taste for politically conscious rap, social-media mavens positioned at the toll gates between news and entertainment. Within minutes of tweeting #BringBackOurGirls, thousands of likes and retweets began cascading into Simmons's phone. One hour later, the rapper Common, weeks short of signing a record deal with Def Jam, and on his way to a meeting in midtown Manhattan, repeated the hashtag to his own millions of followers. "We must help with this! #BringBackOurGirls." The post brought tens of thousands more retweets. Minutes later, the Atlanta-born rapper Young Jeezy joined in: "#BringBackOurGirls!"

Under each influencer's tweet, a crowd of followers began to rally one another, amplifying the hashtag and pivoting their anger onto the news media. This was a moral battle that needed to be joined. The hashtag #BlackLivesMatter was less than a year old, on a website whose user base was more racially diverse than the US population at large. Some of the same unanswered frustrations seemed to be at play in the case of the kidnapped schoolgirls: "WHY DO WE HAVE TO BEG THE MEDIA TO COVER THE ABDUCTION OF CHILDREN WHO LOOK LIKE US?!?" asked one tweeter with hundreds of shares. With a turn of the lens— "And no one is talking about it"—the obscure Boko Haram conflict suddenly had something to say about America. Thousands began to ask how US media would react if these girls were white, many tweeting the hashtag #234WhiteGirls. One tweeted, "There'd already be a Lifetime movie in the works."

#BringBackOurGirls mentions tripled in the hours after @UncleRush weighed in. But the metrics were about to be supercharged by one of the world's most beloved recording artists. At 1:27 p.m. from her estate in Saddle River, New Jersey, the queen of hip-hop soul, Mary J. Blige, addressed her twenty million followers: "It's been two weeks since the kidnapping of 234 Nigerian girls and they still aren't home #bringbackourgirls," she tweeted, adding an evocative black-and-white photo of a young school-age girl wearing a forlorn stare, a tear running down her left cheek. The teenager in the photo was not even from Nigeria, but the

West African nation of Guinea-Bissau, two thousand miles from
Chibok, yet more than sixteen thousand people shared Mary J's
post, broadcasting it to their own followers with furious narra-
tions appended. "We have to do something!" "WHERE THE HELL
IS THE MEDIA!!! WHERE IS THE OUTRAGE!!"

By the time the day closed in New York, sixty-two thousand
people had tweeted #BringBackOurGirls—and millions had seen
it. Suddenly, the tone of global media coverage began to change.
The story, feeding off itself, now had a new protagonist for an online audience: us.

WATCHING THE ANALYTICS SKYROCKET DURING A DENTIST APPOINTMENT
in San Francisco, Jack Dorsey was struck by how quickly the
hashtag was moving. The founder of Twitter had seen people fire
off tweets to coordinate protests in Iran in 2009, and the next
year, the website had galvanized crowds of millions to overthrow
autocratic regimes during the Arab Spring. Closer to home, he'd
watched tweets carry images of antiracism protests in Ferguson,
a mostly black suburb outside his birthplace of St. Louis.

But this, he decided, was different, a story that immediately
resonated. It was about a group of innocent girls being snatched
from a school. Here was that rare universal issue that transcended
the partisan divisions his website was accused of stoking. "Extremely powerful," he thought. "It touches every single person on
the planet."

It was Twitter's eighth year, and the website had never entirely settled an argument Jack and his business partner, Evan
Williams, had had at length: Was this a website to talk about yourself? Or the news?

"If there's a fire on the corner of the street," Evan had said,
"You're Twittering, 'There's a fire on the corner of Third Street.'"

"No. You're talking about your status," Jack had replied. "'I'm
watching a fire on the corner of Third Street.'"

Instead of resolving their dispute, the men had taken their

website live and had let users decide. By 2014, it was transforming into a wildly popular vortex of news, political invective, and
chatroom jabber, with an undercurrent of dangerous subcultures,
including a new terrorist group recruiting foreigners to fight in
Iraq and Syria, all of it powering the rise of a company with a stock
market valuation of $25 billion. "It's as if they drove a clown car
into a gold mine and fell in," Mark Zuckerberg had deadpanned.

Angry condemnations of the media and its collective failure
to cover the events users believed important became a potent
Twitter refrain: "Why is nobody talking about the fire on Third
Street?" A few months before #BringBackOurGirls, Justin Bie-
ber, Oprah Winfrey, and Kim Kardashian had tweeted a YouTube
video tagged "Stop Kony," part of a plan to make Joseph Kony, a
Ugandan warlord, so famous on social media that he would stop
kidnapping children. The video, YouTube's first to top a hundred
million views, felt like a novel phenomenon.

But if the technology was brand new, the themes were not.
Some of the most successful popular culture products ever sold had
supplied Western audiences the same emotional release: African
children are in need, and a celebrity needs *you* to help save them.
Up to 40 percent of the entire world's population had watched
pop acts from Tina Turner to David Bowie play the 1985 benefit concert Live Aid to feed starving Ethiopian youth. That same
year, Michael Jackson wrote "We Are the World (We Are the Children)," a charity release that became the fastest-selling pop single
in history. Ironically, this strange collective ritual had been born
in 1960s Nigeria, when the country's southeast declared itself the
independent Republic of Biafra. As the federal government began
to starve the region into submission, John Lennon, Joan Baez,
and Jimi Hendrix all raised voices and money on behalf of the
hunger-stricken Biafran youth. After two million deaths, Biafra
surrendered in 1970, but the modern aid industry, complete with
its uneasy codependency on rock-star activism, was born.

Nearly half a century later, the same experiment was being

run on a smartphone. If regular people, tapping Retweet from their privileged homes in America, could elevate an African injustice onto the global news agenda, maybe they could summon the collective energy required to resolve it? By May 1, websites were cranking out articles about the Chibok kidnapping, but this time through a new lens. Pop-culture icons were driving a movement that anyone could join with a click. Singers Chris Brown and Wyclef Jean, actresses Alyssa Milano and Mia Farrow were among the first celebrities to demand that news organizations "keep this story in the news," reminding them, "You can save lives here."

Their pleas were activating a feedback loop between pop stars and the news. This was a very different story that was now catching fire, and its focus was no longer the teenage girls taken from a small town in northern Nigeria. It was about the power and promise of the Twitter crowd. America was using the story of the missing students of Chibok Government Secondary School to talk about America.

TWITTER WAS TURNING ITS FIREHOSE ON NIGERIA AT A TERRIBLE MO-ment for the country's president. Back in Abuja, Goodluck Jonathan's administration was putting the final touches on a summit that he hoped would redeem the country's image and potentially bring in billions of dollars' worth of business. The World Economic Forum was normally held in the Alpine resort of Davos, Switzerland, affording captains of industry and heads of state a venue to pontificate on panels about the future of the global economy.

But from May 7, it would take place at the Abuja Hilton, the gigantic 667-room modernist hotel complex that was Nigeria's unofficial seat of money and power. Its bombproof gates shielded an expanse of palm-lined parking lots where chauffeurs on wooden benches waited for calls to collect their VIPs at the glass doors of the smooth jazz–soaked lobby. The hotel's tenth floor, guarded at the elevator and blocked off to ordinary guests, opened into an executive lounge with a panoramic view of the skyline: Inside, government officials, Western oil company executives and the fuel smugglers who stole crude from their pipelines, foreign arms traders, and lobbyists from neighboring African states all hobnobbed at the bar. Jonathan's office had spent months finalizing a security plan to ensure that the summit was a public relations triumph.

And yet across the street, on the dilapidated highway median that cut in front of the Hilton, another more compelling drama was now unfolding. Oby Ezekwesili's protests had swelled from a tiny band of women wearing red T-shirts and berets to a thousand-strong march of fist-pumping protesters, waving placards with evocative messages: "PLEASE DO SOMETHING!" "THE GIRLS DON'T DESERVE THIS." Journalists from major networks who had jetted in to cover the World Economic Forum were walking over from the Hilton, recording striking visuals with their camera crews or phones. The coverage was inspiring parallel protests in Nigeria's largest cities.

"What do we want?" Oby would chant, in a rousing call-and-response.

"Bring back our girls!" would come the answer. "Now and alive!"

Shuttling between meetings inside the Hilton, President Jonathan was increasingly convinced that the kidnapping and the online uproar stemmed from a conspiracy of northern politicians looking to unseat him. Security forces began to arrest protestors and would later deploy armed police and water cannons onto the median. That scene would only provide more arresting footage to help drive the story further up the Twitter algorithm.

Jonathan's officials weren't going to silence Oby easily. Quietly, she was deploying other, less public strategies to outmaneuver them. As the story gained traction in US media, she typed out a series of messages to John Podesta, a colleague she had known from her time working in Washington who was now a senior counselor in the White House: Do you think Hillary Clinton could tweet about this?

Podesta forwarded the request to Cheryl Mills, the former

secretary of state's longtime adviser. At 12:05 p.m. on May 4, in Chappaqua, New York, @HillaryClinton posted a tweet: "Access to education is a basic right & an unconscionable reason to target innocent girls. We must stand up to terrorism. #BringBackOur-Girls." Twenty-five thousand retweets quickly followed.

Jonathan was bringing analog weapons to a digital fight. After Clinton's tweet, some of America's most influential political figures joined the campaign. House Speaker Nancy Pelosi held a moment of silence on the steps of the Capitol, joined by Republican senator Tim Scott. "What has happened in Nigeria is outside the circle of civilized human behavior," Pelosi said. Civil rights hero John Lewis added his voice. Inside the chamber, Miami congresswoman Frederica Wilson, famed for her collection of sequined cowboy hats, convinced dozens of members of Congress to dress in red every Wednesday in solidarity until the schoolgirls came home. For President Jonathan, the host of a summit now ruined before it began, the story could not get any bigger or worse. And then, shortly after one p.m. on May 5, a video appeared on YouTube.

"I ABDUCTED YOUR GIRLS," SAID ABUBAKAR SHEKAU, GRINNING IN COMbat fatigues, his chin raised triumphantly. Standing in a forest clearing, flanked by six masked gunmen, Shekau had finally captured the world's attention, and he was elated. The lens zoomed in to show his rifle swinging in front of his bulletproof vest as he chopped his arms in the air. "I will sell them in the market, by Allah! I will marry off a woman at the age of twelve. I will marry off a girl at the age of nine."

This was petrol for a Twitter fire.

The story now had a ticking clock and a clear villain. Within minutes of Shekau's image beaming onto Twitter on May 5, it was trending, sparking new convulsions of outrage and millions more endorsements for #BringBackOurGirls. Footage of Shekau played on loop on cable news, a pantomime TV villain that helped highlight a six-year-old insurgency as a contest of good versus almost unfathomable evil. Anchors choked up as they presented packages on the madman who had taken *our* girls.

In London, hundreds of demonstrators thronged the Nigerian embassy, holding matching NO CHILD BORN TO BE TAKEN placards. Another crowd swarmed the United Nations in Manhattan. British parliamentarians wrote to Prime Minister David Cameron to ask what action the UK would take in its former colony. In a letter to the White House, twenty women in the Senate wrote, "The girls were targeted by Boko Haram simply because they wanted to go to school and pursue knowledge. The United States must respond." None of them seemed aware that the insurgency had been trying to steal a brickmaker.

Condemnations of Boko Haram were morphing into the language of a military intervention. Conservative talk-radio host Rush Limbaugh was lambasting the White House for being soft on terrorism, asking, "Is the United States really this powerless? And then if you answer yes, we are really this powerless, then isn't Obama to blame?" Senator Dianne Feinstein was demanding that US special forces stage a rescue mission, and if the Nigerian government didn't approve one, then ordinary Nigerians should lift their voices and demand that their leaders open their soil up to American troops. Whatever it took, she said, to locate, capture, and kill the men who'd done this.

IN A MEETING IN THE WHITE HOUSE'S JOHN F. KENNEDY SITUATION ROOM, Deputy National Security Adviser Ben Rhodes had been watching the hashtag's metrics and asking what precisely the endgame was. "Even if it hits a billion tweets, do we think Boko Haram is going to throw up their hands and say, 'OK, we let the girls go'?" Within days of landing in America, #BringBackOurGirls had vaulted onto the screens of some of the Obama administration's most senior figures. Rhodes and other officials for foreign policy, intelligence, and defense held hourlong Chibok sessions in the underground complex where the president had once live-monitored the killing of Osama bin Laden. Other meetings took place in the nearby gray-paneled Eisenhower Executive Office Building. Diplomats in West Africa tuned in on a video conference line, juggling the time difference.

The question was how the world's most powerful government planned to help find and free the schoolgirls. With the hashtag migrating from Hollywood to Capitol Hill, it was inevitable that the kidnapping would appear on the president's morning intelligence summary, compiled at the CIA's Virginia headquarters. The president would want to know what policy options were available.

Formulating an answer fell to the National Security Council, the academics and intelligence officials who sketch out strategy for the White House. Al Qaeda allies in Africa had become a rising concern for the NSC; just as Twitter had also begun to shape its preoccupations. Two of its staff had recently been appointed the NSC's inaugural Twitter monitors, tasked with pasting five or six interesting or popular tweets into daily e-mails to relevant officials. As those monitors watched the analytics on #BringBackOurGirls soar, the NSC staff began summoning top officials at defense, state, and other departments, to discuss the kidnapping. What could America do? Each department needed to reach into its capabilities and present assets it could offer.

The Justice Department could send FBI agents with experience defusing hostage crises. The State Department had medics and psychologists it could fly in from neighboring embassies to treat the girls upon their release. But nobody around the table had anything approaching the resources of the Department of Defense, whose $614 billion discretionary budget for the preceding year was significantly larger than the entire Nigerian economy. Their answer to the problem was drones.

The US had satellites and propeller planes that could scan the Sambisa area. But for daylong flights of continuous intelligence gathering, they could use the "workhorse," an RQ-4 Global Hawk drone, gunmetal gray with a bulbous head like a dolphin, which could surveil a South Korea–size acreage daily. A few weeks later, the US might be able to rotate in another drone, named the Predator. Keeping those machines in the air for extended missions would require deploying troops in the region.

All told, the room was debating the deployment of more than a quarter of a billion dollars of sophisticated matériel over some of the world's poorest farmland, looking for teenagers held by a group that had never attacked the United States, on a mission ordered up by Twitter. "Why are we looking for some schoolgirls as opposed to looking for al Qaeda?" a Defense Department official protested in one meeting. Another asked if the US was going to do this every time a group of girls was kidnapped, and if so, "What's your threshold? . . . Is it five? Is it fifty?"

But this was a chance to dispatch military power for humanitarian ends. The US would send about forty personnel to staff a rescue mission from Abuja: intelligence officers, aid workers, and law enforcement. To service drones, it would base some eighty troops, mostly air force, in the dictatorship of Chad. Foreign allies would follow America's lead. The British would send a spy plane, while intelligence officers from France would scour the former French colonies along Nigeria's borders for leads. China pledged to send satellite coverage. Canada would send special forces, and Israel would offer counterterrorism specialists. John Kerry would sell the plan to Nigeria's beleaguered president in a twenty-minute phone call. On May 6, Goodluck Jonathan agreed.

There was only one other person who needed to sign off. He had two daughters of his own, girls he sorely wished to spend more time with, and he had been following the news from Chibok on an encrypted iPad in the Oval Office.

ARRIVING AT THE WHITE HOUSE IN A BLACK SECRET SERVICE SUV, NAtional Security Adviser Susan Rice brought a stack of reading material to the president outlining the proposed US intervention in Nigeria. On top was a single piece of paper, with two checkboxes: one to approve the mission, a second box if the president had hesitations to clear up. The rest lay out the backstory behind the kidnapping and how America's deployment to resolve it might proceed, wonky and detailed information the president's staff expected him to carefully consider until the last letter on the last page.

Barack Obama checked Approve.

"Do everything we can," he said.

Within hours, the first American personnel were on planes to Abuja, with barely time to pack their luggage. It was May 6. Within a week, #BringBackOurGirls had gone from a small, roadside protest in Abuja to an eighty-one-character tweet from Russell Simmons that would help propel a stack of paper onto the president's desk authorizing US drones over Nigeria's Sambisa Forest. It was a study in the capricious and unpredictable flow of political power in the twenty-first century, the moral authority bestowed on celebrities channeled through the firehose of new media and carrying the force of an evocative story that satisfied the West's faith in its own technology.

"The line that came down from Obama," said an intelligence official, "was, 'Do everything you can to get those girls.'"

The NSC began scheduling regular video conferences to demand results from their respective agencies. They needed updates to brief the president and his staff. The Senate was also set to hold a hearing: "#BringBackOurGirls: Addressing the Threat of Boko Haram." White House officials drafted plans to rent the Jim Henson studio in Los Angeles to record a charity single featuring top rap stars. And in case anybody missed how closely this story hit home with the president, another update went out the following morning.

The first lady was about to tweet.

PART II

An Open-Air Prison

9

The Blogger and
the Barrister

Ahmad Salkida was working at a supermarket as a security guard in a suburb of Dubai when he got a call from his government. Nigeria needed a go-between to free the Chibok girls and the most immediate candidate was him, an exile seeking vindication from a country that he felt had rejected him. Thirty-nine years old with sharp cheekbones and restless energy, Salkida was a blogger and freelance journalist, raised in Nigeria's northeast, who had broken some of the best scoops on Boko Haram, a sect he had covered professionally since its earliest years before the war. At the grocery store, he was easy to miss, a night owl on the graveyard shift, the glow from his phone illuminating his wiry beard. But on the ungoverned space of Twitter, he was one of the most important voices in the Boko Haram war. His unverified account, @ContactSalkida, was followed by tens of thousands of Nigerians, a rare source of information on a sect whose aims, structure, and funding sources were shrouded in conspiracy. He had built such a rapport with the insurgency that before it turned violent, the group's founder, Mohammad Yusuf, had asked him to run its newspaper. But Salkida wasn't interested in the job or in being

anyone's mouthpiece. His business card said INDEPENDENT JOUR-
NALIST.

Then, after the war erupted, his work became dangerous, and
his loyalties came into question. In the violence that followed Yusuf's execution, two police officers had told Salkida to lie on the
floor, then argued over who should kill him, until a government official intervened and the journalist slipped away from Maiduguri.
Salkida reappeared online, but some members of the security services doubted whether he even existed, and if he did, whose side
he was on. Anonymous callers would ring his phone to tell him he
was a terrorist, part of Abubakar Shekau's movement. The blogger kept an exact tally of the number of times he'd been detained
by the military: sixty-one. And he could no longer visit his family
village, where neighbors had promised to kill him. By 2014, he
had been so inundated with threats that he'd been forced to flee,
landing in a working-class neighborhood of Sharjah, a coastal city
made up almost entirely of immigrants, where he moonlighted as
a night watchman in the shadows of Dubai's shimmering skyline.

Now his phone was ringing with a number from Abuja, asking
him to come home. The government official on the other end of the
call had a direct line to Goodluck Jonathan, a president who was
scouring for help from everywhere at once. Nigeria didn't seem to
have the means to meet the moment. Intelligence gathering was
thin, troops were on the verge of mutiny, and it didn't lift their
morale to see celebrities pointing out their failure to protect a
group of innocent schoolgirls. The government had no choice but
to reach out to Shekau, and Salkida, despite all the mistrust and
controversy, was their best choice.

By the first week of May, just as some of the world's most
famous people started tweeting #BringBackOurGirls, the kidnapping reached a critical stage. Early into a hostage crisis, authorities face a dilemma over whether to negotiate with the abductors.
Many of the world's most powerful governments, spearheaded by
the United States, have long proclaimed an immovable public po-

sition against making concessions to terrorists. "We don't negotiate with evil; we defeat it," Vice President Dick Cheney liked
to say. But behind the scenes, the purity of that policy is hard to
maintain. Military raids to rescue hostages fail as often as they
succeed, which means governments often have no alternative.

In practice, many nations choose whether to negotiate based
on public opinion. It is a delicate art, and too much media scrutiny
can be as harmful as too little. In the unusual case of the Chibok
girls, the first tweets and protests within Nigeria had proved sufficient to goad the government into attempting hostage talks, but
just as that discreet, offline effort began, the call to #BringBack-
OurGirls was becoming a global sensation. The students' cause,
once contained to Nigeria, was being supercharged by a Twitter
crowd of millions worldwide, a chaotic new energy as unstable as a
freshly discovered element. It was an inspiring story, for so many
faraway people to lift their voice on the girls' behalf. Yet to what
degree was any of this celebrity attention helpful? The months to
come would test whether international awareness would help end
their captivity, or extend it.

One thing was for sure: the popularity of the hashtag had
started a race to free the women. Abuja would soon fill up with vol-
unteers, each representing a different philosophy. While American intelligence officers downloaded imagery from drones circling
the Sambisa Forest, a bizarre cast of have-a-go heroes and foreign
misfits would fly into the capital, chasing money and fame. The
government of Goodluck Jonathan, clobbered by the torrent of
social-media demands, authorized countless competing negotiating efforts. But the very first was by a freelance journalist guarding a grocery store in the Middle East.

Ahmad Salkida had the connections Nigeria needed, but he
was a complicated person to ask for help. His blog, *News beyond
the Surface*, and his Twitter feed had been a thorn in the side of
a government that had come to see him as in league with its enemies. He was an avowed nonconformist down to the five-fingered

toe shoes he wore under his Muslim robes. Nigerian officials were leery about his fixation with social media. "Everything he does has to be in the public domain. He has to tweet about it," one official complained. "He is such a *millennial*."

But the man on the phone needed the blogger to work his contacts in the sect's leadership to help free the Chibok girls. This was an inversion of the Nigerian social order, the presidency asking for the services of a self-educated emigrant, and yet the situation was desperate. Salkida knew there would be risks, but the Chibok kidnap had become an international sensation and could be a chance to disprove his critics before a global audience. He agreed to help but insisted on one performative precondition: he didn't want a first-class plane ticket. People would think he was doing this for money. A Ugandan co-worker agreed to cover his shifts. "I'm going to meet my president," Salkida said.

THREE-THOUSAND MILES WEST ZANNAH MUSTAPHA FELT THE MODEST rumble of a Swiss train carry him through the mountains. The Barrister had been thinking about the vast differences in temperature you could experience in just a few hours of air travel. Maiduguri had been 99 degrees when his car made its way through traffic and sandbagged military checkpoints en route to the crowded airport. Now he was looking out through thick carriage glass, passing snowcapped Alpine peaks and slopes graced with A-frame cabins that hinted at a life of peaceful seclusion.

Around the same time Ahmad Salkida flew back to Nigeria, the Barrister headed to a hotel on the shore of Switzerland's Lake Thun to begin a series of invitation-only courses by the Swiss government. Pascal, the Swiss diplomat, had arranged for Zannah to participate in this program, a training normally reserved for a global elite. His destination was a century-old hotel, the Park Oberhofen, high on an Alpine hillside studded with tiered fountains that cascaded toward the glacial waters of the lake. Zannah was greeted by a host who introduced himself as an officer from a government division that liked to operate away from the limelight. It was called the Human Security Division, but some in the know still referred to the unit by its earlier name, Political Affairs Division IV.

In the hotel's restaurant, where a 12:30 lunch was scheduled, the other mediators in training were gathering, among them Colombians, Sri Lankans, and several from Southeast Asia. Their tables looked out through arched antique windows to reveal one of Europe's tallest peaks, the Niesen. The longest staircase on Earth was carved into the side of that mountain. Waiting for the class was a stack of required reading to prepare for the first day's session, including a briefing paper titled "Pawns of Peace."

Class in the Park Oberhofen began the next morning in the banquet room, where students gathered awkwardly in front of a pudgy-faced man dressed more like a geography teacher than a diplomat. Julian Hottinger had mediated conflicts from Sudan to Sri Lanka and had once managed a disarmament deal with the warlord Joseph Kony. In Burundi, he had personally reported to the great twentieth-century peacemaker Nelson Mandela. He spent all but ten weeks of the year traveling to untie the knots in faraway peace talks. This, he liked to say philosophically, was the life of a mediator: "We are constantly surrounded by people yet always alone."

Hottinger would be teaching the opening class in a ten-day boot camp designed to explain the art and science of peacemaking, a discipline the Swiss felt they had absorbed into their DNA over centuries of practiced neutrality. The students would examine the strange alchemy through which wars just as terrible as Boko Haram's had reached a conclusion, and if all went well, they would return home to apply their knowledge in their homelands troubled by conflict.

Zannah sat on a padded metal chair, peering through his

designer spectacles at the lecturer. Since the end of the Second World War, 148 ceasefires had been signed, each one unique in its stipulations. But to Hottinger, they were alike in one way: none of them were easy. The public had a deceptive two-dimensional image of how peace talks came together. The leaders of governments and guerrilla movements sat around a table, bartered over terms, and then shook hands. In reality, the effort was infinitely more complex, conducted out of sight by unnamed people who toiled in secrecy, risking their lives and liberty. These "inside mediators," as the Swiss called them, almost never appeared in the photographs of smiling political leaders and their Western counterparts, which capture only the concluding moments of a process that often took years. To accomplish their aim, Hottinger said, the insiders present would need to adopt a careful mind-set of rigorous impartiality. They would need to look into the eyes of killers and see their humanity. "There are no absolutely good or absolutely bad people," he said. "There is always a solution to the conflict."

To bring these concepts to life, Hottinger asked his students to take part in a role-play called the fence game, wherein students would play the parts of two neighbors quarreling over a fence between their properties. To resolve this simplest of squabbles took more than an hour, as Hottinger showed the class how to structure a mediation, narrowing the terms of debate at each step, and allowing both sides to feel heard. It was a cautionary tale of just how hard it could be for two opposing parties to reach an accord. "You don't rush for the solution," said one of the other Swiss specialists helping Hottinger. "There is a series of steps you take in order to reach a mutually acceptable agreement."

The fence game taught an important lesson. Mediations don't always proceed in a straight line. Furious arguments can be a sign of progress, and deadlocked talks sometimes signal that the two sides have identified the issue that holds them back. Smiles, on the other hand, could be a forewarning that one side didn't understand what it was agreeing to or that it couldn't deliver. As a deal comes closer, factions on both sides may try to sabotage it. Hottinger had a mantra he repeated to the class: "Never trust a breakthrough!"

THE BLOGGER AHMAD SALKIDA FLASHED HIS NEWLY ISSUED GREEN ACcess badge to a soldier at the gates of Nigeria's presidential villa. Twenty-four hours earlier, he had been an exile afraid to visit his family village. Now he was authorized to enter Aso Rock, the executive estate secluded from downtown Abuja on a campus of lush lawns. To access the building, he would have to pass through layers of metal detectors, surrendering his phone to the conveyor belts of two X-ray machines. Inside, he would present his plans to President Jonathan.

The night before, the blogger had been picked up from the airport by a senior security official known as Honourable, an erudite specialist in military affairs who sported a beard and a puckish grin. Honourable and Salkida had driven directly to the lobby of a dimly lit midprice hotel where they leafed through a Quran to find a passage that might compel Shekau to negotiate. The two men settled on *sulhu*, an Islamic concept of reconciliation that they believed the fundamentalist could accept. Salkida had been texting his sources in Boko Haram for days, trying to set up a meeting, but to clinch a deal, he needed the approval of the president.

Salkida walked into a carpeted lounge in Aso Rock's residential quarters to pitch his services. It was full of presidential aides wearing skeptical expressions. Some in government suspected Salkida, despite his protests to the contrary, was a Boko Haram sympathizer who should be in jail. In the minutes before he entered, one of them had been insisting that there was "no such thing as Salkida," that the blogger's online identity was a front. But with global pressure building online, their president was prepared to trust this enigmatic exile. Jonathan sat casually in a blue shirt and leather slippers, listened to his guest, and gave him approval, placing his hand on the blogger's shoulder.

"I believe in you," the president said. Salkida could see that the portrayals of Jonathan as an apathetic leader unmoved by the Chibok abduction didn't give the full picture. The president seemed emotionally strained and eager to find a quick resolution. He had a daughter of his own, he mentioned, married just two days before the kidnapping. He also made a comment on Salkida's outfit: the blogger was wearing five-fingered toe shoes.

As the meeting concluded, Salkida walked back down the chandelier-lit hallways of Aso Rock and out through the security cordons. Now that he had the president's approval, he would head straight to his rendezvous point with Boko Haram, a bus station along Nigeria's border with Cameroon, and wait for a member of the insurgency to find him. Before he left, his contacts had insisted on strict security protocols: if he brought a phone, they'd kill him.

10

Thumb-Drive Deal

On Tuesday, May 13, 2014, the first images from a Global Hawk drone circling the north of Nigeria began to arrive at an air force command center in western Germany. The footage had bounced off a satellite and into a receiving antenna that sat neatly behind a barbed-wire fence running through a beech-and-pine forest. Nearby, on a baseball field, local expat kids from the Ramstein American High School team were in spring training.

Some of their parents were inside the airbase, a sprawling campus where military personnel tracked targets in America's drone wars. Ramstein had been born as a Luftwaffe airstrip during the Second World War but was now America's vital perch in a global aerial surveillance program. Noncommissioned officers in battle uniforms stared into nearly two thousand Dell monitors, streaming live footage from the coast of Greenland, the southern tip of Africa, or the mountains of Pakistan. New missions could initiate swiftly. A week after President Obama checked Approve on America's rescue plan, Global Hawk footage of Nigeria started showing on Ramstein's screens.

But the enormousness of the base and its complex of sophisticated antennas cloaked a fundamental shortcoming. America's problem was no longer getting information, but knowing where to look. The officers watching the screens worked in pairs for eight-hour shifts, while other intelligence analysts seated in US facilities elsewhere on the planet watched simultaneously, chatting in small text boxes. They could zoom in or out in a Nigerian search zone the size of West Virginia, but much of what they saw was blank land. They had been through months of coursework to be able to read such imagery and discern whether a shadow, measured against the sun, revealed the height of a target. Was it a child? These could be life-or-death distinctions. Fatal mistakes were part of a desk job in which PTSD was not uncommon.

Yet the air force had never embarked on a search mission like this, for two hundred missing schoolgirls on whom there was almost no biographical information. Many of its analysts had been monitoring individual terrorists for months or years and could recognize them by the way they walked. They knew the times of day their target slept or called his loved ones; they knew the names of his children and which building he called home, and eventually watched on a monitor when it collapsed in a plume of dust.

Paradoxically, looking for hundreds of students involved challenges more complex than the painstaking work of hunting a single well-known fugitive. From above, any group of young women in the vast search zone could include one or many of the Chibok girls, or they could all just be ordinary village teenagers. Days after the first images began to beam across a bank of screens, footage showed pickup trucks driving in a chain through wilderness in northeast Nigeria, but no one could say for sure if it was a convoy of militants or local people on the move. Maybe there were girls inside the trucks? There was scant information on the girls and hardly any digital data trail. There were conflicting accounts of the number missing and discrepancies in the spellings of their names. The girls had never joined social media or left any personal data online. To the Internet, it was as if they hadn't existed until their kidnapping.

On other screens, active conflicts played out, in Libya or

Ukraine, but the Nigerian feed was a frustrating abundance of brown flatland streaked with dried riverbeds and ghostly dirt tracks. A sludge of shallow ponds curled into mudflats sliced by the tire treads of motor scooters that drew lines between villages with thatched roofs that all looked the same. Marshlike groves erupted like small oases from desertlike plains. Scattered trees overlooked tufts of neon-green grass or corn or okra farms tended by lone figures miles from the nearest house. "Empty landscape. A lot of empty territory," said an official tasked with watching it. "For a long time, there was not much to see."

BY THE SECOND WEEK OF MAY, THE BLOGGER AHMAD SALKIDA HAD MADE it into the heart of Boko Haram territory. The journey there had been staggered with obstacles meant to shake off any government minders following him. The blogger had been picked up at the bus depot by the first of a series of contacts who didn't identify themselves but handed him off to other men until he was put on a motor scooter and driven for nine hours over winding dirt tracks deep into the Sambisa Forest.

He arrived at dusk, his back bent and sore, and was ushered into an air-conditioned tent to meet a small delegation authorized to speak on behalf of the Imam. Abubakar Shekau would consider a deal, they said. The men argued through the night, negotiating over the noise of a generator. Hours in, the generator coughed up a loud noise, startling one of the militants, who jumped up and accused Salkida of wearing a GPS locator, then threatened to kill him. His comrade restrained him and Salkida watched in relief as the men calmed each other down. The militants were nervous and seemed paranoid. It was hours before they started to make progress on the operational details of an agreement.

By sunrise they had a deal. Salkida was exhausted as he lifted the flap of the tent to begin his journey out of the forest. But as he stepped out into the daylight, he immediately stopped. There, under the shade of a tamarind tree, were more than two hundred

young women staring back at him. He was face-to-face with the
Chibok girls.

The hostages were gathered onto a patch of dirt, in rows,
knees out or cross-legged. Their clothes were dirty, and they said
nothing. Around them stood masked fighters, some dressed in
army fatigues, others in polo shirts and flip-flops. Salkida could
see that a few of the hostages were still wearing their blue-and-
white checkered-print school uniforms. They had been moved at
gunpoint at dawn so that the blogger could have them filmed.

The militants unfurled a white tarp, stretching it as a backdrop for a makeshift studio, then invited the girls to walk toward
a man holding a camcorder who began to ask questions. Salkida
asked for two tapes—one for the president as proof that the captives were alive, another as a scoop to give to the media, if talks
stuttered and he needed a dose of public pressure. Boko Haram
gave Salkida both videos, loaded onto a thumb drive, and after
another day of disorientating travel winding his way out of the
Sambisa Forest, he was back in Abuja, and the video was playing
on a screen in the presidential villa. Goodluck Jonathan watched
it on a couch in a small room darkened by heavy drapes while a
translator sat on the carpet.

One after the other, ten students appeared on the screen,
hands clasped in their laps, as a man off-screen asked them questions in Hausa with a strong Kanuri accent.

"Are you being fed?"

"Yes."

"Did we touch you?"

"No."

The terrified students spoke into the camera, their voices
breaking as they recited scripted lines: "We are begging the president. We are his daughters."

The video ended, and Jonathan was visibly moved. "Every life
that is lost pains me," he said. He gave a green light to the second
stage of the negotiations and asked to be briefed on the results. In

the final shot of the video, more than one hundred hostages could be seen guarded by masked men with Kalashnikovs. Most of the schoolgirls stared ahead vacantly or looked at the ground, trying to avoid eye contact. But one could be seen staring defiantly back at the lens. Naomi was wearing her school uniform.

DAYS LATER, ON SATURDAY, MAY 17, AS THE FINAL DETAILS OF SALKIDA'S plan came together, President Jonathan arrived at the Élysée Palace in Paris, walking through a row of silver-helmeted guards into the gilded Salle des Fêtes ballroom to meet with presidents of four West African countries and several Western allies. The leaders, summoned by French president François Hollande to coordinate an international military campaign against Boko Haram, took their seats under a ceiling painted with angels said to represent guardians of the peace.

The men around the banquet table were all determined to see an end to the insurgency and its kidnapping phenomenon. Chad's president, Idriss Déby, a former paratrooper in power for a quarter of a century, boasted that he could deploy his troops into Nigeria to mop up the insurgents. Paul Biya, leader of Cameroon since 1975, promised "total war" against the sect. Europe had been paying ransoms for more than a decade to free citizens kidnapped by terrorists in West Africa, including Boko Haram. Months earlier, the French government had approved a ransom of just over €3 million to free a family of seven tourists abducted by Shekau's men. It had taken Paris and the Cameroonian government only six weeks of secret, offline negotiations to liberate the family, but it was an episode Hollande didn't want to see repeated. The British and American position, stated repeatedly before and after the Chibok abduction was clear: make no significant concessions to a terrorist group. "If you pay ransoms, you're making this into a business, a successful business," one American diplomat warned. "I don't want to wake up tomorrow morning and hear that more girls have been kidnapped."

Goodluck Jonathan, wearing his trademark black fedora and matching kaftan with gold trim, looked tired and distracted as he fumbled through a barrage of questions on why Nigerian soldiers hadn't found the Chibok girls. In the preceding twenty-four hours, another bizarre coalition of celebrities, politicians, and Western institutions added their voices to the campaign. Archbishop Desmond Tutu and conservationist Jane Goodall held up placards, while the supermodel Irina Shayk, girlfriend of soccer star Cristiano Ronaldo, posed topless with a #BringBackOurGirls sign. Thousands had marched in the streets in Washington and Boston while students at public schools from Kansas to Connecticut held vigils calling for the Chibok girls' safe return. That night, Harrison Ford, Wesley Snipes, and Mel Gibson would hold #BringBackOurGirls signs on the red carpet at the film festival in the French Riviera town of Cannes.

Nobody in the French presidential palace except Jonathan and his closest advisers knew that back in Abuja, Ahmad Salkida was sitting in a top-floor office in Radio House, a towering building looking out across downtown Abuja, trying to finalize a hostages-for-prisoners swap over ordinary text messages with his sources in Boko Haram. The deal that was coming into shape seemed to be as good as Nigeria could hope for. At first, Shekau's negotiators had demanded that Nigeria cede territory to Boko Haram, mentioning the example of the Swat Valley in Pakistan, the region where Taliban gunmen had closed schools and shot student Malala Yousafzai on her way home from an exam. Nigeria's government rejected that request until the two sides settled on a lower cost solution. If government prisons released a hundred of his members, Shekau would free the girls. It would be a swift symbolic victory for the warlord.

A swap was set for Sunday, May 18, with Jonathan still in Paris, but the bureaucracy had been a problem. By that morning, Salkida and the security official named Honourable had managed to get only sixty-seven prisoners cleared for release through the state prison system. At five a.m., Salkida began driving in a six-car convoy toward the staging ground for the exchange: Damaturu, a small town eighty miles from Maiduguri known as a militant stronghold.

As Salkida's convoy rumbled toward the rendezvous point, Honourable drove to the home of Alex Badeh, the chief of defense staff, to confirm when the prisoners would be airlifted to meet it. But Badeh was relaxed and about to leave for church, complaining that he had not been briefed on the operation. He stepped into another room to make a phone call, and returned with news that a C-130 Hercules turboprop transport plane was available. It was strangely easy.

Reassured, Honourable drove home to see his newborn daughter, who had arrived just two weeks earlier. He had been so busy with the Chibok file that he'd barely seen her. But shortly after he arrived, his phone rang. The call was from a cabinet minister, who spoke slowly and deliberately. The deal had been canceled. No exact reason had been given. Jonathan's military chiefs didn't trust the Twitter personality who had become the middleman in the war they'd been fighting, and most of all they simply couldn't stand the raging, celebrity hashtag movement that they felt had embarrassed Nigeria abroad and *politicized* these girls. To the president's inner circle, the foreign calls to #BringBackOurGirls had crescendoed into what now felt like an intrusion on Nigeria's sovereignty. The very fame intended to free the students was complicating their release.

When Honourable finally reached Salkida over the spotty network, he spoke slowly into the phone: "We are men of faith," he began. "The deal has been canceled." Salkida screamed into the handset; then the call cut out. A few minutes later, Salkida called back: "I need to get out of the northeast now." The journalist was in danger. Until that phone call, Salkida had hoped to be the rare individual who could connect both sides of the conflict. Now he'd fallen off the tightrope. Boko Haram had every reason to believe he was planning to lead them into an ambush, and if he stayed in the country, he would certainly be detained or killed.

The blogger paid a taxi to drive him to Abuja, then walked into a meeting room full of Nigerian officials arguing over who was at fault for the botched deal and whether Salkida should stay. "Fuck it, I'm leaving," the blogger said. Within hours, he was on a commercial flight arcing over the Sahara, headed home to Dubai. Ten days later, @ContactSalkida returned to Twitter. "Its enough chalenge to be Nigerian," read one of his first tweets, "& extremely distressin to luv Nigeria."

LYING ACROSS THE BED OF HIS HOTEL SUITE, AS THE CHAMPAGNE-colored drapes softly filtered the light crossing the shores of Switzerland's Lake Thun, the Barrister, Zannah Mustapha, was nearing the end of his two-week course on making peace. Since arriving from Maiduguri, he had immersed himself in the vocabulary of mediation. Over long dinners, he'd listened to classmates from Sudan and Colombia, countries whose wars had seemed just as bleak as Nigeria's until, suddenly, they had lurched toward resolution. There was hope. Holed up in his room, he pored over texts by Kofi Annan, the Ghanaian diplomat who had become UN secretary-general, studying how other nations learned to put their violent disagreements to rest.

His course literature described how some wars created obstacles so fundamental that peace talks couldn't proceed until those "boulders in the road" were removed. South Africa couldn't resolve its struggle over apartheid until Nelson Mandela, the subject of a global campaign, had been freed. Now the Chibok students were becoming the human riddle at the heart of Nigeria's troubles. The schoolgirl hostages were the lock to the conflict, he decided. "I am trying to pick the lock," he would say.

The Swiss invited dozens, if not hundreds, of people to courses like this each year, and the lawyer sometimes questioned how much of their theory applied to his country. One night by the fireplace, as students gathered to listen, instructor Julian Hottinger compared Boko Haram to Palestine: Both were *identitarian* movements of people seeking their own state, Hottinger argued. Zannah was impressed at first, but later the Maiduguri lawyer realized that he disagreed. Nigeria wasn't much like Palestine at all and would require its own solution. The course came with a certification that Zannah placed neatly in his suitcase, picturing the wall of the office at his school where he would hang it.

After one of his classes, instructors told the group to layer up, as they would be taking a field trip into the cold. Zannah changed from his starched white kaftan into a light-blue hoodie. The class boarded a red, two-carriage train that wound its way up mountainsides for hours to the highest railway station in Europe. Stomping out through snow and ice in a pair of black tennis shoes, Zannah followed the group. There, spilled out in front of them was a vista of the Swiss plateau, the low-lying land of lakes and rolling pastures carved by tectonic plates crumpling over millions of years. Sporting wraparound sunglasses, Zannah snapped selfies with his fellow mediators, posing next to a sign, TOP OF EUROPE.

"When you make peace," Zannah remembered one of the instructors saying, "it should feel like this."

11

Malam Ahmed's Rules

Naomi dragged herself to her feet, her clothes, hair, and skin soaked with rain, and followed her classmates as they gathered for the first of the day's five prayers. It was the third week of May, and the rainy season had arrived over the Sambisa Forest. Even the thick, leafy branches of the tamarind tree weren't dense enough to prevent downpours from drenching the hostages, and the rock-strewn ground where they'd been sleeping. All night, they had been pelted by a ferocious storm that left ankle-deep pools of muddy water and streaming rivulets. They had hollered for their guards to bring them plastic bags, or a tarpaulin sheet to cover them. Naomi had heard one girl scream, "We are dying!"

It was still dark when her captors had barked orders to rise for the morning head count. Naomi was hoping the sun would dry her sodden clothes. She was now wearing a long dark *mayafi*, a kind of flowing, floor-length hooded shroud that the militants had told her to pin under her chin to cover her whole body except the face and feet. "We don't want to see your bodies," the man distributing the garments had said. "They will make us sin." The militants had demanded that the girls hand over their school uniforms, but Naomi had decided to wear hers surreptitiously

underneath. Now both layers were soaked through as she walked
to an adjacent tree with more than two hundred mostly Christian
hostages to perform the first of five daily Islamic prayers. Each
day, they were learning to clean their bodies in a special sequence
before worship, washing hands, mouth, noses, and arms in ablu-
tions monitored by their guards. This was part of becoming a good
Muslim, the men would say. The hostages would sit on their legs,
raise their palms to the sky, then bend forward. But as she went
through the motions, Naomi told herself that she wasn't going to
perform the ritual with conviction. "We should just pretend," she
told her cousin Saratu. "Not take it for real."

More than a month into their captivity, the Chibok students
were adjusting to an almost unimaginable new normal. Days
and nights were spent under the branches of the tamarind tree,
guarded by men with guns who watched them eat and sleep and
enforced the laws of the "true Islam." The breezy reassurances
that they would be going home "soon, soon," had stopped.

The fundamentalist leading the young women in prayer was a
bearded figure who walked with a stoop and carried a thin, reedy
switch in his left hand. Older than the guards, the man spoke in
broken Hausa and walked around the camp reciting passages of
the Quran to himself in Arabic with a single index finger raised.
He had introduced himself as Malam Ahmed and said he had
been chosen as the hostages' official "guardian," replacing Malam
Abba, the military man in the red cap who had led the abduction.
"I will be the first man you see each morning and the last you see
each night," he had said.

Malam Ahmed was a religious leader who would be responsible for their care and spiritual education. His calling, he explained,
was to mold them toward the "true religion," so they could enter
paradise. "You are daughters of infidels," he would say, flashing a
smile that revealed a row of stained teeth. His hair was jet-black
save for the few errant strands of gray that betrayed his use of

dye. He was their chief jailer and their shadow, eating near them, teaching them, and sleeping next to them. Most important, he would make sure they did not try to escape.

Dressed in civilian clothes, usually a brown or black kaftan, the Malam walked with a limp that stemmed from his time in the Giwa Barracks, when Nigerian prison guards had planted a hot electric iron on his spine. It made him look frail, but the hostages would disobey him at their peril. He had been a fighter once and still carried a rifle over his shoulder. Each morning, he organized the students into what he called Islamiyah school, a Quranic class held under another nearby tamarind, with a blackboard leaning against the trunk. The hostages would sit on the ground and listen as the Malam enforced the rote memorization of scripture. He had a teaching assistant, a scrawny teenager younger than most of the girls, who would leer at and taunt them: "I must marry a Chibok girl." He would write passages on the blackboard, which the students had to commit to memory. During class, Malam Ahmed would circulate, smacking a student on the thumbs with his switch if she missed a letter or mispronounced a word or if her hood was slipping from her head. "It seems you still have *boko* in your hearts," he would say. Of course Naomi hated the class and would pretend to be sick, holding her stomach to avoid being called on. The first lessons had covered the creation story of Adamu and Hauwa in the Garden of Eden. "It is you women who brought sin," Malam Ahmed said.

In these daily hours-long classes, the girls were being indoctrinated into Boko Haram's beliefs. Democracy was a sin, and their minds had been polluted with lies about the world and its weather by the "geographers" who had taught them in their school. "The world is flat, and it is Allah, not evaporation, that causes rain!" the Malam would say. Now that the hostages were following God's teachings, they would have to obey strict rules. They should refrain from speaking loudly or laughing, which was unbecoming of a woman. They were to never wander off, except when escorted by a guard, and would be required to wear their long dark apparel at all times, in a specific color that identified them as the Chibok girls. They should never speak in their pagan language, Kibaku, which none of his men could understand. And they should instead learn the vocabulary of Boko Haram's new civilization. Young fighters, often in mismatched uniforms or baggy shalwar kameez, shod in flip-flops, were the *rijal*, foot soldiers. Commanders were *qaids*. A *jasus* was a spy. A girl who had embraced the struggle could become an *emira*, or "princess," a leader among the women. She would be expected, the Malam said, to serve a fighter as his wife, and in return, would be given a home and as many dates or cuts of dried meat as she desired. In time she would bear children. "If you get married, you will become part of our religion," he said. "And if you have children, it will be more people for our cause."

LIFE IN CAPTIVITY DRIFTED BETWEEN DEPRIVATION AND TEDIUM. THE hostages spent almost all of their time sitting under the Tamarind, waiting to be given orders to pray or study. They had come to think of that tree as almost a person, its wrinkled and rutted arms outstretched and shielding them from the elements in the wilderness. Their lives had been stripped of every comfort and all autonomy. Days became hard to remember, and endless afternoons of inaction made them grateful for sundown and the thin veneer of privacy that darkness brought.

They noticed that the young men guarding them would rotate as they went on military operations, then returned days later recounting grisly stories of their bravery on the front. Some of them were frightening, but others were childlike, practicing clumsy martial arts or lifting each other up to build muscle for the war. Under their turbans they wore their hair shaggy and long, cutting it only once a year during the holy festival of Eid or for "spying in Nigeria" when they cropped it in a trendy style and donned colorful T-shirts and canvas shoes to blend in with civilians. Sometimes they would gather in a small group to sing a *nasheed*, one of many

Islamic victory songs chanted a capella in Arabic, with lines like "Nigeria today, tomorrow conquered!" and "Only one side will go to paradise!" The guards talked loudly about which of the school-girls each would marry, calling in their direction, "I see beautiful girls!"

The girls called this being *viewed*: "They viewed us."

These guards were responsible for monitoring the captives' most basic activities, transforming everyday chores into surreal experiments. In the mornings, the militants would escort a small handpicked group of girls to walk for hours to forage for their own food supply. The hostages would climb into the branches of bao-bab trees to pluck its taut green, vitamin-rich kuka leaves and to snap off kindling for a fire. When they returned to the Tamarind, another group designated as cooks would pound the kuka into a powder to be mixed with oil in large cooking pots they recognized from the Chibok school pantry. They lit a fire with the kindling and sent ribbons of smoke rising through the branches and into the sky.

This was a strange reversion to muscle memory for teenagers who had been following this recipe their whole lives. Back home, the students had grown up preparing kuka soup on their parent's wood-fired stoves, part of the chores young women in Chibok were expected to perform, alongside their schoolwork, cleaning, and caring for younger siblings. The ritual of pounding the leaves and the peppery smell from the pot could sometimes, just fleet-ingly, transport Naomi home. She had been boiling up kuka soup since age five under the watchful gaze of Kolo. Now she and the other students had to repeat the steps in the open-air corner of the camp the insurgents had designated as a kitchen. The ingredients were thin, and the soup wasn't as filling as it would have been in Chibok. And it would get thinner as the months wore on.

Malam Ahmed would often join the hostages as they ate, loudly chewing squares of spiced beef from a bowl of noticeably better rations. Naomi saw that the Malam was trying to ingratiate himself with some hostages while scowling at or threatening others. "I know what it's like to be imprisoned," he would say.

At dinner, he would hand out rewards for the girls who had pleased him in Quranic class or by showing some sympathy for the sect. It could be a piece of meat or some fresh fruit. "That is why I like you," he would say, conspicuously handing them a larger portion. It was a tactic to create friction among the classmates. When other hostages would cry or call for their parents, his lips would curl into a smile as he offered them chilling updates from home that Naomi wondered whether to believe. "Chibok is now flying our flag," he said one evening. "Your parents are infidels, and many of them are dead."

BY LATE MAY, THE MOST TERRIFYING MOMENTS OF CAPTIVITY WERE COMing from above. While the girls were learning to endure Malam Ahmed's bullying and the deprivations of thin meals, the sky was filling up with aircraft sent to find them.

Some of the planes would screech into earshot out of nowhere, shaking the ground as they blasted low over the treetops. The sound was nothing they had heard before, like a gigantic piece of paper being ripped from top to bottom, the students would say to one another, their hands still trembling. At the first sign of an incoming aircraft, the hostages were ordered to shelter as close as possible to the trunks of trees, cloaked from the eyes of the jets by the branches the men called a *sitara*, an Arabic word meaning "curtain" or "cover." This was for their own safety, explained the guards, who shared harrowing recollections of the warplanes that fired off rockets before the girls' arrival. The deafening howl of the engines caused acute distress, sending the hostages huddling together, grabbing for their classmates' hands or lying on the ground shaking in panic. One of the older guards assured the girls there was no need to fear, at least, not about the planes they could hear or see.

"If you can hear the plane, it isn't on you," he said. "It is the plane you don't hear that will hit."

The young fighters, the *rijal*, who were supposed to watch the girls, were now spending hours of each day studying the sky, trying to divine a flight schedule that could help them anticipate a strike. They had told the captives to remove any item with a bright color—a blue washbowl, a red plastic bag—all should be buried in the ground. The aircraft could even see the silver glint of a Maggi soup-stock cube, they said. Leaning against the trees, a flat hand shading their eyes, the men watched for aircraft, all of which they had given nicknames. A Hilux made growling, grinding sounds, like a Toyota pickup. Pink Plane came in the morning, and the men were sure it took pictures. Triangle, a drone, could be heard buzzing in the clouds but was almost never seen. The *rijal* had recently become interested in the arrival of a bulkier, beige aircraft that had been seen cruising above. They called it Chibok Girls Airplane, and took pains to hide their captives from it. "Chibok Girls Airplane is looking for you!" they would say when the lumbering craft juddered overhead.

Naomi was frightened by the planes but also intrigued by what their appearance meant. These flights were the best clue the girls had that there might be a rescue mission seeking to free them. One day, a plane dumped leaflets that floated to the ground or got snagged in branches. They were black-and-white fliers with stenciled images of tanks and helicopters. Hannatu Stephen, at sixteen one of the youngest captives, was able to look at one long enough to read what it said: THE GOVERNMENT IS COMING TO RETRIEVE THE GIRLS THAT WERE ABDUCTED ON APRIL 14. She was puzzling over the crude drawings of tanks and guns on the paper before one of the *rijal* snatched it from her. The men built a fire and she watched as letters offering a note of hope turned into the acidic smell of burnt paper, billowing outward.

Malam Ahmed had promised to lash any girl not taking shelter under the *sitara* when the jets flew near, but they wondered if there was some way to signal for help. One day, before anybody could understand what was happening, Rhoda Emmanuel realized a plane was approaching, and sprinted out from the shade of the tree into the open, her arms flailing in the air. She had tossed off her full-length *mayafi* covering and was wearing her blue-and-white school uniform, jumping up and down to catch the attention of the pilot and shouting with all her strength, "Greet our parents! Greet our parents!" Other girls watched in astonishment, then began to stand and chime in, a chorus of teenagers yelling, "Yes! Say hello to our parents!" The guards began to shout and one dashed after Rhoda and dragged her back under the branches. She knew she would be lashed by Malam Ahmed's switch, or worse, beaten with an entire branch cracked off from the Tamarind. A guard screamed a warning in her face: "We are not the ones that will kill you," he said. "It is the airplane that will kill you."

12

The Mothers' March

Kolo Adamu looked in the mirror and watched strands of her gray and black hair fall to the floor, settling on the dust of the porch. Her third daughter, Philomena, was silently working the razor across the top of her head, edging closer to the scalp. The removal of hair was a mourning ritual in Chibok usually reserved for the newly widowed. But Kolo had made herself a vow. She would not grow her hair back until Naomi was home to tie and braid it.

"All the way back to the scalp," she told Philomena. "Like a man."

Two months into her eldest daughter's captivity, Kolo was struggling to hold a conversation or a thought. She could sense her daughter's presence everywhere. Naomi's sandals lay tangled in the pile of shoes on the porch. Her clothes still held her smell. When Kolo ate, she would wonder if her daughter had food and when she went to farm, she found no energy or motivation to tend her crops. At the local clinic, a nurse had taken her blood pressure and warned her it was perilously high. "Whenever I remembered her, I became depressed from heart to stomach," she said. Sometimes, while fetching water from the well, Kolo would stare down and consider throwing herself in.

One day, when she had summoned the energy to go to buy groceries at Chibok's open-air market, she saw her daughter's face staring back at her from a newspaper's front page. The picture showed Naomi leaning forward into the camera with a confident half smile, wearing a yellow dress and a silver pendant, against a blue sky–like backdrop. It was one of the wallet-size photos she had taken for fun at the portrait studio in town with her friends. Kolo had given it to the authorities in the hopes it would serve as a useful clue. Instead, it had become a stock image the papers used to illustrate the latest update or false lead on the Chibok girls. When Kolo caught sight of it, she felt dizzy and nearly dropped her groceries. A bystander had to help her carry them home.

Kolo felt like a "human shadow." To perform daily errands, she would start down the dirt road outside her home, then look up an hour later only to realize she'd forgotten where she was headed. One afternoon, she heard a car driving up, blasting a joyful and cathartic song that seemed familiar as it drew closer. It was the bubbly, syncopated gospel of Mama Agnes, Naomi's favorite singer. "Instantly, tears started pouring from my eyes," Kolo said. Frequent reports and rumors of sightings or rescues were aggravating the pain. Nigerian troops were advancing into the forest, hoping to extract the hostages, Kolo would hear one day. The next, she would hear speculation, repeated through town, that their daughters had been drugged and were headed back to Chibok as Boko Haram bombers. She had begun to have nightmares of her daughter swaddled in a suicide belt.

THE IDEA TO STAGE A MARCH CAME FROM SARATU'S MOTHER, RIFKATU Ayuba. She and Kolo had sat listening to the same radio in May, the two mothers sharing anxious glances as the BBC reeled off the names of celebrities blasting their children's plight into hundreds of millions of smartphones around the world, before the British and US militaries deployed spy planes and drones into the skies above them. It was a surreal twist. The mothers had heard fathers and uncles enthusiastically placing their hopes in the idea of American special forces rescuing their daughters, or in a prediction of the preacher of a Nigerian megachurch, T. B. Joshua, who had prophesied to his congregants that the girls would be released. The rush of online attention had left Kolo with conflicting emotions. She and Rifkatu agreed it would pile up pressure on the Nigerian state. Yet Kolo had felt a pang of fear over how a messy international movement on a website she'd never heard of might alter the calculations of Abubakar Shekau. They worried the fame could increase the duration or hardship of their daughters' captivity, and worst of all, it was out of their hands.

Yet by June, the wave of global activism had begun to recede into other concerns, and the pair of moms wanted action. Oby Ezekwesili was still leading daily marches to the Abuja highway median, but the number of mentions of the hashtag had fallen precipitously to just a few dozen pings a day. #BringBackOurGirls had been replaced as a symbol of celebrity-backed global solidarity by the #IceBucketChallenge, a campaign to generate awareness for motor neuron disease. The coverage on Kolo's radio had shifted from the Chibok girls to other battles and attacks by Boko Haram. By month two, Kolo and Rifkatu had decided that if they were going to keep pressure on the government, they would have to do it themselves.

Together they had recruited Mary Ishaya, whose daughter Hauwa had been abducted, who then recruited three more neighbors, all of them mothers of missing daughters. Struck by guilt and unable to think of another way to confront the pain, they had gathered in each other's homes and charted a march route they intended to take. "We want the world to know that the abduction is real," Kolo said. The district head of Chibok, approached by some of the mothers, had declined to approve this idea. "Let's just do it anyway," Rifkatu had proposed. "The district head won't stop us."

The women walked alone, six mothers, each of them dressed in black, along a dirt road to the outskirts of town. They carried their daughters' school uniforms and charred textbooks, halting at the walls of Chibok Local Government Secretariat, a single-story office whose zinc roof shimmered under the baking sun. None of the officials inside stepped out to meet them. The only witnesses were local women occasionally pedaling by on shiny bicycles, turning to gaze at the small throng of mothers pumping their fists.

"Bring back our girls!" the women chanted, their voices carrying over the walls of the secretariat. "Bring back our girls!"

Their cries were a shout into the void. Not a single journalist, foreign or local, covered the march. There was not a single mention on the Internet. Just a small band of mothers marching through the broiling heat of the rainy season, from the government office to the torched remains of the Chibok Government Secondary School for Girls, where they faced the torched shells of the hostels in which their daughters once slept. There they chanted in tears, "Bring back our girls!" "Bring back our girls!"

By sundown, Kolo and Rifkatu had returned to Kolo's neighborhood. The mothers sat in a courtyard and discussed the march. They hadn't attracted much attention beyond the well-wishes and cries of solidarity from local residents, but they had made their voices heard, and Kolo felt that the action had strengthened their resolve. Their daughters would be proud of them if they knew, she thought. The mothers decided to protest again, and in the days and weeks after the march Kolo, Rifkatu, and dozens of other relatives began to fast, hoping their hunger would create a positive spiritual energy. They would gather under the tin roof of the local branch of the Church of the Brethren, pacing among plastic chairs, their voices rolling over each other in a plea to the heavens. "Free at least one girl," Rifkatu would ask God, "so that we can get to know the situation of our missing daughters."

"God should touch the heart of the Boko Haram terrorists to release our children," Kolo would pray. "Make their captors repent."

WERE THE SCHOOLGIRLS ALIVE? FOR KOLO TO BELIEVE SO MEANT IMAGIN-ing their continued suffering in a forest that the parents of the town were powerless to enter. Elsewhere in Chibok and surrounding villages, some had found it easier to presume their daughters had died. Hope could be cruel. The kidnapping had not only robbed them of their daughters but left them tortured by the uncertainty. Local families would meet in church halls and homes for emotionally charged gatherings where the pain and anger often found a voice. At one such meeting, a small group of desperate fathers concluded that they should take a request to the Nigerian Air Force: "Please bomb the forest where our daughters are held." At least that might deliver a measure of justice to the men who did this. "Even my own people have given up," a leader of the parent's group said one afternoon. "People are saying, 'Bring back the bodies so that we can bury our dead.'"

In helplessness, other parents had begun attempting to sustain some thin connection to their children. One of Kolo's friends, Yana Galang, had started washing and folding her daughter Rifkatu's clothes every month, then laying them out neatly on her bed. She wanted them to be ready for her when she came home. Birthdays were rolling around, and when Lydia John's eighteenth arrived, her parents held a small and somber gathering. A few families were trying to get pregnant, hoping that a baby might bring some joy into a heartbroken home. But like many parents who give birth after losing a child, they would eventually find that the arrival of a newborn didn't resolve their grief, but complicated it.

The families' psychological traumas were mutating into physical problems that left the families unable to work, and many grew poorer. Stress-related illnesses, like stomach ulcers, hypertension, and heart problems, were becoming common. Crops went unharvested. After the kidnapping, one mother stopped eating, fell into a feverlike stupor, and after two months, discovered she could no longer lift her left arm. Some of the families fared worse still. Within three months, it was certain that eleven of the parents would never see their children again: they had died. One father kept repeating the names of his two missing daughters on his deathbed until he passed.

13

"You Must Write One Too"

Under the Tamarind's branches, hidden from the prying eyes of the aircraft circling above, the Chibok girls were laboriously copying down Quranic verses from the blackboard to the sound of Malam Ahmed's instructions. One day, before class, the Malam had asked his teaching assistant to hand each of the girls a pen and a small floppy booklet of thin lined pages. He had told them that memorizing his lessons would help save "daughters of infidels" from the *arna*, the pagans. "There will be tests!" he had warned. Naomi's exercise book wasn't like the ones she had used in Chibok. It was bright yellow, forty pages stapled into a glossy plastic cover printed with an image of Disney's Cinderella standing in front of a horse and carriage. Naomi filled out the box at the top right-hand corner, apostrophe included.

Name: Na'omi Adamu.
School: Sambisa and the Chibok.
Subject: Islamic.

As the daily Islamiyah classes ground into their third month, Naomi had found herself sitting next to a girl she had only vaguely known from the overcrowded Gana Hostel, one the school's star pupils. Lydia John had been such a diligent student that she would fall asleep with a book in her lap. At home, her mother, Rebecca, would check on her near midnight to turn off the lamp and close what was often a Bible. Lydia had arrived in Chibok at the start of her senior year, fleeing her hometown of Banki, near the border with Cameroon, after an attack by Boko Haram, and in her new school, she immediately rose to the top of the class. The new girl spoke the best English at the school and wrote in bold, loopy cursive.

At CGSS, Naomi and Lydia had moved in different circles. Naomi hung with her cousins and the girls from her neighborhood she had known since childhood, while Lydia sang in the school choir, the Glorious Singers. While Naomi had struggled with health problems through high school, Lydia breezed through, absorbing the required reading from the second half of senior year, which included *A Woman in Her Prime*, by Ghanaian novelist Asare Konadu, and *Lord of the Flies*, by Britain's William Golding. When she graduated, Lydia felt certain she'd win a place at a university and had already begun to dress the part, wearing Western skirts she found more fashionable than the local wrappers her mother bought. She was tall and elegant and seemed to wear her clothes more like a girl from the city than a high-school senior from small-town Chibok. When she was a child, a bout of measles had left her with an infected left eye that drifted to one side and had made her feel self-conscious. When she laughed, she would cup a hand over her mouth. But by senior year, she had been thinking of herself in new ways. Visiting the local photo studio, she leaned toward the camera with a self-assured smile, a hand on her knee, in a white V-neck dress with red pendant earrings. The backdrop she chose was a mansion with plush furniture and golden curtains. This was her future.

But now, in captivity, Lydia was finding it harder than most of the other hostages. She was struggling to sleep, missing meals, and seemed to be lost in the haze of her own fears. Some of the students had found ways to occupy their minds. They braided their hair using a pin they'd carved with a razor blade from a piece of wood, or they fashioned thorn-studded sticks to pull the seams of their *mayafi*s and readjust them so they didn't drag on the dirt or trip them as they walked. Lydia had been spending a lot of time alone. When she spoke, she talked about escaping, whispering about elaborate plans she was concocting in her mind. Most of her school friends were just trying to hold on. "You're thinking too much," several told her.

Lydia began to confide in Naomi during class and under the Tamarind, leaning into the older classmate for moral support. The two talked about music, church, and the gospel singer Mama Agnes. Lydia told Naomi that before she moved to Chibok, she had been a prayer warrior at her church, one of the evangelicals who banished evil spirits from the community. And Lydia had the talent for writing that Naomi had always wanted for herself.

One day after lessons, when they were back behind the fence under the Tamarind, Lydia said she wanted to tell Naomi a secret. She checked to be sure nobody was looking, then carefully opened the cover of her blue exercise book and began to slowly flick through the pages. Naomi watched as the mostly blanks toward the front gave way to pages covered in wide-looped handwritten English. She pulled the notebook from Lydia's hands and began to read, her eyes following each curl of the hidden script. It was a written record of their ordeal, starting from the night of April 14. Lydia was writing a diary.

"You must write one too," she said to Naomi.

It began:

First of all on 14th April 2014 on Monday. 11 oclock. We heard the sound of gun and then some started phoning their parents, brothers, sisters, uncles. And we started praying.

Before they enter into the hostel, that moment we heard the sound of the first Gun.

If we ran they are going to shoot us. So we get out of the hostel gradually. As we left the hostel quarters, the principal's house and the rest is burning. So we went behind the staff office. Then they collect our clothes our torch light and our dresses.

Then they burn the staff office. Then one of them said that we should sheeft a little out behind the office, then one of them said no that we should stay there. Then the other one said no they should sheeft out of the hot side. he said don't you know how the heat is Incrising. then we sheeft a little under one tree.

they said that anyone who deside to run they will shoot her. so we went under a big tree. As we sit down there then they said that we should pray our final prayer.

then they asked us have we finish praying then we answered them yes. Then they said since you have finish praying we should lie down. As we lie down many of us are saying lie on me, lie on my head.

One girl picked up her shoe and put it on her chest and said even if they shoot her the bullet will not enter.

Then the BH told us to get up and enter into the car.

Soon after, a small club of secret diarists began huddling together in the afternoon shade. They shared paper and pens and ideas about how to tell their story. Naomi would sometimes write and sometimes dictate her reflections to Lydia, who wrote them in her rounded handwriting on the thin, blue-lined notebook pages, hidden under pages of halfhearted notes they took from

Malam Ahmed's class. Each diary described the same events from
a slightly different perspective that evolved as the schoolmates
read one another's work.

At first they wrote about the kidnapping, disjointed and fragmented recollections from their journey into the forest and the
terror of their first weeks in captivity. As weeks dragged on, however, other material went into the pages. They copied passages
from a small Bible one of the teenagers had stashed in her clothing on the night of their abduction. One schoolmate would ask
for permission to go to the bathroom so she could find a bush
to crouch behind and copy the verses. She would return with a
shorthand-replicated section. A popular choice was Luke, chapter
2, verses 1 to 20, the story of Mary giving birth. Several of the
girls memorized it as a way to pass the time. They copied the Book
of Job, the biblical account of a man tested with endless suffering
who refused to renounce his faith. Sometimes they wrote down
psalms:

Oh my God I keep calling by day
and You do not answer.
And by night.
and there is no silence on my part.

They took turns writing down family phone numbers, or love
letters to boys they liked back home. Sometimes, Naomi wrote
notes to Andrew, other times to her other crush, Baba. "heart to
heart my only lover my only darling," she wrote Andrew. "I love
you like [our] government loves America."

The hidden diaries gave them a private space where they could
be themselves again. It helped them keep track of time and bond
over their shared experience while storing and sorting through
traumatic memories. They had discovered a survival mechanism
many prisoners of war would recognize. Nelson Mandela wrote
his autobiography at night on scraps of paper he had buried in

a vegetable garden by day, and dozens of Holocaust victims kept similar diaries. Like other captives, the teenagers were keeping a record of injustice that they believed would ultimately come to light. "We were hoping that we would eventually be released," Naomi said. "We wanted the world to see what we saw."

Amid the bleak expanse of captivity, the diaries could briefly transport them back to Chibok, a portal to their family kitchen, the Gana dorm, or the schoolyard where they used to chatter and sing the latest Afrobeat hits under the fragrant shade of the guava tree. One day Naomi and Lydia wrote down the lyrics to "Shake," a pulsating Nigerian dance-floor hit that had been on heavy rotation in the months before their abduction. They laughed as they read the verses and chorus aloud, bobbing their heads in a silent rhythm.

> *Mr Flavour on the Microphon*
> *baby you are so fine*
> *baby you are so fresh*
> *muh baby, muh baby, muh oh oh oh baby*

THE DIARIES WERE A SUBTLE FORM OF REBELLION THAT WOULD SOON record more daring acts of mutiny. One morning at the beginning of Ramadan, before the dawn prayers, Naomi and her classmates noticed Malam Ahmed speaking animatedly with his guards, several of whom hurried away and boarded motorbikes that screeched into the bush. The Malam's daily head count had come up short. A pair of girls had made a run for it during the night, disappearing without informing their classmates. The news sent ripples of hushed excitement around the Tamarind. Naomi hadn't heard anyone leave. The hostages were looking around, asking who had gone? It was Patience and Tabitha. The whole group seemed to be considering the same thoughts as tidbits of information traveled through the crowd.

It was time for prayers, and the Malam organized them into

rows. Naomi bowed her head in front, but as she pressed her face to the floor she was thinking about her classmates and prayed they could make it out. The Malam had always warned the girls that mounting an escape was futile. Boko Haram had spotters and lookouts posted all around their camp, and even if they managed to make it to the edge of the forest, the military would shoot them. As the hours went by, Naomi could hear her classmates debating whether to run, speaking in Kibaku so their guards wouldn't understand. One small group of hostages, ordered to cook a pot of spaghetti, were discussing their options, a few footsteps away from the oblivious fighters lounging nearby. Would they survive? What would happen to them if they were caught?

One of the students seemed ready to flee. "Me, I am ready. I am going home. I will get some girls and we will plan," she said. Others were too afraid. "I wish you good luck," responded another. "Maybe God will help you."

The plotting ended with a commotion of gunshots fired into the air and screams announcing that the two breakaways had returned. Patience and Tabitha's hands were tied with rope, and they were shoved forward by an escort of more than a dozen guards. The girls' faces and garments were caked in dirt. Malam Ahmed called for all the hostages to gather and watch the runaways' punishment. He ordered the girls to kneel and then gestured to another tall, scowling militant who approached them carrying a large tamarind branch.

His name was Abu Walad, and he brought the gnarled branch down hard, across their shoulders and their back. They screamed and buckled, and yelled at him to stop. "I'm sorry Malam, I'm sorry." Naomi and the hostages watched, wincing, some sobbing or looking at the ground. "Do not look away!" Malam Ahmed said. Abu Walad beat them twenty times each, counting the blows, their bodies splayed out on the ground.

Another guard arrived with an assault rifle. He hauled the escapees back onto their knees and placed the barrel next to one of their heads, and screamed: "Let me open your ears." Then he fired it into the air.

"You will listen," he said.

Malam Ahmed addressed the crowd to issue a final warning. "Anyone who tries to flee will be beheaded." When the guards had left, the two young women explained how they had hiked through the forest in the darkness. Then by sunrise, they'd made it into the open countryside, hiding behind trees or taking cover behind tall grasses at any sign of figures on the horizon who could be militants. They had asked a group of villagers for help as they approached a small settlement hours away. But these men were loyal to Boko Haram. Worse, they recognized the girls from a YouTube video, released in May, in which they'd pledged to follow Islam, a clip that had been played on loop on Nigerian television and social media for weeks. The two escapees had been wearing the same distinctive dark-colored coverings. The fame meant to free them had also made it harder for them to escape.

Naomi and Lydia briefly recorded the episode in the diary:

There is another day that some girls ran they try to escape but they couldn't . . . They tied her two hands and we are begging them, we are crying. Then they said If we ran again they are going to cut off our neck . . . Malam said we should better stay and pray . . . then he said we should be careful.

In the forest, the Malam said, "Boko Haram is everywhere."

14

Tree of Life

CIA STATION, ABUJA
MID-JULY 2014, THREE MONTHS INTO CAPTIVITY

Three months after the kidnapping, a photo landed on a bank of monitors in the CIA station in Abuja, a group of sealed drop-ceiling rooms concealed behind the blastproof walls of the US embassy. It was the spy agency's best clue in months as to the location of the Chibok girls, a picture forwarded over an encrypted channel from the airbase in Ramstein, Germany, where drone and satellite footage had been collected. On the screen glimmered a full-color image of about eighty young women in full-length *mayafi*s milling about in the sun, just far enough beyond the branches of a gigantic craggy tree to be visible to the high-definition cameras peering down at them. They were stationed near a handful of men who appeared to be their guards. Pickup trucks were parked near a small mudbrick structure with a shiny tin roof.

For months, American intelligence officers had been staring into computer monitors at the rolling landscape beaming in from Nigeria's northeast, of vacant desertlike terrain and ordinary villages, hoping to find some hint of the schoolgirls. The CIA had kept two agents on payroll in the region, locals the agency considered reliable but who had proved unable to report back any indications of the students' whereabouts. The British had trained an informant close to Boko Haram at England's elite Royal Military

Academy Sandhurst, but then Nigeria's military had arrested him and he'd since vanished. All told, analysts had received nothing as promising as this lead now glowing on the screen. The image traveled over subterranean cabling and eventually landed in the Pentagon, where officials flagged it up the chain of command. It was unusual to see so many young women in the open, military analysts agreed. One of the officials who reviewed the photo said it was clear the women were hostages. "We're assuming it's not a rock band of hippies out there camping," he said. The photo quickly worked its way through the US national security architecture all the way to President Barack Obama's morning briefing. In time, it would gain a nickname: the Tree of Life.

OUTSIDE THE CIA STATION, THE AMERICAN RESCUE EFFORT IN NIGERIA had been stuttering in the months since the president had ordered that "everything be done" to bring the Chibok girls home. US drones had been flying missions over a vast area of the Sambisa Forest and beyond to the borders of Cameroon, Chad, and Niger, but on the ground below, Boko Haram was continuing its advance. Every day seemed to bring grisly new headlines of a murderous attack on a military base, the bombing of a church or mosque, or the sacking of a village, followed by the summary execution of fighting-age men and the forced conscription of their sons. More young women had gone missing.

The death toll was breaching new thresholds in the now five-year-old war. More than fourteen thousand people were known to be dead, and about half a million were on the move, fleeing the villages and towns for the safety of more heavily defended cities. Army convoys were so routinely ambushed in the countryside that soldiers had begun to retreat, leaving villages and smaller towns to the insurgents. By July, the militants were encircling the city of Gwoza, one of the largest in Borno.

As the insurgency expanded, its fighters swept up hundreds more hostages, sending fresh reports into Abuja of boys who had been forced into combat and of girls forced into slavery. Almost none of those abductions were noticed beyond Nigeria. In May, while some of the most famous people in the world were tweeting about Chibok, the militants had seized six villages in the nearby mountains and rounded up children from each, with virtually no coverage outside the Nigerian press. In June, the group had raided a village only a few hours from the school and kidnapped another sixty women and girls, along with thirty-one boys. By the end of the year, after the sect had grabbed three hundred more boys from an elementary school, the Chibok abduction was no longer the conflict's largest kidnapping.

Nevertheless, it would remain Nigeria's most famous abduction, the one set of victims that the American Interdisciplinary Assistance Team (IDAT) deployed by the White House in early May had been ordered to rescue. The thirty-eight officials of the Assistance Team included two top FBI hostage specialists, military intelligence analysts, and psychosocial therapists. The crew had arrived in Abuja with their own office-support clerk, who set up a front desk in a spare area of the US embassy.

Their lodgings would be in the 667-room Hilton, the sprawling hotel where Goodluck Jonathan had held his ill-fated World Economic Forum months earlier. By July, it had filled up with other guests coming from around the world to help with Chibok. On July 13, Malala Yousafzai spent her seventeenth birthday in the hotel, hosting a charity dinner in aid of the missing schoolgirls. Washington lobbyists stayed in a tenth-floor executive suite hired by Nigeria's government on a $1.2 million contract to "change the international and local media narrative." Around the same time, on the same floor, in rooms booked by the government, four men who claimed to be senior Boko Haram members were holding hostage talks that went nowhere.

On another floor was an Anglican priest who was friends with the archbishop of Canterbury and had flown himself in from western Australia bringing screenshots from Google Earth that he believed showed Boko Haram camps where the girls might be kept. President Jonathan had given the priest unlimited dollar funds and the use of a military helicopter, whereupon he vanished into the countryside, then returned with sordid but inaccurate stories about their whereabouts and treatment. The government had also given money to his partner, a mysterious Maiduguri woman who called herself Mama Boko Haram and boasted of having once cooked thick, oily plates of *egusi*, a protein-rich stew thickened with ground seeds, for Abubakar Shekau himself. The administration was throwing money at the problem, and the Hilton was filling up with con men, foreign and local, who wanted to catch it. Meanwhile, on the highway median across the street, Oby Ezekwesili gathered with protesters every day at 4:00 p.m. "What do we want?" her fist-pumping supporters chanted. "Bring back our girls! Now and alive!"

The American Assistance Team left the hotel regularly to meet at the US ambassador's gated hilltop residence for tea with delegations from the governments of the UK, Japan, Australia, Canada, and France. Even Scotland Yard had an official present at the table, where diplomats and military personnel shared magnifying glasses to study printouts of American drone footage. Senior diplomats from Western allies were convinced they could retrieve the girls, and early on they had assured each other that at a minimum, they would rescue at least *some* of them.

And yet the Tree of Life photo was not just a snapshot of the hostages, but a window into the mission's ethical complexities. The ambassador's table debated whether they could trust the Nigerians with the photo. The British ambassador urged caution, warning that the air force "would simply bomb the baobab tree." Their host country was fighting a brutal war, and considered defeating Shekau a bigger priority than rescuing the schoolgirls that the warlord had turned into human shields. Western allies, on the other hand, were responding to public pressure to rescue one particular group from the hundreds and hundreds of children kidnapped by Boko Haram, a victory that might not change the tide, but would provide a feel-good moment for a superpower with little to show for its trillion-dollar counterterrorism adventures.

The hashtag was refracting the war through two different and incompatible lenses. Western news articles were identifying Boko Haram principally as the group that had kidnapped 276 school-girls, not as the killers of thousands of Nigerians. An insurgency that opposed *Western* education was understood by Americans as opposing *girls'* education. The abduction of the Chibok girls had originated as an example of what is called the identifiable-victim effect, a phenomenon wherein a single incident allows people to care about a larger tragedy, much as Anne Frank helps young readers connect to the Holocaust. But somewhere in the social-media storm, the dynamic had inverted, and to Western public opinion, the missing schoolgirls had become more important than the war.

All through July, cities as big as New York and as small as Bradford, England, held candlelight vigils, and in Washington congresspeople dialed the State Department demanding updates. Michelle Obama's chief of staff had continued to ask about the Chibok girls in the president's morning security briefing. The military was under pressure from the Senate to review its options. Senator John McCain, himself once held hostage in Hanoi, had been privately lobbying the Pentagon for special forces to rescue them. When the US ambassador went home for his son's gradua-tion, he overheard a distraught young woman in a Chicago sand-wich shop complaining to her parents: "These girls are missing and nobody is doing anything because they're Africans."

THE NIGERIANS WHO SAT AROUND THE LONG TABLE AT THE NEWLY formed Fusion Center, a sealed conference room for sharing infor-mation at Abuja's Office of the National Security Adviser, knew that the US wasn't trusting them with the full intelligence pic-ture. Americans from the Assistance Team would arrive with new leads, but they needed clearance from Washington to share any significant piece of intel, a process that took up to seventy-two hours. The photos the Americans brought in were lo-fi, grainy printouts that the Nigerians knew had been purposely degraded, and they felt condescended to.

The Americans were withholding information because they didn't entirely trust the Nigerians and because they didn't want to reveal the extent of the imagery, phone calls, and private messages their drones could capture. The distrust stretched beyond the drone footage. Intelligence officers posted to the CIA station in Abuja were listening to intercepted calls in Nigerian languages they couldn't translate or in some cases identify, but they didn't share these with their hosts. The US had hacked into e-mails, texts, and phone calls from Nigeria's military that included incriminating details on the number of prisoners who'd died in Maiduguri's Giwa Barracks. At one point, a US drone took high-resolution photos of a Nigerian military base, and an official bringing the latest stack of photos to the Nigerian government forgot to yank the printout from the pile until the last minute.

The spying went both ways. Assistance Team members would return to the Hilton to find their room "tossed," their belongings conspicuously rearranged, a warning that they took to mean they were being watched. Eventually, arguments erupted at the two tables where the allies were meant to coordinate their rescue efforts. The Nigerians felt the Americans were too prescriptive, as if Africa's most populous nation was subservient to the US. "Obama is not our president! You're not in Washington now!" one had shouted across the Fusion Center table. Privately, Assistance Team members griped that the Nigerians didn't seem interested in resolving the kidnapping.

Other international efforts were also unraveling. The British Sentinel R1 spy plane sent to scour the search zone broke down and had to return to a base in Senegal. French intelligence struggled to confirm London's suspicion that the schoolgirls had been trafficked over the border with Cameroon, or possibly Niger. No Nigerian officials ever recalled seeing the satellite images China promised to send. After a few weeks in the Hilton, the FBI hostage specialists flew home, and their teammates began to feel unwelcome in the cramped embassy back office where they were working long days writing briefing papers and proposals that the Nigerian government didn't act upon. The Nigerians were losing territory, soldiers were going AWOL, and many of their helicopters had been destroyed. Some simply hoped the Americans would solve the problem. Eventually, Goodluck Jonathan himself asked John Kerry to send combat troops.

The American and British discussions about a possible special-forces rescue never gained traction. It was hard enough under the best conditions to deploy helicopters to such a remote area and free two hundred hostages simultaneously, but Nigeria also lacked the infrastructure for the mission. The government had spent hundreds of millions of dollars building airports, but in the morass of corruption, key radars barely functioned and runways lacked lights, meaning that any operation would have to take place by day or from another country. Moving enough aircraft or armored cars into the area risked being noticed by a bystander or someone breaking the news on Twitter. In other words, the intense global focus on whether the girls would be rescued was part of why they couldn't be. By August, the Assistance Team that had ridden into Abuja on a wave of optimism was struggling under a cloud of impotence. Nursing sundown drinks at the Hilton, they were philosophical about the limitations of American power. "Everybody thinks, oh you're the United States, you can just fly some troopers in there and go in and rescue them," one said. "It's not that simple."

In August, officers in the CIA's Abuja station noticed that the women under the Tree of Life had been moved. There was nobody there, and the trail had gone cold. All told, US drones spotted clusters of young women in the search zone about five times, but no group as large as those under the Tree of Life. It was never certain that the individuals in the photos were the Chibok girls, and by August, the drones circling Lake Chad were redeployed over Iraq and Syria, where the Islamic State was holding Americans hostage. By the end of the year, the only members of the Assistance Team left were a handful of military intelligence officers, staring into a bank of monitors, studying the occasional picture beamed in from a satellite.

15

The Senator's House

Naomi stepped through a high-framed doorway and into the hallway of an enormous white mansion, surrounded by towering walls, and entered through a black iron gate. Dozens of her classmates were already inside, exploring the cavernous interior of an eight-bedroom villa, their flip-flops smacking against a cool tiled floor that reflected the light from a vaulted ceiling. Around her, confused students were anxiously peeking into spacious bedrooms emptied of furniture. Only one of the rooms had a bed, and there was a master bathroom where a woman's underwear hung near a sleek showerhead. Fanning out, the girls toyed with light switches and opened bathroom faucets to watch tap water pour into the sink. In the kitchen, a toaster and an electric water kettle came to life at the flick of a button. The refrigerator hummed with chilled air. Behind them, armed guards had filtered into the house and were also excitedly inspecting the rooms. More were outside in the courtyard and stationed beyond the gate. This vast empty property, far larger than any house in Chibok, was to be the captives' new prison. It was the first time they would sleep with a roof over their heads since the night of April 14.

Naomi had no idea where they were or why they'd been brought to this place. Other hostages were posing the same question to the guards making themselves at home in the living room. "Where did Boko Haram get this house?" asked Mary Dauda, one of the young women who had joined Naomi's secret club of diarists. "A nice, fine house?" When she asked one of the guards whose home it was, he conferred with the others, then replied, "We don't know."

In the first week of August, almost four months after their kidnapping, the Chibok students had again been spirited away by their captors. Roused before dawn under the Tamarind, they had been ordered to walk for hours, trekking from the camp called Timbuktu through the tangled underbrush until they reached a group of large flatbed trucks that were normally used to transport cattle. The guards had ordered them onto the vehicles but said nothing about where they were headed. The trucks had barreled through the forest until the foliage got thinner and they reached a dirt road that eventually gave way to blacktop. As night fell, they saw fluorescent-lit shops and the white lettering on Nigeria's green road signs alongside a highway, and some of the girls had wondered aloud to each other whether they might be going home. Many of the students had already drifted to sleep when the convoy began to weave past the shopping plazas and cement homes along a central boulevard of what appeared to be a sprawling city. The trucks halted by the high walls surrounding an enormous villa. "Enter!" a guard ordered, and the iron gate creaked open to reveal a palatial white mansion.

The hostages had not all arrived at once. Boko Haram had moved them in batches, probably to avoid being seen by search planes, Naomi thought. The first group had been scared to sleep in the mansion's vast empty halls. The lights turned on and off by themselves, they complained, each making a clicking noise when it happened, sparking panic. The students decided that they would sleep in the garden. The house might look like a palace, but it was a "ghost house," they had told Naomi when she arrived.

When the guards couldn't cajole the captives to sleep inside, they had accused them of ingratitude and beat them with heavy rods until they ran through the doorway screaming. The teenagers still had no idea where they were.

About a week into their house arrest, a militant walked in with an answer to the girls' questions about who owned the building and all the electronics, furniture, and clothing inside it. "Don't you know?" he asked. "This is Senator Ndume's house." Naomi and several of the classmates had never heard of Mohammed Ali Ndume, a businessman and veteran politician who had gone to college in Ohio, represented Chibok's district in Nigeria's legislature, and risen to the rank of majority leader. Ndume had fled Gwoza before Boko Haram had captured the city and stolen his home. Most of the students didn't recognize his name and had no way of knowing that they had ended up imprisoned in the residence of one of the country's most prominent politicians. The guard shrugged off their questions with a dismissive and befuddling remark. "He is your father," he said. "You know him."

THE CHIBOK STUDENTS HAD BEEN TAKEN INTO WHAT HAD BECOME BOKO Haram's new capital and the heart of Abubakar Shekau's state-building experiment. They were in the city of Gwoza, a centuries-old grid of low-slung homes on neem tree–shaded dirt roads centered around the ornate palace of the emir, the traditional ruler whose family had governed the town since the 1980s. Located on the northern side of the Sambisa Forest close to the border with Cameroon, the city was a strategic prize. Its suburbs included the training academy for Nigeria's riot police, and all around it stood the terrace-farmed mountains known as the Gwoza Hills. Boko Haram's members had been hiding in these hills for years. Some of the girls' parents had sold their crops in Gwoza—the hostages were only sixty miles from Chibok.

Boko Haram had stormed into the city two weeks earlier in a hundreds-strong column of gunmen on motorbikes swarming around trucks and armored cars they'd looted from the army. Within hours the 350 soldiers guarding the city had fled and the streets around the emir's palace were littered with hundreds of bodies. Terrified families had gathered whatever belongings they could carry and scurried into the hills. Panicking policemen had thrown off their uniforms and joined the retreat. The emir's brother, trying to escape in his flowing traditional white-and-red robes, was shot dead as he ran.

The residents, scampering up the mountainside to hide, had expected Boko Haram to loot the city and then withdraw into the forest, but in the days after the attack it became clear the sect had a detailed plan for their town. They watched as an endless stream of trucks, cars, and motor scooters ferrying thousands of followers poured into deserted city streets. Boko Haram military units began to patrol neighborhoods that Nigerian security forces had abandoned, sweeping from house to house to arrest and kill government loyalists. Fighters occupied the emir's palace and decorated it with their own black-and-white flags. The militants' wives, with children in tow, arrived on trucks and began to select which of the prime houses would become their new homes.

Insurgents fanned out into the hills to hunt civilians they found taking shelter. Residents who remained in the city were required to swear allegiance to the group and pray in mosques now graffitied with battle slogans and sketches of Kalashnikovs and tanks. The *hisba*, the religious police, punished sinners with public amputations, stonings, and beheadings. Commands were blasted from the speakers of the mosques' minarets: Gwoza was no longer part of Nigeria, a godless nation, but was now the seat of a new society built on the "true religion." Women would be fully veiled at all times, and men's trousers would be hemmed above the ankle, the fashion worn during the time of Prophet Mohammad. Men could not walk with any woman who wasn't a relative. All subjects had to accept that the sun rises because God wills it and not because the Earth is spherical. Anyone carrying a government ID would be killed. Refugees who escaped the city would later recall how local women were forced at gunpoint to dig shallow graves for those who'd fallen afoul of these orders. Others were taken to the city's abattoir and slaughtered there.

Three months earlier, another jihadist group, this one calling itself the Islamic State, had swept across Iraq and Syria and named the Syrian city of Raqqa as the capital of a new caliphate. Its leader, Abu Bakr al-Baghdadi, had called on Muslims throughout the world to "pledge allegiance." To pay his respect to the new caliph, now the most wanted terrorist on Earth, Shekau dispatched an emissary with the first in a series of handwritten letters that pledged his *bay'ah*, oath of allegiance. In time, al-Baghdadi would accept his pledge. The Iraqi terrorist group had admired what Boko Haram had accomplished in Nigeria and had applauded how Shekau used the kidnapping of schoolgirls to jump from obscurity to global infamy. On the same day that Boko Haram captured Gwoza, Islamic State fighters stormed into the Iraqi city of Sinjar and began to kidnap thousands of women of the Yazidi religion. Later, Islamic State's *Dabiq* magazine would say the idea to mass-kidnap women came from Chibok.

The fall of Gwoza was a major turning point in the war. Months earlier, Boko Haram's fighters had been living in thatched-roof homes in hidden forest encampments, under the branches of trees. But as the military melted away, it surrendered most of Borno State to the insurgency, which was now moving into concrete houses in large towns and small cities. By the end of the year, Boko Haram controlled some twelve thousand square miles of Nigeria's northeast, a territory the size of Belgium. Nearly everywhere the militants went, they kidnapped children. Some communities fled to Maiduguri for refuge, others were abducted at gunpoint or merely swallowed into the insurgency's expanding borders.

On the eve of the Chibok kidnapping, the estimates of the number of minors abducted by Boko Haram had lay in the low hundreds. By August, as the group captured Gwoza, more than two thousand girls and ten thousand boys had been reported missing. Kidnapping became a self-perpetuating cycle as boys forced into combat swarmed new towns and dragged others into their ranks. The boys were useful to the insurgency. They could be deployed as front-line fighters, sent into battle after being plied with opioids like tramadol, tactics that echoed the cocaine-fueled child soldiering in Liberia or Sierra Leone a little more than a decade earlier. The young unmarried women represented a financial and theological problem, and as Boko Haram's camps filled up with girls, reports of an epidemic of sexual abuse began to filter out of Nigeria's northeast.

ALIYU, THE MILITANT LEADER WHO HAD FIRST ADDRESSED THE GIRLS IN their earliest days of captivity, threatening to decapitate unbelievers, was now seated on a chair in the living room of the senator's mansion, giving a long speech about the evils of government schools, which he claimed taught "lesbianism," a word he said in English with emphatic disgust. Malam Ahmed stood next to him, encouraging the Chibok hostages to listen closely. Aliyu was working himself into another rage, arguing that Nigeria's government kept women in class during years when they could bear children. As he concluded, he declared that some of the girls were already too old to remain single. It was time for these women to be married.

If they agreed, they would be rewarded with a luxurious home of their own and all the meat they could eat, but if they didn't, he warned, they would be among the first of the Chibok girls to become "slaves." Soon the hostages would all be sorted into married women and slaves, Aliyu explained. The slaves would fetch water, cook, and clean for Boko Haram's wives. They would carry sacks of rice, unload trucks, or work as midwives. Some of them would work for Aliyu himself, in the mansion he occupied nearby. This would be degrading work, stressed Malam Ahmed, adding that the first to be assigned would be the oldest of the Chibok hostages.

It was sinful in the eyes of God for a woman in her twenties to remain single when she could start a family, he said. When Aliyu had left, the Malam asked a small group of teenagers in the living room to bring Naomi Adamu. He wanted to speak to her on the veranda.

WHILE THE CHIBOK STUDENTS WERE SEQUESTERED IN THE SENATOR'S mansion, fragmented stories of mass rapes and conversions underway in Boko Haram's territory had been reaching the rest of Nigeria. In August, the scope and scale of the abuse was still unclear to the wider world, but inside the group's caliphate, hundreds, perhaps thousands, of women had been forced to make a choice. They could be slaves, or they could enter what the sect called marriage. This "marriage" was a form of rape and sexual slavery, but the men who perpetrated it referred to themselves as husbands and called their victims wives. They saw their arrangement as legitimate and held weddings in mosques, with catered meals and fired guns to celebrate the vows.

The system had a displacement of responsibility at its core. Rape was impossible within marriage, the group's theology held, and thus any woman who "agreed" to marry could not be the victim of sexual abuse. If they didn't accept, then they were "slaves," and could be denied food, beaten, and punished with hard labor until they relented. In some cases, women were locked into small spaces until they capitulated. And once a woman gave in, she consented to be the property of her husband, who could never be deemed guilty of abuse. The theology absolved the men's conscience and their guilt in front of their peers and God, but left the captives with no easy choices. Often, in the tumult of war, men would take multiple wives and engage in short marriages that could be consecrated and ended quickly by a religious authority.

Armed with a theological justification for forcing women to become their wives, male members of Boko Haram considered themselves more pious than the Nigerian army, whose soldiers faced their own credible accusations of rape. Insurgents were banned on punishment of death from partaking of sex outside of marriage, whether rape or simple adultery. The fighters taking towns would address their new subjects through loudspeakers to proclaim that their harsh enforcement of fidelity was God's law. Those pronounced guilty were "apostates" and would be punished in gruesome public executions filmed by the insurgents and disseminated on social media. In practice, of course, Boko Haram members often committed rape without even the pretense of a forced marriage.

Though women are often portrayed solely as victims of Boko Haram's insurgency, many joined willingly and believed in its cause. Some smuggled weapons, spied on the movements of military convoys, and were among its most zealous supporters. In some cases, female members felt they had just as much freedom as at home in the remote corners of northeast Nigeria, where most teenage girls marry before they turn eighteen in weddings arranged by their fathers. Additionally, many of the men who committed grave crimes against women had once themselves been victims, kidnapped or coerced into joining. The moral universe of child soldiering was not as black-and-white as the Boko Haram flag.

In their first months of captivity, the Chibok girls had been an exception to Boko Haram's rules. They were prized hostages, a symbol of the insurgency's ascendance and preserved as currency to be traded. But as negotiations broke down, social media moved on to new topics, and the government's attempts to liberate them lost steam, their captors began to explore another way to exploit them. The students had begun to lose some of the special status that had protected them.

MALAM AHMED WAS STANDING ON THE VERANDA OF THE SENATOR'S house, waiting for a conversation with Naomi. He had a proposal for her that he said she should take extremely seriously. The hostages were not going back home, ever, and the most senior of them should do the correct thing and marry. If Naomi got married, it would please Allah, help her on the road to salvation, and be an example to her younger peers.

The Malam started with a sales pitch. "If you get married, you will have access to air-conditioned vehicles and luxury houses," he said. The unmarried women would become slaves, he continued, and their meager rations would be cut even further. Naomi said nothing at first. The Malam had been smiling through stained teeth, advising her to make a decision that would "please God." A refusal would have consequences, he warned, and they would be unimaginably severe. If she didn't marry, he wouldn't be able to protect her from the guards and their animal instincts. "There are seven guards. . . . We will have to lock the gate and release them on you."

Naomi had heard this kind of violent threat before. She could see that the Malam was struggling to find an argument that would intimidate her into submission. She responded with a suck of her teeth: "You can do whatever you want, since it is not with our consent." The body language between the two tightened. "Maybe you should bring twenty?" she added. Malam Ahmed pushed on, but he was losing face. He began to yell.

The argument was growing louder on the veranda. A cluster of students filed over to watch as the usually composed Malam Ahmed erupted in anger at this diminutive young woman. At one point, loud enough for dozens of schoolgirls to hear, Naomi herself started to shout: "Even if heaven and earth come together, I will not marry!" She clapped her hands to emphasize her resolve. But she could feel her heart galloping. Malam Ahmed looked stunned, as if he had been slapped hard around the face. He was about to abandon his self-control.

"I will kill you." he replied.

"If you kill me, I don't mind. No problem."

Malam Ahmed lifted his Kalashnikov in the air and slammed it down onto Naomi's back. The rifle butt hit the curve of her neck, lacerating the skin. Her knees buckled and her face collapsed in

Naomi Adamu was one of the oldest seniors set to graduate from Chibok Government Secondary School in 2014. Her friends called her "Maman Mu." *(Courtesy of Naomi Adamu)*

Naomi and her friends would dress up to have their portraits taken at Chibok's photo studio—a way for a town unconnected to social media to share pictures. *(Courtesy of Glenna Gordon)*

Kolo Adamu, Naomi's mother, converted between Islam and Christianity several times in an attempt to keep the peace in her multifaith family.
(Courtesy of Mohammed Bukar)

Naomi (left), Hauwa Ishayah, and Palmata Musa were schoolfriends who became the core of a group of holdouts that refused to convert to Boko Haram's creed.
(Courtesy of Mohammed Bukar)

Maryam Ali was a bookish senior certain to win a place at university. She was set to graduate from high school in the same year as her elder sister, Sadiya.
(Courtesy of Maryam Ali)

Until April 14, 2014, few people had heard of Chibok, a farming community of tin-roofed homes and whitewashed churches built beneath a hill studded with jagged boulders.
(Courtesy of Glenna Gordon)

By tradition, women in Chibok are each given a bicycle on their wedding day, a means of mobility.
(Courtesy of Gbenga Akingbule)

The Church of the Brethren, the heart of Chibok's Christian community, where the Adamus attended Sunday service. *(Courtesy of Glenna Gordon)*

The Adamu family home on the outskirts of Chibok, where Naomi and her six siblings grew up. *(Courtesy of Kolo Adamu)*

Chibok lays on the southern edge of Nigeria's Borno State, the second-largest state in the country and the crucible of the Boko Haram war. *(Courtesy of Glenna Gordon)*

For half a century until the night of April 14, 2014, families across northeast Nigeria sent their daughters to the Chibok Government Secondary School for Girls to give them a better life.
(Courtesy of Getty Images)

The examination hall was ransacked and torched during the Boko Haram attack. Only the blackboard at the front remained undamaged.
(Courtesy of Getty Images)

Posters of the missing girls were circulated to law enforcement agencies and printed in national newspapers. Naomi Adamu is on the second row, far right.
(Courtesy of the Bring Back Our Girls campaign)

Oby Ezekwesili, a Harvard-educated former government minister nicknamed "Madame Due Process" for her no-nonsense approach, became a leader of the #BringBackOurGirls movement.
(Courtesy of Getty Images)

Abuja's dilapidated Unity Fountain, built as a symbol of Nigeria's national project, became the scene of daily protests to free the Chibok hostages. (Courtesy of Glenna Gordon)

Goodluck Jonathan, the former zoology professor from Nigeria's deep south who became president, struggled to make sense of the war in the country's Muslim north. (Courtesy of Goodluck Jonathan, official Facebook account)

A video released by Boko Haram in May 2014 showed most of the Chibok hostages chanting the Shahada, or "proclamation of the faith," and claimed that they had converted to Islam.
(Courtesy of Boko Haram video)

Michelle Obama's May 7th tweet, from the White House Diplomatic Room, became one of the most shared posts of all time.
(Courtesy of Michelle Obama, Twitter)

Our prayers are with the missing Nigerian girls and their families. It's time to #BringBackOurGirls. -mo

23:03 · 07/05/2014 · TweetDeck

73.3K Retweets **95K** Likes

At the Cannes Film Festival, movie stars, including Harrison Ford, Kelsey Grammer, Wesley Snipes, Sylvester Stallone, Ronda Rousey, and Mel Gibson, held placards before the screening of their movie *The Expendables 3*.
(Courtesy of Getty Images)

pain. She managed to stay on her feet, but the strike would leave a permanent scar.

"God will judge you," she said.

Malam Ahmed stormed off. Naomi stumbled back into the house, and a group of schoolgirls who'd watched the scene rushed to her aid. The pressure to marry was now violent. The group decided on a new strategy: for the next three days, they would fast. Boko Haram couldn't starve or lure them into marriage if they *chose* to go hungry. They had grown up believing that fasting was a way to seek God's intervention and find clarity of mind. Naomi's mom, Kolo, had fasted to seek strength during Naomi's worst bouts of illness, and when she recovered, they had fasted together to give thanks. Naomi believed that if fasting had saved her before, it could do so again.

Naomi's defiance stuck with Lydia John, her fellow diarist, who had been mulling the marriage proposal out of despair, reasoning it could be a way to get more food and perhaps escape. During the brief hunger strike that followed, Lydia gathered some friends out of earshot of the guards. She wanted to make a vow. Naomi watched as she placed her hand down flat on an imaginary Bible, then swore in Kibaku that she would never marry one of these men.

Around this time, Lydia and Naomi snuck off to the courtyard at the back of the mansion where they could open their notebooks in secret. Naomi needed Lydia's help to write a letter to her mother.

Dear my lovely Mum,

I'm so worried about you. I love you so much mum when i remember how you suffered to raise me, I felt so sorry. You took nine months of my pregnancy and you suffered to take care of me when i was born, my growth and my health you take care of me by the grace of god. Now Ive grown to help you but the BH took me away from the school. They are telling me words of nonsense. God is with us. When God says yes, nobody can say no, because god is the creation of the faith. Have a nice day mum.

> *i wish to*
> *stop*
> *here.*
> *Your lovely daughter*
> *Naomi Adamu.*
> *From sambisa to Gwoza.*
> *From Gwoza to Chibok, city of life.*

Naomi signed the letter and circled it with a large heart sketched in pencil. She crested it with flowers blooming to the top of the page, then closed the book.

16

Maryam and Sadiya

Maryam Ali and her sister Sadiya had grown up in a single-story thatch and mudbrick *gidan tabo* home on the outskirts of Chibok. Their father, a civil servant, would say they were "like peas in a pod." Born less than a year apart, the siblings did their chores together, their complementary natures making each other feel whole. Maryam, the elder, was reserved and bookish and liked to escape into her economics textbooks. She wanted to be a business-woman and was focused on graduating. Sadiya was an extrovert, more interested in friendships and having fun. Whereas Maryam kept her head down, Sadiya could be dramatic, delivering so many passionate speeches that her family joked that she would run out of tears. They affectionately called her Biba Problem, after a movie character famous throughout northern Nigeria for creating a lot of trouble for her husband. Sadiya's mother had to implore her to finish her studies: "Your father paid your school fees," she would say. "Don't waste his money."

The Alis were a pious Muslim family, and in their conserva-tive household, surrounded by mostly Christian neighbors, both daughters worked during school vacations on the family's small farm. The days started before dawn, the sisters each gathering a reedy basket and walking the twenty minutes to their father's

plot, where peanuts and beans were planted in bushy rows. They
sang Hausa pop songs as they stooped over the sandy soil to yank
up the crops, shake off the dirt, and place them in the basket, each
taking a verse. When their father built extensions to the house
as more children arrived, the sisters shaped and dried the mudbricks. To make the time pass, Sadiya encouraged Maryam to pretend they were baking cakes.

As the sisters neared the end of high school, however, their
contrasts had begun to edge them apart. Maryam was one of the
top students at Chibok, eager to leave town and study at the college of her father's choice, the Federal University in the northeastern city of Yola. She saw education as a way to give herself
more choices, to be a "respected person" who could stand tall in
the world. Sadiya had found a boyfriend, a construction worker
named Baba, and the two chatted constantly, meeting in secret and
making plans to marry. Why, Sadiya wondered, would anybody
want to leave Chibok?

At school the sisters shared a bed in the same dorm, Gana Hostel. On weekend evenings at home, Maryam would wake up before
dawn to read her schoolbooks by flashlight, only to hear her sister
whispering on the other side of their dark room, holding a cell
phone to her ear. Sadiya used those hours to talk to Baba, when
the phone network offered cheap calls and she knew her father
would be sleeping. The sisters imagined they would likely end up
in separate places after graduation, pursuing separate lives.

By August 2014, all of that had been left behind. In their senior year of high school, the sisters had been taken together, their
fates entwined in ways neither could have ever foreseen. They
had spent more than three months as captives of Boko Haram
and were about to become the first Chibok students to be forced
into marriage.

INSIDE THE REFLECTIVE TILED HALLWAYS OF THE SENATOR'S MANSION,
the early days of captivity in Gwoza passed with a rhythm simi-

lar to that of their days in the Sambisa Forest. The students were locked inside the compound, venturing onto the veranda or the small courtyard only for Quranic lessons or to eat meals they'd prepared for themselves. They were not allowed beyond the metal gates, heavily guarded on both sides by armed militants.

A couple of weeks after their arrival, Malam Ahmed addressed the captives to announce that some of them would be allowed to briefly walk on the street outside the complex, escorted by the guards. It was a historic occasion, he said, and the hostages needed to witness it. On this day, Gwoza would be renamed Dar al-Hikma—House of Wisdom—and declared the official capital of Boko Haram's new caliphate.

Maryam and Sadiya stepped onto the road in the middle of a long line of Chibok girls to a riotous chorus of car horns and celebratory gunfire. They could see dozens and maybe hundreds of vehicles graffitied with Boko Haram's black and white flags tailgating each other in a parade of tanks, armored cars, and Toyota pickups with fist-pumping fighters manning truck bed–mounted machine guns. Car stereos were blasting the religious songs of the *nasheed*: "Nigeria today, tomorrow conquered!" Women in niqab veils, covering the entire face except for the eyes, stood in clusters along the street, shaking their arms and cheering wildly. Maryam noticed that one of the vehicles, a gleaming white Hilux with blacked out windows, was flanked by a phalanx of bodyguards. An ecstatic crowd hollered in praise as it passed. She knew it was Shekau, the man they called the Imam. "He didn't show himself in case there was an assassination attempt," she said.

The Boko Haram warlord was at the peak of his powers. Shortly after, he would appear in a mosque in the flowing gray-and-white robes of a cleric to officially declare the city the capital of a new society.

"By the grace of Allah we will not leave the town. We have come to stay," Shekau said in the twenty-five-minute video published on YouTube, his sermon cutting to images of militants gunning down men who had tried to flee. One piece of footage showed Boko Haram fighters using shovels to beat to death two men caught in women's clothing. Gwoza, Shekau said, "now has nothing to do with Nigeria."

On the streets outside, Maryam could see that Malam Ahmed and the guards were jubilant. They were embracing one another to celebrate their victory, and singing in Arabic: "We took Gwoza, we took Timbuktu!" For the hostages, the spectacle reinforced the demoralizing sense that the war was lost. Back in the senator's mansion that evening, Malam Ahmed gave another speech claiming Boko Haram was creating a new civilization that would rule over all of Nigeria. The hostages would never see their parents or their hometown again. "It is time for you to marry," he said.

ONE DAY SOON AFTER THE PARADE, MARYAM, SADIYA, AND TEN OF THEIR classmates were called by name and instructed to gather before Malam Ahmed in a backroom on the mansion grounds, sequestered from the other hostages. When they arrived, the Malam was smiling. He had assembled them to make an announcement of the utmost importance, he said. The dozen hostages had been *chosen* to "fulfil God's wishes." The group were a mixture of Muslims and Christians who had agreed to convert in the early weeks of captivity and would now be given mattresses and extra rations the Malam said. The reason was that Shekau had ordered that they be wed.

Maryam looked at Sadiya, neither saying a word. A grim silence hung over the group. They knew they had already been identified as more compliant than some of their classmates; the insurgents had given them jobs teaching Quranic verses or managing the food supply. Some of them had privately discussed with one another whether marriage could make captivity more bearable, or even provide a means of escape. Their parents were dead, they'd been told, and they would never return to Chibok. And maybe these men were right that God was on their side? How else had they conquered swathes of Nigeria from the Sambisa Forest with little more than a few guns? But now the Malam was standing before them announcing that they would be wed to a fighter whether they liked it or not. They should be happy they would no longer be living in sin, he said, then turned to walk out. The realization took time to sink in. In the hours that followed, Sadiya sobbed inconsolably and refused to eat. Maryam was numb at first, but then she told herself that if she could deal with the abduction, she could deal with anything. Whatever happened was "the will of God," she would say.

A few days later, Malam Ahmed returned, this time beckoning Maryam to follow him. She had a visitor, he said. In an adjacent room, sprawled on a prayer mat and cradling a rifle, was one of the guards. Maryam recognized the tall, taciturn young man. He said little and never smiled. She had once seen him beat several of her classmates on their backs with a tamarind branch. Maryam sat down on the carpet with her back to the stranger, staring at the ground. This situation seemed so unreal, almost dreamlike. The room was dark, and it was hard to see his features. Her eyes focused on the metal of the gun on his lap. He seemed cold and uncaring.

After some time, the man spoke: "Malam Ahmad sent me to see you," he said, staring straight ahead. "If you do not like me, I will see another girl."

Maryam said nothing. After a while he got up and left. Malam Ahmed returned smiling and asked excitedly if she would accept the fighter as her husband. His name was Abu Walad, he said.

Maryam said no and the old man's jaw clenched. He slapped her hard across the side of the head. "You idiot! How dare you . . . I introduce a suitor to you and you reject him?" Maryam had seen Malam Ahmed striking other hostages before but she had always managed to avoid it. "Do you mean to embarrass me?" he screamed. "You will marry him whether you like it or not."

As he stormed off, Maryam was shaking. What she had just

experienced was not a proposal but an ultimatum. Sadiya, she soon discovered, was to be married to a fighter from another unit. One of the twelve, Ummi, would be married to Malam Ahmed himself, a man about thirty years her senior. It had all been decided.

THE BRIDES WERE NOT INVITED TO THEIR WEDDINGS.

Maryam was locked in a corner bedroom in the mansion, thinking about what the ceremony would have looked like if she were at home. In Chibok, a suitor would have first sought permission from the men in her family, bringing them bowls of kola nuts, cookies, and chocolates. On the wedding day, her father, or maybe an uncle, would have represented her at the mosque alongside the groom and an imam. There, after reciting the first chapter of the Quran, the marriage would have been sealed between the two families, and in the eyes of God.

Maryam would have stayed behind, in a bedroom with her best friends, chatting in anticipation as they covered her body in perfume, scented flowers, and intricate henna patterns. Women from the groom's family would have dropped by with stacked boxes of *kayan lefe*, or bridal-shower gifts, perhaps a new home-made dress, shiny, fashionable sandals, jewelry, a prayer rug, and always a copy of the Holy Quran.

But she found nothing familiar or festive about the Boko Haram nuptials. Maryam was in a gilded prison. Her mother and father were either dead or far away, with no idea that their eldest daughter was alive and about to be "wed." Their absence plunged her into an even darker place. *This is a totally alien wedding*, she thought to herself. *Even a corpse is more important than you, since nobody is with you.*

Instead, Malam Ahmed represented her at the ceremony, which started shortly after sunset prayers at the house of one of Shekau's deputies, an emir, or senior commander, who was wanted by Nigerian and US authorities for planning the 2011 bombing of a UN office in Abuja. The twelve classmates chosen for marriage stayed behind, gathered in a room of the senator's house. Outside the compound, decisions were being quickly made on their behalf. Maryam was in a daze.

"We are the first to be married, and nobody is giving us any information or advice," she said. "I am just worried about what kind of life it will be. How am I going to live with them as a wife?"

Maryam decided to ask for advice from an older classmate. She wrote a quick note to Naomi Adamu. Maman Mu would know what to do. The letter asked, "How should I live a married life?"

The reply from Naomi left Maryam in tears: "What do I know about being married? Just be faithful."

Naomi felt little sympathy for Maryam and thought she hadn't protested enough against the proposal. Maryam was abandoning the group for better treatment, Naomi believed, and she was doubting whether her friend's loyalties lay with her schoolfriends or their captors. The thought made her feel cold. The guards had tried to sow suspicion among the group to divide them and it was working. Maryam folded up Naomi's note, feeling alone in the room full of other anxious young women being forced into marriage. Her sister was staring into space. At one point shortly before nightfall, they heard celebratory gunshots, and soon after, Malam Ahmed was back.

This time he reclined on the mattress next to Maryam to deliver another message. She was married now, and so was Sadiya. The young women would now officially become emiras, fighters' wives with new status and privileges, he said. As a gift, he had dowries for each of them—15,000 naira, around $80—a handful of cash he put Maryam in charge of distributing. Their husbands were preparing to welcome them, he said, as he handed out a box containing eye shadow, lace, makeup powder, and perfume. Maryam knew she would soon be asked to put it on, and the thought made her feel ill. The Malam reassured her that he would come back in a few hours.

Maryam was in shock. "I could not believe it. It was like a play."

She turned to her sister and her friends, looking for reassurance. "It is a forced marriage and the man does not laugh; everyone is saying he is cruel," she said. "How can I live with him?"

"The one I am going to marry is not polite either," another girl replied.

"We should be patient and take it as our fate from God," said a third.

Malam Ahmed handed a final gift to Maryam, a bag of presents from the man now calling himself her husband. Inside was brocade fabric, new shoes, and some jewelry—an earring and a necklace. "Put them on," said Malam Ahmed. "You will now be taken to your homes," he said.

Maryam found the prospect of being taken from the others frightening and one of the other girls was starting to panic. "I don't know where I will be taken to," she said. They hurriedly began to discuss where they were likely to live, sharing dread at the prospect of being alone. "I am afraid we will be separated," Maryam said.

Shortly after ten p.m., Abu Walad arrived for Maryam, his gun slung over his shoulder. The pair walked in silence through a neighborhood of tall cement houses adorned with satellite dishes, arriving at the door of a well-appointed home. Maryam could see the possessions the family who owned it had left behind. There were suitcases full of clothes and perfumes in one room and a flatscreen television in another. She wondered who the people were who lived in this place, ate their meals on the kitchen chairs, and wore the floral-print dresses in the wardrobe. Where were they now? Were they still alive?

Abu Walad was sick with a fever, coughing uncontrollably, and rushing to find the bathroom so he could vomit. He returned and slumped on the sofa, then made a demand: "Maryam, where is the money?" It wasn't hers to keep, he said.

17

"These Girls Are Global Citizens"

Oby Ezekwesili sat under a painting of George Washington at war while waiting in the West Wing of the White House for a meeting she hoped would provide some answers on the missing school-girls. The Nigerian activist had come to the American capital with a set of questions. How could it be that the administration had deployed some of its best surveillance technology—spy satel-lites, drones, and an unparalleled intelligence network over the Sambisa Forest—and yet for months there had been no news on the whereabouts of the Chibok girls? America's rescue effort, an-nounced with so much fanfare, had faded away without so much as a press release. Neither the international press nor social me-dia seemed aware that the drones sent to find the Chibok girls had been repositioned to Iraq and Afghanistan, or that most of the American personnel sent to help free them had returned home.

What Oby did know was that the hostages' families were des-perate for information. At night, she was still receiving calls from parents looking for an update or just for someone to listen to their call for help. Each afternoon at 4:00 p.m., protestors would gather outside the Abuja Hilton, but the crowds were waning. To keep the cause alive, she'd led a march through the capital on October 14, the six-month anniversary of the abduction. That same day, R&B star Alicia Keys and her husband, the beatmaker Swizz Beatz, had waved #BringBackOurGirls placards at a parallel demonstration outside the Nigerian consulate on Manhattan's East Side, briefly returning the hashtag to Twitter's list of trending subjects. A few days later, Oby had conducted another march, titled The Need For Speed, to the front of the US embassy in Abuja, to ask why America hadn't found the girls. "The administration remains engaged," a US diplomat had assured her, but the protest hadn't seemed to make any difference. Oby had resolved that if she was going to get America recommitted to the cause, she would have to go to Washington and personally petition her administration contacts.

John Podesta, President Obama's senior counselor, was in his West Wing office when Oby walked in for their meeting, perching herself on a red couch under his wall-mounted collection of Catholic iconography. His forthright old colleague didn't seem herself, uncharacteristically melancholy as she lamented how the world had abandoned the kidnapped students. She asked for an update on the US rescue effort, but Podesta had little to offer.

More than six months after they had disappeared, even the world's solitary superpower seemed to have stopped trying to solve the riddle of their captivity, Oby complained. She felt the girls had ceased to be a priority. Her frustrations bubbling over, she kept repeating the same unsettling point: If these schoolgirls were American children, there is no way you wouldn't activate everything to find them. "You are leaders of the free world," she said. "These girls are global citizens."

Inside the White House, the effort to free the girls had frozen into a kind of bureaucratic limbo. There had been scant new information, and the issue had faded from priority lists. At one late 2014 meeting with the secretary of defense and the national security adviser, John Kerry had plowed in, trying to convince officials around the table that the United States could still mount a rescue mission. But the secretary of state's eagerness was checked by skeptical voices at the table more convinced than ever of the difficulties of operating in Nigeria or partnering with its government. Still, an election was coming up in Nigeria, and senior officials agreed that the moment was right to sketch another package to help defeat Boko Haram and free the Chibok girls.

What they needed was a partner they felt they could trust.

18

Christmas

GWOZA

DECEMBER 2014, EIGHT MONTHS INTO CAPTIVITY

Dear my lovely dad,
I miss you so much in this moment . . .

To celebrate Christmas, Naomi and Lydia snuck into an empty room of the senator's house, a quiet space where they could crouch next to a cement wall and write letters home. Malam Ahmed had told them the holiday, a day for the *arna*, or pagans, was near, and that the militants had chosen the day to launch an attack across the border on Christian communities in Cameroon. The two students already knew their holy day was approaching. They had been keeping a December calendar in the pages of their secret diary, crossing off the days of advent. On Christmas Day in Chibok, the town would flood into the Church of the Brethren for the morning service. Pastor Enoch Mark would lead the hall in four hours of prayers and Bible readings interspersed with praise songs belted out by the *zumunta mata*, the women's choir. At Naomi's home, Kolo would prepare a pot of beef, and the family would visit neighbors with bowls of fruit and sweets to share. Naomi was in tears as she dictated a letter to her father, Musa Adamu. No other students were around, and to reassure her friend, Lydia placed her hand on Naomi, and repeated, "Calm down, don't cry."

"I miss them," Naomi said.

They were writing in a new notebook given to them by the Malam after they told him they had filled the originals with Islamic verse. But this booklet looked very different. It was a dulled sepia color, caked in dust, and had begun to split along the spine. Naomi had found a hair needle and black thread a girl had used to fix her veil, and stitched the book back together. With its fraying and crease-riven cover, crisscrossed in dust-caked brown thread, the notebook looked like an ancient manuscript. On it Naomi had scrawled "my mummy my blood," in honor of Kolo, and "Talatu," her mother's endearment for her. The notebook seemed to have been stolen from a Nigerian soldier; there was a half-torn sheet inside that read "PRINCIPLES OF ATTACKING TERR," the rest of the word cut off. The first principle was the element of surprise, it stated. Another page featured an ad from 1999 offering iMac computers in five colors.

Inside the notebook, Naomi and Lydia had been drawing pictures of each other or jotting down the phone numbers they'd most love to call, but weren't able to. "0816558 . . . —Talatu Adamu. 0803887 . . . —My Angel." They listed the names of family members and wrote the names and corresponding capitals of all thirty-six Nigerian states. They continued copying down verses from Psalms or the New Testament. "Psalms 22:1 My God, my God, why have you left me?" Naomi sketched the image of a woman's face from the discarded wrapper of a bar of soap, her ballpoint pen shading a carefree expression from the outside world alongside the tag line "NEW: Ivory Complexion Care." A few times, Naomi lent the notebook to other hostages, who wanted to read the Bible passages.

All told, Naomi and Lydia had now filled three exercise books with letters, reflections, and sketches, a record of the trauma of captivity. Naomi swaddled them inside her *mayafi*, tying them around her thigh in a makeshift pouch. Lydia volunteered to hide one of the diaries, but Naomi insisted on stowing them all. She learned to sit and sleep in a position where the pouch, made with strips of fabric torn from a veil and tied together, would not rub her leg uncomfortably. Keeping the pages pressed against her body was a way to feel a connection to home.

> *Dad, I want to see you, i'm so worried about you and*
> *mum and the rest of the people at home. I wasn't aware*
> *that this could happen to me, none of us who the BH*
> *kidnapped realized that. By the Grace of God dad, i*
> *missed you so much. I want you to help me in prayer*
> *all the time so that i will defeat the devil each time he*
> *comes to torment me. So dad, i will like to stop here. I*
> *missed you so much. Goodbye have a nice day. Your lovely*
> *daughter Naomi Adamu.*
> 　　　　. . . *Wish you a merry Christmass.*

BY DAY, NAOMI WAS NOW WORKING AS A NURSE AT A MAKESHIFT CLINIC that Boko Haram had established inside a small cement house, a short walk from the senator's mansion. So long as she refused to marry, she would dress wounds and boil needles in a dirty room where pregnant women would regularly miscarry. Children would die from disease. Fighters would arrive with horrendous injuries, and the hostages would have to administer anesthetic or help hold the patient down during operations or amputations.

This was a purgatory Malam Ahmed hoped would weaken their resolve. "You are slaves now," he had told Naomi, her cousin Saratu, and eight others of the most disobedient hostages on the day they had been given the assignment. Their first task, not strictly medical work, had been to carry twenty-pound bags of guinea corn, or sorghum, and wedge them into the potholes of a nearby road. This was state-building in the caliphate.

"It is better to carry a bag of guinea corn on our head than to marry Boko Haram," they told each other in Kibaku, as they staggered forward.

Other students had been sent away to perform drudgery that the Malam considered less punishing. Lydia John had been ordered to sweep out the houses that Boko Haram members had stolen, mend their wives' dresses, or look after their children. Others were to carry bundles of firewood on their heads to a neighboring camp for hours under the scorching sun. Each group was called a *halqa,* a cell, or a link in a chain, and the more pliant they were, the easier their chores.

Naomi's *halqa* was supposed to endure the worst punishments, but their sentence was lightened by the doctor tasked to look after them. Ali Doctor, as the ten students called him, had himself been kidnapped from the driver's seat of his jeep while traveling not far from Chibok, and forced to become a Boko Haram field medic. Somehow, along the way, he'd sworn allegiance to the group, started carrying a gun, and had risen to a position of influence. Nevertheless, this tall bearded man quickly became a sympathetic figure who took pity on his fellow captives. When the young women arrived in the clinic, he saw that they were at best halfhearted apprentices, and he understood. He quietly began letting them shrug off the grislier medical tasks, which the Malam had promised would be their lot. When other militants ordered them to do chores, Ali would chastise them. "They are not slaves!" he yelled at one woman who ordered the hostages to roast her some peanuts. "They know their worth."

Ali Doctor patiently explained to Naomi how to sterilize and use needles and how to conduct pregnancy tests—and how to console the women who looked crestfallen when the result was positive. He noticed that Naomi tired easily and sometimes would stand crumpled against the wall, holding her side. After he had examined her, he began to give her palliative treatment for the growing throb in her kidney. In his clinic, she often felt like more of a patient than a servant. He had a soothing bedside manner and would check her vitals in a ritual that sometimes reminded her of a previous life of endless medical visits. "Be calm and don't worry," he would say. "Hopefully you won't be here for too long."

Outside the clinic, other women on the streets of Gwoza called Naomi and the other hostages who refused to convert "pigs of infidels" and spat on the floor as they walked past. But inside, the group felt free to express themselves. Ali Doctor told Naomi he had a family back in Maiduguri, and two adult children who had graduated from college, and he prayed for them every day. He had taken a new wife in the forest, a plump, kindly woman who would visit the clinic and greet the hostages. Naomi never asked this well-educated medic how he had transitioned from captive to believer, but she felt he was different from the other men in camp. "He would always be there to help us," she said. "He wasn't interested in anything else."

Life with Ali Doctor had made the daily grind of captivity a little more bearable. Naomi could pass a day without being noticed by Malam Ahmed or subjected to his constant pressure to marry. Ali Doctor worked alongside an elderly nurse with round cheeks and an attractive smile. She would ask the Chibok girls about their lives back home and liked to share a joke with Naomi. They had laughed when she said she knew the songs of Mama Agnes. Before she'd been kidnapped, she'd been a church leader and chorister in her village. Boko Haram had taken over so many of the towns and villages she'd known, she complained, that there was nowhere left to go for her, or for the thousands of others who weren't members of the group but were now ruled by it. She'd had several children, she said, all boys. They, too, were missing.

ONE DAY, NAOMI AND A GROUP OF STUDENTS HAD BEEN SITTING UNDER A tamarind tree when a masked visitor approached flanked by the *rijal*. He was wearing a white tunic, trousers cut short above the ankle, and a turban that covered much of his face but not his eyes, which were narrowed into a scowl. He said nothing, nodding first at the captives, then a guard, before walking away. As the stranger moved from earshot, one of the *rijal* seated near Naomi and her friends asked them a question: "Did you recognize this man?" The hostages shook their heads.

"It was your teacher."

The classmates looked at each other as the realization sank in. The masked militant was a teacher from Chibok, the men said. The girls debated who the man in a white robe with trousers hemmed high above the ankle could be as the *rijale* erupted into laughter. One turned to the captives and said, "Even your teacher is a member!"

PART III

A Global Conflict

The Foreign Fighters

MAIDUGURI
JANUARY 2015, NINE MONTHS INTO CAPTIVITY

A strike force of South African mercenaries, more than a hundred men with leathery skin, many in their fifties and sixties, landed on the cracked tarmac of the bomb-damaged airport that would be their operating base for the next three months. The blue fiberglass letters welcoming visitors to MAIDUGURI INTERNATIONAL AIRPORT had all come off the beige concrete walls, leaving behind the bare exterior of the arrivals hall. By the runway sat the charred shell of a burned-out helicopter facing a row of four empty flagpoles. The airport, besieged by militants firing rocket-propelled grenades in the preceding months, was still closed to commercial flights.

But by January 2015, its runway was bustling with military cargo planes bringing a new wave of foreign fighters, mercenaries, and Western military advisers into Nigeria's northeast, further transforming what had begun as a remote local struggle into a global conflict. At the forefront of this cavalcade were the South Africans, grizzled veterans who had once fought as special forces for the apartheid government's Defense Force, tracking Nelson Mandela's allies across semi-arid terrain not too different from the land outside Maiduguri. They had arrived with sacks of dry food, utility belts, medical kits, and South African–made armored

gun trucks, a gush of supplies unloaded in front of Nigerian soldiers who were mainly armed with cheap Kalashnikovs. The South
Africans had come to Nigeria on a mission to rescue the Chibok
girls.

The country was now only weeks away from nationwide elections, and President Goodluck Jonathan, in a desperate gambit to
win a vote he looked set to lose, had invited some of the world's
most controversial foreign soldiers and mercenaries to free the
girls and beat back the men who held them. Outside Maiduguri,
more than two thousand troops from neighboring Chad were driving across the border in a column of Toyota pickups, supplied with
intelligence and weaponry by the French security services. The
Chadian Army, which had only recently stopped recruiting child
soldiers, had fought civil wars and ruthlessly quelled years of re-
bellions in the similar landscape of their own desert country. They
were joined by Ukrainian helicopter pilots, hired to fly nighttime
raids. The British Royal Air Force had sent three Tornado surveillance jets with thermal-imaging cameras to help find the girls
and track Boko Haram's movements. Vladimir Putin's Russia and
Angela Merkel's Germany had offered weapons or assistance as
well.

From its base in Syria, the Islamic State was working to in-
ternationalize the fundamentalist side of Nigeria's conflict. Boko
Haram had started a Twitter account, posting images of the boys
it had conscripted, and Abubakar Shekau was on YouTube, calling on foreign fighters to join his jihad, as they'd done in Iraq and
Syria. Islamic State leaders were endorsing and amplifying that
summons, encouraging their followers to enter Nigeria's battlefield. "Oh Muslims, this is a new door opened by Allah," an ISIS
spokesman announced in an online message. But few fighters managed to reach northeastern Nigeria, which was much less acces-
sible than Istanbul, whose bustling airport allowed thousands of
recruits to enter Turkey then cross the border into Syria.

Shekau's military success had appeared to contain a fateful miscalculation. The Chibok kidnapping and Boko Haram's alliance with the Islamic State had elevated his movement into a problem too big for the international community and the countries near Nigeria to ignore. By the end of 2014, his army had claimed a swath of the northeast the size of Belgium, but as the new year began, Boko Haram was starting to retreat back into its network of camps in the Sambisa Forest, pursued by an awkward coalition: Chadian shock troops, aging South African mercenaries, Ukrainian helicopter pilots, and Nigerian soldiers, following local wildlife hunters as guides, who chanted as they rode out: "Where is the hardship? The hardship is in Sambisa!"

The most contentious figure in this new offensive was Colonel Eeben Barlow. The blond, blue-eyed pensioner leading the South African mercenary force had started his career fighting for South Africa's apartheid government, corked in Black Is Beautiful face cream during locate-and-kill missions targeting freedom fighters along the dry riverbeds of Angola or during operations to secure diamond mines in Namibia and Botswana. As white minority rule crumbled, Barlow had looked for other battlefields, founding Executive Outcomes, a mercenary firm that recruited men from both sides of southern Africa's fading conflicts. On his Facebook page and on *Eeben Barlow's Military and Security Blog*, he bridled at accusations that his men were relics from a racist regime, countering that he had achieved the rank of major general in three African armies, and had been hired for dangerous work by sovereign governments, but the brand was hard to shake.

Nonetheless, by early 2015, he was leading a "mobile strike force" of veteran mercenaries and handpicked Nigerian soldiers into Sambisa to flush Boko Haram from its camps. Just a few weeks after his arrival, the Jonathan administration had dramatically expanded his contract, widening his mission from "finding and freeing the Chibok girls" to confronting the insurgency and reversing the course of the war. None of the details of the deployment would be made public. The president was on a tight timetable and needed results fast.

Barlow's team, with men they recruited from a Nigerian unit called the 72 Strike Force, moved quickly, driving into insurgent territory supported by attack helicopters, using a tactical concept called relentless pursuit. The strategy was designed to make Boko Haram hungry, tired, and disoriented. Barlow and his men would fight around the clock, in hours-long shifts, wolfing down packaged food. "Never stop to cook," went one guideline. The fighters tracked trails from retreating militants to see if they were carrying heavy weapons, walking fast or slow, exhausted or full of verve. Disrupted soil signaled land mines or improvised explosives. If the target stopped to snack, clues in the dirt would give away what, if anything, the target was eating.

As the men pushed deeper into Sambisa, they were looking for anything that could betray the girls' location. Barlow badly wanted to be the man to show the West how "an African bush war" was won, and thus redeem his name. "The Western response was hashtag-save-the-girls," he later said. "That type of nonsense does not save girls."

NAOMI HAD BEEN SLEEPING IN A TIN-ROOFED SHED, JOLTED AWAKE IN between terrifying dreams. In one nightmare, she saw a warplane circle over an enormous mosque where Boko Haram held its evening prayers. Hundreds of worshippers were crammed into the building unaware of the jet taking its aim above. As the bombs began to fall toward the mosque, Naomi realized that many of her classmates were praying inside, about to die.

Dreams of airstrikes had become recurrent for Naomi, now nine months into captivity. Weeks earlier, she had been moved with her classmates away from the senator's mansion to a new camp called Sabil-Huda, or Path of Guidance. It hadn't been safe to stay in the city of Gwoza. At odd hours, the rumble of airstrikes had begun to shake the surrounding countryside and increasingly hit targets in the city, sending hot shards of shrapnel and the screams of neighbors over the walls of their prison compound. By the time Malam Ahmed had told them they were leaving, some of the *rijal* had become afraid to stay inside the mansion but had banned their hostages from stepping outside. When the sound of a helicopter or jet passed over, the girls would lie flat on the tiles and pray.

Sabil-Huda was far larger than "Timbuktu," the encampment where they had spent their first months of captivity under the Tamarind tree. It was a bustling town hidden in the forest, home to hundreds of fighters and their families clustered in tin sheds and small mudbrick houses built flush against the trunks of red-leafed thorn and tamarind trees whose long, knotted branches looked as if they were sprouting from the roofs, cloaking them completely from the sky. *Boko Haram are everywhere, like maggots in fruit*, Naomi thought. These cramped metal structures, just large enough for a few bodies to lie down inside, would be the hostages' new homes. It was a return to a life in the forest, but this time at least they would have a roof over their heads. Malam Ahmed had set up home in a larger house next door from which he could monitor them and the *rijal* who were guarding them. From the doorways of their tin shacks, the girls could see elaborate thatched stick fences that zigzagged through the forest camp and up toward a nearby mountain. Boko Haram's black-and-white flags fluttered from the trees and had been spray-painted onto the sides of pickup trucks and vehicles. The ground was hard, parched dust that somehow gave life to clumps of pale, waist-high reeds. The earth under Sabil-Huda housed a new threat that scuttled underneath them as they slept, scorpions that lay in the thin soil and emerged after dark.

Naomi and her *halqa* had been transferred with Ali Doctor, who was now running a new grass-roofed clinic in the heart of the camp. During the days, they would help tend to the sick or

wounded, then walk accompanied by guards to forage for food or
to bathe in a nearby river. By late afternoon, they would be returned to the shacks under the trees. There were radios scattered
around, tuned to BBC Hausa, and when Naomi was close enough,
she could hear news that Nigerian soldiers, backed by foreigners,
were advancing into Boko Haram's territory. She and her friends
discussed whether the army sent to rescue them might acciden-
tally kill them in the crossfire.

Ali Doctor was becoming increasingly nervous. He would spend
much of his mornings staring at the sky, trying to divine clues as
to the impending attack, noting a distant rumble, a vapor trail, or
a tiny shimmer of dark metal on the horizon. One morning Naomi
saw him squinting. "I think the war might start today," he said.
"We should prepare for war."

NAOMI WAS WATCHING LADI SIMON STIR A POT OF THICK MORNING *KUNU*,
a rice-and-peanut porridge, when they heard a popping sound.
Naomi had been school friends with Ladi, one of Chibok's more
style-conscious students who shared the fashion tips she cut out of
Hausa gossip magazines like *Fim*. Naomi called the younger student "my child," and in turn, Ladi called Naomi "school mother."
They were both in the same *halqa*, exiled to the clinic.

The two friends turned to each other. "It's like there's an attack today," Ladi said to Naomi. In a flash, the distant crackle
of gunfire, now growing closer, was eclipsed by the roar of two
helicopters, huge metal shadows approaching the camp from the
horizon. For a moment, the friends stood spellbound as the gun-
ships glided over the tops of trees, sucking the air from their ears
and making the ground tremble under their feet. They were still
staring when the forest floor began to light up with flashes of
fire, chased by billows of smoke and a series of deafening booms.
This was more than an airstrike; it was a coordinated military
assault.

Ali Doctor screamed for the students to clamber into a nearby truck. They ran toward it, hauling themselves inside, as he cranked the ignition and sped off. The vehicle lurched over the lumpy ground toward the trees for cover. The ground was being strafed by some sort of rockets or possibly heavy gunfire, ripping through vehicles and the mudbrick homes. "Kai!"—the doctor shouted. "These helicopters will hit us and nobody will survive. It's better we come down and start running."

Doors flew open and Naomi and Ladi began sprinting in a scrum of insurgents, women, children, and other Chibok hostages. Naomi's ears were ringing, yet the sounds around her were muffled as if her head had been dunked under water. Another crackle of gunfire cut across her path, and a woman in front of her fell, her leg bludgeoned into a bloody stump from the shin down. Naomi recognized her face. She had been a member of the insurgency.

Other bodies were collapsing as she sprinted past, hiking up her *mayafi* to lengthen her stride. Armored gun trucks were now heading toward the camp as well, rolling over the reedy plants in their path. At one point, Naomi thought she saw white men in military uniforms. The camp was in chaos. Strange sounds were mixing in the air, the thuds and crackles of weapons and a chorus of panicked screams.

Anybody running in the open was at risk. It was better to carve a path through the trees a short distance away. Naomi darted toward them, but there too, gunfire was ripping through branches. Another person fell down, too fast for Naomi to see if it was a man or woman, and she jumped over the body. Behind her, the attack helicopters were firing into homes, which were erupting in flames. Tied around her thigh was the piece of fabric holding her school uniform, her phone, and the folded pages of the diaries, slapping against her leg as she ran. Her mind was willing her body forward, her feet careening over slippery dirt and jagged clumps of grass. She could hear herself panting, and the sound of it eased her into a kind of metronomic trance as she pushed toward the horizon. A voice jolted her back into consciousness as she sprinted past a Boko Haram member who was shouting in her direction. "The soldiers are looking for you, the Chibok girls!"

It wasn't until the gunfire seemed distant and her lungs were heaving with exhaustion that Naomi allowed herself to slow down. Hundreds of the camp's residents had sprinted in different directions. As she stood there catching her breath, she saw that she had become separated from her classmates and her guardians in the melee. For the first time in nearly a year, she was alone.

She kept walking, the greens, yellows, and browns of the forest melding with with passing hours into a foggy blur. It was oddly silent now, except for the sound of her trudging sandals and the ringing in her ears. She felt parched and disoriented, and noticed that her legs were trembling. Walking through the unfamiliar landscape, her mind replaying scenes from the attack, she was suddenly completely overwhelmed. With tears streaming down her cheeks, she began to sing the chorus of her school's warm-hearted end of year song, "Departure Is Not Death My Sister."

Goodbye my friends,
Departure is not death my sister,
Farwell my sister, it's time to leave
Goodbye my sisters,
Being apart is not death . . .

Naomi continued until the sun set, then took shelter against a tree trunk. Her leg was swollen and pierced with thorns, which she yanked out in the dwindling light. She couldn't gather the strength to keep walking and wasn't sure where to go. At times, she could hear what sounded like helicopters and gunfire in the distance. The moon was large and bright enough that she worried a helicopter would be able to see her. She lay in the dirt to camouflage herself, listing the dangers in her head. Boko Haram could also find her. They might think she was a *jasus*, a spy, and execute her.

Naomi's thinking turned down a dark path. She was losing her sense of balance, and now felt, for the first time, that she would die. She fell in and out of fitful sleep, physically exhausted but alert to every rustle. She stayed flat and still on the forest floor, until at some point she dreamed that something was coming at her and she needed to get up, *now*. She opened her eyes and saw a black snake winding its way toward her through the dirt and hissing. Startled, she grabbed her things, lifted her body off the ground, and hurriedly limped away.

DAYS LATER, NAOMI WAS HAVING HER BLOOD PRESSURE TAKEN AT ANother makeshift clinic that Boko Haram had created from yet another grass-roofed structure. Outside, the insurgency had rapidly assembled a hidden set of modest dwellings that looked much like the last one, nestled into a similar forested landscape. Naomi was still disoriented and weak, but the aging nurse's round face and her welcoming smile were a comfort as she held her arm.

Ali Doctor, driving around in his truck, had found Naomi and had brought her here to a new camp for treatment. Her schoolmates were just outside the doorframe, and her cousin Saratu pledged that the next time the military came close, she would grab Naomi "and hold her so she doesn't go missing again." The nurse was taking Naomi's vitals and asking for details on her ordeal. She was lucky to have survived the attack, said the woman. The two traded jokes and observations about their predicament.

Then the nurse lowered her voice. She had heard something on a camp radio, interviews with some of the parents of the Chibok girls. The parents had been denying themselves food for days on end, believing their hunger would attract God's favor.

"You don't know," she said, "but they are praying and fasting for you. You too should pray and fast."

If there was a fast, Naomi knew for sure that Kolo would have joined. The news from home filled her with a new resolve. She could picture her mother praying in the Church of the Brethren, under the blue cross on its white cement wall. Naomi reached a decision. She too would fast and encourage her classmates to do the same. It was one way she could take back control.

As the nurse finished Naomi's checkup, their conversation returned to their shared love of the gospel songs of Mama Agnes. Quietly, alone in the clinic, they began to sing the melody of one of their favorites, "Dogorana," meaning "Trust."

It is a must for trials to come
But if we stand firm to the end
We will receive the crown
And be like the angels

20

Agents of Peace

Under strip lighting in a conference room on the second floor of the Swiss embassy in Abuja, the Barrister, Zannah Mustapha, was staring into a projector screen showing a large color-coded organization chart of Boko Haram's leaders, contacts, and messengers. At the center of the spidergram was a smiling Abubakar Shekau, a spaghetti crosshatch of red, blue, and green lines linking him to jihadists and weapons smugglers from Nigeria to Libya and Syria. Zannah scanned the names and recognized only a few of them from Maiduguri; the rest were a mystery. They were known by noms de guerre and had never been photographed. Across from Zannah sat Pascal Holliger, the Swiss diplomat who had become his close partner, posing questions in his pitch-perfect northern Nigerian accent. Dressed in a blue suit with a frumpled tie, he was asking which of the Boko Haram commanders in the diagram were the most influential. But while the diplomat was studying the screen, Zannah had been studying him.

Almost eighteen months had passed since Pascal had first traveled to Zannah's orphanage in Maiduguri, and the Barrister had been gleaning clues to his partner's backstory. He had been born in Brazil but spent his adolescence between a watchmaking town in Switzerland and apartheid-era South Africa, where his

mother sometimes sheltered local black families from the authorities during the state of emergency. He would offhandedly recall
conversations with political prisoners who had served time in the
same Robben Island jail as Nelson Mandela. It was hard to know
what to make of him or his way of mirroring people. Zannah had
the habit of muttering "That's good," in the middle of conversation, and recently, the Barrister had noticed his Swiss partner
had begun to bounce his own stock phrase back at him.

For months, while troop battalions and mercenaries had been
fighting their way into insurgent territory, trying to kill the leaders and free their captives by force, these two men had been trying
to decode the structure and the mind-set of Boko Haram. Their
assignment, funded by a discreet Swiss government agency, was
to pursue dialogue with the terror group in the hopes of starting peace talks. But to achieve that end, they needed a top-to-
bottom understanding of an insurgency that had been reduced
to a ghastly caricature in the public eye. For all the thousands of
people left dead in the war, for all the drones in the sky and the
reward money on the group's heads, there remained some basic
questions that no government had managed to sufficiently answer. How was Boko Haram led? Who, if anybody, would Shekau
listen to? Who *was* Shekau? Was he one man, or a chain of secret
successors? And whoever he was, what exactly did he want with
the two hundred teenage girls who had finally given him the no-
toriety he craved?

The Swiss diplomat had neither enormous resources nor special access to classified intelligence. Pascal's home government,
which had an almost entirely volunteer military and a tiny spy
agency, had dispatched exactly one person to deal with Boko Haram, and that was he. What he did have was Zannah and the
budding team of Maiduguri men the Barrister had helped bring
on board, Nigerians who had the contacts and skills the embassy
was looking for. These were Zannah's friends from the northeast,
often eccentrics from the fringes of the conflict, and they had

learned to survive the war by artfully managing their relationships with both sides.

One was a former soccer goalkeeper turned Quranic scholar,
Tijani Ferobe, whose father was a prominent cleric. This gregarious man knew how to track down contacts close to the insurgency,
living as far away as Egypt. Another was an elderly disciple of
Mohammad Yusuf who'd refused to follow Shekau into war but
had helped families slip messages to Boko Haram suspects in the
crowded cells of the Giwa barracks. His name was Tahir Umar,
though the prisoners and their guards alike called the unassum-
ing old man Ba, or Father, as a sign of respect.

Together, these men became the core of a unit the Swiss diplomat called the Dialogue Facilitation Team: a neutral party for
both sides to confide in. In time, Pascal would allow himself a
more idealistic term—Agents of Peace.

All of the men were signing up for work that would be politically sensitive and sometimes dangerous. They would keep their
day jobs and their normal lives. Zannah would continue to run his
orphanage. Each of them had to assume that their phones would
be tapped and their movements followed by intelligence agencies.
Some of them would be asked to sneak into insurgent territory
to meet contacts and deliver messages, assignments that risked
death or imprisonment. All agreed to abide by the one condition
the Swiss held as an article of faith: absolute discretion. This was
a low-profile effort, and the less said, especially on social media,
the better. "Everything is being done transparently but quietly,"
Pascal e-mailed one new recruit, "away from the 'fanfare of the
other efforts.'" It was his subtle dig at the clamorous but ineffec-
tive parade of mercenaries, Nigerian troops, social-media influencers, Americans drones, and celebrity politicians that Twitter
had inspired to try to free the girls. If his time Nigeria had taught
him one thing, he would say, it was this: "The people who talk the
most have the least to contribute."

By January 2015, his Dialogue Facilitation Team had begun

to coalesce, trading thoughts over encrypted phone apps or during hours-long sessions around the embassy's conference table. Just reaching the embassy could be tedious for the Maiduguri men, who had to travel two days over deadly roads through army checkpoints or talk their way onto the governor's private jet. Zannah would try to inspire them by daydreaming aloud about giving an acceptance speech for the Nobel Peace Prize.

At the turn of the year, the Swiss diplomat had found another recruit who would ultimately become one of his most indispensable advisers. The man was a razor-sharp analyst of jihadist groups who walked with a limp and spoke with a nervous energy that sent his eyes darting from behind Coke-bottle spectacles. For several years, he had been what he called an "associate" of terror organizations throughout Nigeria and the Sahara loosely allied with al Qaeda, forming relationships with some of the region's most influential Islamists. He went by the nickname Nasrullah, meaning Victory of Allah.

Fulan had several phones, used separately and packed with contacts, some of whom knew him by different names. A former migrant who had braved the dangerous journey through Libya, he was fluent in the language of Nigerian militancy and wasn't afraid to travel into the countryside to sneak across battle lines. When he had first heard about the prospect of joining a Swiss-run team, he'd snorted at the idea, which to him reeked of white privilege. What business did Switzerland have jetting into Abuja to tell Nigerians how to pursue peace? Westerners had caused enough problems with their meddling in Muslim lands, he felt. No foreigner could ever hope to understand this conflict.

But Fulan was tired of life on the margins and wished to reinvent himself as a respected analyst on counterterrorism who could contribute to regional peace and security. He wanted to correct the dangerously simplified Western narratives about Nigeria's insurgency. In face-to-face meetings, Pascal had convinced him that he was different from other Europeans and was building a team worth joining, built on the expertise of Nigerians who understood the insurgency.

In time, the diplomat would learn to keep the bespectacled analyst and his other team members motivated by drawing on their interests beyond the battlefield. One day, watching soccer in a bar, Fulan seemed to be down, uncomfortable with the boozing going on around him. Pascal asked him if he had any hobbies.

"Tennis," he responded.

Some time later, Pascal invited Fulan into his second-floor office at the Swiss embassy, placed a speakerphone call, then asked an old friend to say hello to a fan. At the other end of the line was Roger Federer, the Swiss tennis champion, who offered a spirited hello. "I'm a fan," the onetime fundamentalist managed to stutter out, cracking a wide smile.

AT FIRST, ZANNAH AND HIS SWISS PARTNER HADN'T SET OUT TO FREE THE Chibok girls. Neither man used a Twitter account or saw the point of one, and the May 2014 hashtag campaign had erupted faster than they could process it. But by 2015, they had come to accept that the fleeting storm of activism online had permanently changed the reality on the ground. Social media had transformed the Chibok hostages into the most potent symbols of the war, and they could hold no broader peace talks until the women were set free. They were the boulder in the road, the key that could unlock the conflict.

In January, Goodluck Jonathan had given the Swiss embassy official clearance to pursue hostage talks to try to free the Chibok girls and use their release as a trust-building mechanism to kick-start a broader dialogue. "You have my permission," the president said. The first problem, as the Swiss diplomat noted to his colleagues, was that the president had given his permission to everybody.

The second problem was that they were all, from the insurgency's point of view, *kafir*, infidels working for a Western government with a cross on its flag. Bringing Boko Haram to the table would be hard and possibly without parallel, Fulan, the analyst, had warned. "Nobody has ever started a mediation from the point of 'We want to kill you,'" he told the team. "The only reason these guys want to talk is because of Michelle Obama." There was fame in it. And yet therein stood one of the central problems to structuring a negotiation: the millions of tweets to free the hostages had turned them into the insurgency's most jealously guarded asset. One of the team's contacts had a way of describing the power the girls had over its Imam: "What the girls mean to Shekau is what 9/11 meant to Osama bin Laden."

To convince him to give them up, the team would need to win his trust. "Trust is the fuel of the mediator," read one of the guidebooks Zannah had brought home from his course in Lake Thun. The riddle, for the Dialogue Facilitation Team, was to understand how they could get there. To solve that problem, the men turned the second-floor embassy conference room into a makeshift classroom for the study of Boko Haram. There they took notes and printed out articles that could help them understand why the insurgency and its like-minded allies waged war. They mapped its leadership structure, from its top decision-making body, the Shura Council, down through the qaids, or commanders, to the associates, and then the *rijal,* the foot soldiers. A whiteboard began to fill up with notes, near framed pictures of Alpine scenes and brochures advertising courses at Lausanne's elite École hôtelière school of hospitality management.

For hours and hours, they waded through Shekau's rambling online videos, trying to find some thin thread of logic sewn into his threats to kill and maim. One of their sources claimed there were clues planted in each video: Shekau would rant about cannibalism or Abraham Lincoln, but these were distractions from the hidden message, which could be a phone placed conspicuously to signal his openness to dialogue. The men carefully studied each violent tweet from Boko Haram's new Twitter account, flush with images of kidnapped boys, whom it called cubs, training with Kalashnikovs.

Western specialists connected to their own governments would visit to share and crib notes. An American from West Point spent several days in the embassy to help chart the sect's connections with Islamic State chapters as far away as Uzbekistan. Boko Haram, he argued, was a part of a wider world of aggrieved local insurgencies who'd stumbled into a shared set of answers to their frustrations. For lunch, the team would break for meals of *jollof*, a one-pot meal of spicy rice and tomatoes, while hanging out on a balcony that overlooked the neighboring embassy of Denmark, whose staff were intrigued by what the Swiss were up to.

In the pile of notes and biographical printouts of scarcely known terrorist commanders, it was easy to forget that this was sensitive and perilous work. Pascal knew Nigeria's feuding security agencies were monitoring him. Zannah, too, received strange reports from friends warning him that he was under constant digital and personal surveillance. The Barrister had abandoned two of his three cell phones and assumed that every call was listened to, every e-mail read. The other members of the team, some of whom were traveling into the field to talk to Boko Haram members, were most at risk. Pascal hatched a strategy, delivered in advance to coming guests whose attendance needed to remain confidential. At the gate, they would meet a security guard with a visitor's log, and the diplomat would warn them in a text, "Don't write your real name."

Pascal was also under orders to be clandestine in his communications. Nigerian officials were beginning to ask why this peaceful and wealthy Alpine nation would want to insert itself in the Boko Haram war, a question with no simple answer. When nobody else was around, he would slip into a sealed room in the embassy to send updates to his superiors. Other governments used e-mail, but the Swiss demanded that he use a cumbersome device only found in their embassies, an encrypted fax machine whose special-purpose product name struck the employees of the Swiss foreign ministry as some sort of wry joke: the 007.

BERN, SWITZERLAND

Four thousand miles north, Pascal's faxed reports were landing on an upper floor of a nondescript government office building opposite a toyshop in the center of his country's capital, overlooking the river Aare. The building was not listed on Google Maps. Curtains blocked the view into the lower floor, and only a small placard indicated the foreign ministry unit based inside, the Human Security Division.

Inside, a staff of about seventy officials, many of them trilingual graduates of Europe's finest schools, were pursuing a grand peace-building endeavor that was starting to seem almost naive in its global ambition. From this cramped gray-paneled office, Switzerland's government had been recruiting and training personnel to become peacemakers. Under the flag of Switzerland, a country that had proudly professed a policy of neutrality for five hundred years, it had sent dozens out into the world to help solve intractable conflicts in far-flung countries.

Many Swiss people had no idea the division existed. It didn't like to advertise its services, although an insistence on discretion left it with a reputation for secrecy that some personnel didn't like either. The shop was set up in 2000 as Political Affairs Division IV, but its name had recently been changed after staff complained that it sounded too Orwellian. The new name struggled to stick. Some in the know still referred to the unit by a cryptic acronym: PD4.

The division had been established at a fleeting moment, "the end of history," when it had seemed possible for America to preside over a worldwide lasting peace. This was a problem for Swit-

zerland. The wars of the twentieth century had made the country
influential beyond its size, as the neutral meeting ground where
warring states could speak to each other. Once the Cold War had
ended, powerful nations were no longer at arms. Moscow could
call Washington; Berlin could call London. Few governments required the services of another country to talk to their enemies.

Switzerland, with a population smaller than New York City,
needed a pivot, a way to retain its influence. It looked south, to
the developing world. There, as the century turned, new wars had
been erupting, not between nations but within them, in broken
states like Somalia, Rwanda, and the Democratic Republic of
Congo, poorly understood conflicts that neither the great powers
of the West nor its postcolonial allies seemed able to end. Resolving them would burnish the public brand of a country infamous
for banking dictators' embezzlements. A former Swiss president,
Micheline Calmy-Rey, told reluctant lawmakers that the peace-
building program was a mechanism for a country landlocked behind the Alps to advocate for itself in the era of globalization:
"You cannot stay anymore in this world hiding behind your mountains because you're not able to defend your interests."

The division had been less than a year old when the 9/11 attacks heralded an immediate opportunity. America's War on Terror brought sweeping new legislation that meant few European
nations would risk engaging with anyone linked to terrorism.
But Switzerland, fiercely independent and officially neutral, had
never joined the European Union and could not be sanctioned
for violating its restrictions on engaging with such groups. The
country's diplomats could issue visas to wanted men whom other
states called terrorists but whom the Swiss called by the less judg-
mental term "armed non-state actors."

Representatives of outlawed groups, including the Tamil Ti-
gers and Hamas, began flying on commercial SwissAir flights to
meetings in Geneva, where officials welcomed them with flowers
in their hotel rooms. For years, the Colombian FARC insurgency

sent Swiss presidents Christmas cards. One government-backed
NGO started inviting wanted militants from around the world to
sneak across borders and gather on Geneva's Lac Leman for boat
trips, where they could talk to each other about human-rights
best practices. "Or perhaps they discussed where to buy a Kalashnikov," an organizer said. "You never know."

These backchannel contacts helped Swiss leaders schedule face
time with US secretaries of state, meetings they often used to push
back against American financial regulations that were targeting
Swiss banks. Thomas Greminger, one of the founders of the Human Security Division, told his American counterparts that they
were "too conventional in their thinking about who was a terrorist." In 2009, the day after Swiss citizens voted in a referendum
to ban minarets, Greminger had his officers call the Islamist
leaders they had come to know, expressing disappointment with
the vote but asking them to discourage any terrorist attacks on
Switzerland. Later, when Swiss citizens were kidnapped in war
zones like Mali and Libya, their foreign ministry knew how to
reach the militants who'd taken them. At one point, the government suggested opening talks with Osama bin Laden.

The Human Security Division had been born into a golden
era for the peacemaking business. From 1988 to 2008, the world
signed some 650 peace agreements, and negotiation brought more
wars to an end in those twenty years than in the previous two
centuries. The number of people killed in hostilities plummeted.
The division's diplomats helped ink more than a dozen accords
from Sudan to Indonesia, crisscrossing battle lines in the margins
of a world order framed by the United States, the solitary superpower. The Swiss model seemed to be working.

However, in the decade that followed, the dynamic flipped and
hardly any wars ended. Libyan warlords, Yemeni militant leaders,
and Syrian rebels came to Geneva, but all left without progress.
Through the 2010s, the peace industry had tipped into a recession. The War on Terror had seeded irresolvable conflicts, first in

Afghanistan, then Iraq, and eventually in dozens of countries, proceeding from an American "never negotiate" logic that had proven more durable and widespread than the Swiss had hoped. The world's richest and poorest countries alike were carving out enormous chunks of their budgets to fight terrorists, an economy of war that was daunting to untangle. The insurgent side was just as intractable. Radicals were using YouTube and Twitter to stage grotesque executions before an online audience. Drones offered a cheap way to sustain war but not to end it. This modern phenomenon was given a name: the Forever Wars.

By 2012, the Human Security Division was looking for a years-long conflict it could help bring to an end. It turned its attention to Nigeria. Neither Abuja nor its powerful Western allies seemed to have a handle on the deteriorating security situation. The division labeled Nigeria a tier-one target and decided to post someone there for the first time on a scoping mission to see if there was a way it could offer its services.

The individual they sent was not a career diplomat. Pascal Holliger had been living in South Africa, running "township tours" in an old Volkswagen bus, driving foreigners to visit poor black settlements and spend their money there. He had brought Roger Federer to the tennis camps he ran for local children. Living along the coast, he had gained a reputation for speaking English with a Xhosa accent, the mother tongue of Nelson Mandela, ingratiating himself to his hosts. A small item in a Swiss newspaper had given him a nickname: the White African.

In Nigeria, the first months of Pascal's assignment had not been filled with optimism. Through 2014, his faxes back to Bern had outlined the tailspin of a country slipping further into conflict. Yet by early 2015 he was reporting a new and encouraging reality on the ground: Nigeria was under new leadership. Perhaps, for the Chibok girls and the country at large, there was hope.

21

The General

Nigeria's new president was staring into a camera through a pair of thick horn-rims, his salt-and-pepper mustache framed under a colorfully embroidered kufi cap, and talking live to Christiane Amanpour. The CNN anchor was posing a question many people in the West had once asked but had by now forgotten.

"What about the girls, the famous Chibok girls who were kidnapped so long ago? Is your government willing to trade those girls?"

Muhammadu Buhari, a ramrod-straight former general who had won the election four months earlier on a promise to vanquish Boko Haram, spoke slowly and deliberately. During the campaign, he had pledged to liberate the girls. Now came the pressure to make good on the stump speech. "We have to be very careful about various Boko Haram leaderships claiming they can deliver," the seventy-two-year-old said, his gravelly voice enunciating each consonant. "But our main objective as a government is to secure those girls safe and sound."

It was July 2015, and Buhari had landed in Washington for his first visit as president on a mission to win support and military aid for his strategy to defeat the insurgency. He knew exactly the standard on which success would be measured at home and abroad. "We can not claim to have defeated Boko Haram without rescuing the Chibok girls," he had said in his inaugural speech. On his US trip, the sharp-jawed septuagenarian was eager to present a fresh face to the world and show the public that Africa's largest democracy was under new management.

Back home, though, the incoming president was a throwback to the 1980s. There wasn't a single voter who didn't know the stories about the ruthless military man from Nigeria's north who had once ruled the country with an iron fist. President Buhari had previously been Major General Buhari, a soldier-dictator, who used his twenty months in power to jail journalists, execute drug dealers, order the expulsion of nearly a million immigrants, and imprison dozens of politicians for corruption, many for life.

A generation of intellectuals had fled the country, and the ones who stayed found plenty to criticize. "The trouble with Nigeria is simply and squarely a failure of leadership," wrote the author known as Nigeria's conscience, Chinua Achebe, the year the dictator came to power. But in Buhari's worldview, the problem seemed to be the other way around. It was the people who lacked discipline and this raucous, dynamic society would accomplish much more if only its citizens were instilled with the kind of regimentation he'd imbibed as a young cadet at an infantry college in Aldershot, England. The trouble with Nigeria was the Nigerians.

To change the national character, General Buhari launched a nationwide War Against Indiscipline. Soldiers used whips to force bus passengers into orderly queues and made civil servants who showed up late for work endure the humiliating punishment of performing frog jumps in front of their peers. The goal was nothing short of rewiring Nigeria's cultural DNA.

Instead, the country's economy collapsed, the wave of emigration sped up, and repressed grievances began to boil over into protest. As General Buhari became unpopular, his close friends in the military seized power for themselves, arresting him in his home at gunpoint, detaining him for two years and replacing him with another army general in a succession of military dictators who would last until the country emerged as a democracy, laden with debt and broken infrastructure, in 1999. All told, Buhari's short and ignominious experiment in statecraft had been a key step toward Nigeria's slow-motion state failure, and he had seemed destined to spend the remainder of his life in the political wilderness. But in an ironic twist, thirty years later, Nigerians returned the general to the presidency to save the same nation he had once driven into ruin.

Buhari had won a resounding victory in the March 2015 elections with an audacious act of political rebranding, casting himself as a democrat who happened to have the military credentials to restore order. Some of his billboards simply read SECURITY. Nigeria's new leader might not be a born democrat, went his argument, but he had spent much of his life fighting insurgencies inside Nigeria, and knew what it took to win the war. Crowds of mostly young men numbering into the tens of thousands flocked to watch him speak across the north, with spectators climbing up lampposts in the hot sun to cast their eyes on the man whose campaign symbol was a broom: a clean sweep to rid Nigeria of its myriad troubles. His election was a signal of just how dire circumstances in Nigeria had become. Boko Haram, though it had been pushed back during the final weeks of the Goodluck Jonathan administration, had left a trail of human wreckage. More than a million people had been made homeless by war. The economy was plunging into what would become its worst recession in four decades. Jonathan, the zoologist who'd become a president, had swiftly and graciously conceded, beaming as he boarded a flight to sightsee wildlife in the Galapagos Islands.

President Buhari had spent his first weeks in office trying to make sense of the strange and murky finances of the government he inherited. His supporters, still optimistic, were becoming impatient for more immediate change, and had begun calling him Baba Go-Slow. Returned to power after three decades, the president knew there was one way he could set a successful tone for his second chance. His advisers understood that everywhere he went he would be asked the same question from reporters and politicians. "What about the Chibok girls?"

Getting them back would be a symbolic victory that would confound his critics at home and be reported all over the world, generating goodwill for his administration. It would be redemption for a military man whose last turn in power had ended in disaster. Two days before his trip to America, after meeting in a conference room of the Aso Rock Presidential Villa with a pair of crying mothers of the missing schoolgirls, President Buhari had gathered his most senior advisers for an important discussion. He was authorizing another attempt at negotiating the girls' release. More than that, he was setting a deadline. He wanted the Chibok girls returned within his first hundred days as president.

The Chibok matter should be ended not one day later than September 5, 2015.

IT WAS 94 DEGREES ON THE MORNING OF JULY 20, WHEN BUHARI ARrived at the White House in a black SUV with tinted windows, walking through a phalanx of ceremonial marines to meet Barack Obama. The Nigerian president and his staff had spent the morning with Joe Biden, the vice president launching into a long story involving a helicopter landing in Afghanistan while his visitors around a breakfast table smiled politely. America's forty-fourth presidency was in its twilight, and some in the administration were leaving with the regret that Africa hadn't gotten more bandwidth.

Before a row of reporters, Obama welcomed Nigeria's new leader as a "man of integrity" who "has a very clear agenda in defeating Boko Haram." When the cameras had disappeared, the meeting began with the two presidents talking about Nigeria's elections. Buhari pushed for help fighting corruption and money laundering. Then Obama turned to the Chibok girls.

It was hard to believe that just over a year had passed since the kidnapping. Nobody could imagine what the schoolgirls had been through during the same time. There had been no recent sign of the girls in the surveillance planes and satellite images of Nigeria's north, and the US had only scattered intelligence reports of them being split up into separate locations. Neither president had an easy answer on how to bring them back.

But America hadn't given up helping Nigeria find them, Obama said. Helping secure their release was part of a dramatic increase in military assistance the US was offering Nigeria's new president and his neighbors, an escalation that the American public barely registered, which would come to include training, weapons, and the deployment of intelligence officers to a military base in Maiduguri to track Boko Haram. Into the countries surrounding Nigeria, the Pentagon was pouring hundreds of troops, who were bringing new tactical vests and night-vision goggles to the local soldiers, whom they would accompany on border patrols, in trucks fueled by American funds. This was the strange and continuing mutation of the #BringBackOurGirls phenomenon. A Twitter hashtag to save schoolgirl hostages had been slowly and covertly morphing into a billion-dollar deployment of drones, intelligence officers, and turboprop planes full of American troops. The State Department had been so inundated by congressional calls during the height of the social-media campaign that it created a new post of special coordinator for the Boko Haram conflict, filled by a veteran ambassador who was now pushing to have much more American intelligence scooped up and supplied to the Nigerians. In Maryland, a contractor for the National Security Agency had just put out a Help Wanted ad for Americans who spoke the northeastern Nigerian language of Kanuri to scan intercepted audio. It appeared in a local newspaper, next to a posting for "Crew at Chipotle."

Buhari wanted to bombard the insurgency from above and had pledged to "crush" it with a renewed ground advance. The

United States, while worried about civilians being killed in the airstrikes, was willing to sell the country four easy-to-fly fighter planes, called Alpha Jets—with the weapons removed. The Nigerians would have to buy their own rocket launchers and attach them to the wings. They wouldn't be flying blind, though. By the end of the year, one of the US Army's most elite units, the Night Stalkers, which hunted its targets after dark, would be flying four Gray Eagle drones over Nigeria, each capable of soaring for two days on end, with equipment to sweep up phone calls, and cameras to map body heat or read a license plate from fifteen thousand feet. This was the unit that had flown cover for the operation that killed Osama bin Laden, now coming to help dismantle Boko Haram—and, it was hoped, locate the girls.

But securing the hostages' freedom would have to flow from a Nigerian solution, Obama argued. Buhari's predecessor had used the last days of his term trying to convince John Kerry that America should send combat troops into Nigeria, which the US wasn't going to do. Meanwhile, America's tech giants had belatedly started to act to remove Boko Haram's violent propaganda, and Twitter had finally deleted the group's account.

As the meeting wound up, Buhari asked Obama if he couldn't squeeze in a trip to Nigeria before the clock ran out on his second term. Obama offered to consider the invite, though he wasn't going to be around much longer. The graying president was on his way out, hoping the American people would elect someone who could carry out his legacy.

"I don't know why anyone would want to stay in a job like this longer than you had to," he joked to his Nigerian guest, an ex-general who had secured an improbable return to power after thirty years. "But as my wife reminds *me*, you volunteered for it."

22

Shadrach, Meshach, and Abednego

When Lydia John was a senior, she auditioned for the class choir, an exclusive club in school. The Glorious Singers accepted only the best. To make the team, each chorister needed to launch her voice into its top register, bending vowels between notes so high that their cheeks hurt after practice.

Chibok's school had no shortage of girls with ability. Singing was a collective release for the town's residents, from grandparents to kids, and almost everybody knew the popular *sabon rai*, "reborn soul," hymns by heart. On slow weekdays, when the girls grew bored in their dormitories, a student would often pull out a drum and beat an intricate, syncopated rhythm. Classmates would rise and gather into lines, then step on the spot, jabbing elbows and ducking heads together in a collective choreography. Their words would build into a wall of sound, dozens of teenage voices all chasing the same impossibly high notes.

"Today is a happy day!" went one praise song. "Everybody, shake your body, thank God."

From this oversupply of talent, only twenty vocalists had the skills to qualify for the Singers. Lydia had been nervous before auditions. The choir performed physically exhausting renditions that tested the limits of the vocal range, featuring nasal call-and-response refrains, but it wasn't enough to reach the upper octaves. The Glorious Singers had to dance and clap in synchronized timing. In school, Lydia wasn't a natural performer, but she did have an exceptional voice and won a place in the choir.

Malam Ahmed loathed the blasphemy that he heard in melody. He had lectured his captives at length on why it was haram for a woman to raise her voice in any context other than a Quranic recital. Music was the preserve of the *arna* and made God quake with rage. Under his rules, any girl caught singing would be beaten with branches cracked from a tamarind tree.

In her early months in captivity, Lydia had been recalling songs in her head to while away the long empty hours. Sometimes, when guards weren't paying attention, she and Naomi would write down lyrics they didn't want to forget. Their notebooks were supposed to be full of Quranic verses, but they had been filling them up with hymns. In the first year of captivity, they had felt it too risky to vocalize the melodies.

By mid-2015, however, Naomi and some of her closest friends were starting to lose their fear. They had witnessed Boko Haram in retreat, fleeing the gunfire from helicopters, and seen its members dying. Their captors, who'd once seemed larger than life, appeared to be weakening. "We had seen so many dead bodies, we were no longer afraid to die," Naomi would later say.

Their new home was another flat, dusty settlement, this one named Ukuba, an abandoned village of thatched-roof homes that Boko Haram had made its own. Dozens of Chibok students, separated in the previous months, had been corralled back together, sharing stories of the hard labor and petty chores they'd been forced to do in other camps. Malam Ahmed was around more often now, living in a nearby home made from air-dried mudbricks. None of the girls could say for sure where the rest of their classmates were being held, including those who had been forced into marriages at the senator's house.

Reunited after weeks apart, Naomi and Lydia began to find their voices. When the guards were out of earshot, the two would hum, then murmur, the words to popular and religious songs. Naomi would choose which songs to perform and dare Lydia to sing louder. The younger hostage was worried about the consequences. "Anything that happens, happens," Naomi told her. But as the months drew on, their songs became loud enough for other girls to hear. Naomi's cousin, Saratu, had also been singing and sometimes joined in.

The main material they drew on was Mama Agnes. As they performed the vocals, they would strive to remember the bubbly Casio keyboard beat and brisk rhythms that would have poured from a speaker at a wedding or church service in Chibok. Mama Agnes had a voice like a flute, which fluttered along at a heavenly register, like a record playing on the wrong speed. Naomi couldn't hit those notes, but she could help Lydia reach for them.

An insurrection was brewing among the captives, and singing was its most provocative expression. At prayer times or whenever guards were dozing or distracted, clusters of hostages would sing through cupped hands while lying on the ground to muzzle the sound. Other times they gathered in a tight circle, heads bowed toward each other, singing into the dirt, their voices bouncing back. One evening, when the guards ambled off for Maghrib prayers, Naomi heard Saratu begin to sing with two other hostages. Within a few minutes, dozens of girls were singing together in hushed chorus.

"Nebuchadnezzar is the king of Babylon, big king of Babylon," one small group sang, the opening lines to a hymn about a trio of Israelites who refused to abandon their faith. "We, the children of Israel, will not bow."

A YEAR INTO CAPTIVITY, BOKO HARAM'S AUTHORITY OVER THE CHIBOK girls was starting to wane. The group had been pursued by the Nigerian army and forced to break camp so often that its fighters seemed disorientated and distracted. The girls could hear the *ri-jal* share rumors that Shekau had been badly injured. Even so, Malam Ahmed would ask them the same question almost every day: When will you get married?

One group of holdouts was becoming bolder and more determined to refuse at any cost. They talked back, shrugged off the slavery work he demanded of them, and suffered beatings for their refusal. Malam Ahmed gave these most insubordinate captives a label. They were the *arnan daji*, "pagans in the forest," an insult Naomi wore with pride.

The group was coalescing around a few unexpected rebels who refused to yield. At school, Hauwa Ishaya had been a bubbly conversationalist with a contagious smile and a big laugh. Saratu was quieter but calculating, a top student who kept her thoughts and emotions private. Palmata Musa had been their mutual friend. Their group of "pagans" grew and shrank as the months wore on. In time, they began to absorb a collective identity, speaking as "we" more than "I." They understood that, as Chibok hostages, they were special and had a value to their captors that afforded them some level of protection.

The result was a new strain of quiet, and occasionally fullthroated, defiance. The girls would act out scenes from the northern Nigerian romantic drama *Bashin Gaba*, "Debt of Hatred," a Romeo-and-Juliet story of two young lovers who eloped after their parents objected. They retold Bible stories, revisiting the notes copied into their diaries.

Eventually, word of the girls' indiscipline reached Malam Ahmed. The girls were singing, he learned, and were hiding a Bible.

He was furious. His guards arrived, a mass of men descending on them all at once, barking orders, and demanding to search the area. The girls stood to the side while the men rifled through the piled-up clothes and kitchen utensils they kept under a tree. But the girls had already buried their diaries and a Bible, marking the spot with a stone. They'd tied another Bible to one of their legs, using a piece of rope. The group had been tipped off; two of them had heard the men talking about the upcoming raid while washing their hands before the afternoon prayer.

The militants confiscated medicine, mainly basic painkillers the girls had been hiding. They found a cell phone. To Malam Ahmed's frustration, no girl claimed it. One guard picked up a notebook, but saw only Quranic writing. Another announced a punishment: "If we see anybody holding any kind of book, we will beat them," one said, holding up a rifle and a tree branch. "Now we can't find it, but there will be a day when we do."

When the guards fell asleep, the girls, who had been pretending to doze off, would turn to one another and begin to pray. Most often they prayed for freedom. They felt guilty for the grief they were causing their parents, so they said supplications for them. They asked God to heal the sick and wounded among them. Naomi prayed for fighter jets to kill Boko Haram.

"One tactic was to pray into a cup of water," she said. "We would sit together. We brought it to our mouths. It would disguise the prayer and also make the water sanctified. We would then sprinkle the water over our sisters who were injured or depressed. We felt it could heal them."

Faith became their language of resistance. The group were embarking on regular fasts, taking turns renouncing food for a few days while others ate, before trading roles, to create a constant spiritual energy.

Malam Ahmed and his guards responded to their insubordination with beatings and more pressure on the hostages to wed a fighter. As he brought the branch down on their necks, he would chastise them with each stroke, "You. Are. Too. Stubborn." One day, he instructed all the students in Ukuba to stand in a line and watched as a guard whipped every single one with an electrical wire. The lashings inflamed the skin on their backs, turning it raw. One girl couldn't stand for a week.

Another time, he ordered ten lashes to a group of hostages

who'd walked away without permission to pick leaves from a nearby baobab, an ingredient for their dinner, kuka soup. The girls promised not to step away again if he let them keep the harvest, but a few days later, they were back at it, snapping leaves from another tree.

"We went anyway," said Ladi Simon. "We were hungry."

Sometimes, the girls had no idea why Boko Haram was beating them. Other times, they suspected their classmates coerced into marriage were informing on them.

One afternoon, shortly after the guards slipped away for the Maghrib prayer, Comfort Bukar began to sing by herself, almost inaubibly at first, then louder, gradually rising to her feet. Singing hymns had become more common but none of the hostages had dared to stand up and dance in such a gesture of defiance. But Comfort seemed to be somewhere else, transported far away by the melody, her eyes shut as she began clapping and spinning to the tune. The girls watched her, lost in the moment, pivoting on the balls of her feet.

Then a guard came rushing back, furious. Without hesitation, he picked up a heavy stick and brought it down hard on Comfort's back. She fell on the ground, and curled into a ball, screeching each time she was struck. The guard kept hitting her. "She was writhing in pain, saying, 'Sorry, sorry.'"

The hymns stopped for a while. Then, slowly they began to restart, through cupped hands, under cover of darkness. "We should just hide our singing," one of the girls said.

TO MALAM AHMED, THE MUTINY AMONG HIS MOST VALUABLE ASSETS could be traced to a few disrespectful malcontents. The face of this group was the stubborn hostage he'd been "forced" to strike with a rifle butt on the porch of the senator's mansion in Gwoza. The chief problem was Naomi Adamu, whom he started to call by a nickname, the Chief Infidel.

At times, he would bring Boko Haram members to come meet

the teenagers and introduce her as the one who most strenuously refused to get married. Naomi felt that he viewed her as an obstacle. One day, he told her that all the girls who had married had gone home. When she called him on the lie, he stormed off. Another day he declared to the group that Naomi's illness was not a kidney disease but the result of a secret abortion. It wasn't true, but he insisted it was to slander Naomi in front of her classmates.

Finally, he came with a pad of paper and took down a small list of the full names of the captives who'd done most to undermine him, starting with Naomi Adamu. The Chibok girls would struggle to embrace the true Islam so long as agitators like them were in the mix. The Malam could see their influence was spreading to more classmates, who were refusing to pray and denying themselves food for days. As he jotted the names of the disobedients, he warned them he was being forced to take more drastic action: he would report them to the Imam.

ONE AFTERNOON, FIVE GIRLS SNUCK OFF TO THE BANKS OF A SMALL stream, under the premise of needing to wash their clothes. The real reason they were slipping away was to test one of the cell phones they'd been hiding.

The plan was to remove the battery from Naomi's battered old Nokia, held together with a rubber band, and see if it fit the camera phone that Yanke Shettima had been keeping in secret. Yanke's battery had died, but Naomi knew hers had power left in it. Maybe Yanke's phone could place a call. Hauwa would be the lookout to make sure the guards stayed a safe distance away.

A small hill shrouded the bank from the view of their guards, and as they crouched beside it, Naomi cracked open the case and clicked her battery into Yanke's phone. Yanke stared into the glowing screen and waited to see if her phone could muster service. The five knew they would be lucky to get a signal. As they stood waiting for a bar to appear, one of the girls had an idea. Posing in the sunlight by the small stream, they took turns snapping pictures of themselves, before breaking into embarrassed laughter. Until now, they had seen their reflections only in the rearview mirrors of motorcycles or trucks parked around camps, and now here they were, on the screen. Naomi was shocked to see how hungry and tired she looked. Some of the hostages saw themselves and started to cry. Naomi told them to keep taking pictures. "When we get home people will see how we suffered," she said.

One of the girls snuck over the hill and called some of their classmates down to the creek to pose for a picture. They took turns, and stopped caring that there was no cell service. "We were so happy," Naomi said. "We felt that we were home. We felt free." Then, after a flurry of selfies, Naomi looked to one side and saw Yanke sitting on the ground, head bowed. Her friend had been flipping through the photos stored on her phone, and halted at a picture from a wedding just five days before the abduction. Standing in front of the Church of the Brethren was her mother, father, and six siblings, all of them gathered around her grandfather, Tama, the family patriarch, smiling contentedly in a red kaftan, his favorite. Yanke began to weep. It was the first time she'd seen her family and friends since she had been kidnapped.

23

Sisters

Maryam and Sadiya were standing close to each other in a large boisterous crowd, their faces concealed by full black niqabs. Hundreds of men and young boys were standing at the front, the women bunched behind into a smaller, separate group. The attendees were excited, jostling behind a rope that sealed off a grassy clearing next to a tin-roofed building converted into a mosque. A young bearded man standing on the grass began to solemnly recite some prayers, then make some announcements into a microphone hooked up to a single black speaker. "You must all pray at the mosque at the allotted times," he said. "If you do not abide, you will be punished."

Behind him stood a group of fighters in masks, some carrying a large black-and-white flag, others holding what appeared to be large knives or meat cleavers. The announcer began to read out sentences for the condemned. Neither Maryam nor Sadiya had wanted to attend this spectacle, but the sisters' husbands had said the *hisba*, the religious police, demanded it.

A group of convicts, hunched over in shackles, stepped forward as their names and crimes were blasted through the speaker. An old man was sentenced to a hundred lashes for using drugs. Three thieves were to have their right hands placed on a log and amputated at the wrist. A couple convicted of adultery would be buried to their necks and then stoned to death. The prisoners stared straight ahead, looking through the crowd but making no eye contact. The man with the microphone pointed to them and asked a question designed to whip up the faithful: "Should they be punished in accordance with God's law?" Voices began to cheer. Some of the fighters fell to their knees in prayer.

Then he turned to the audience with another message that Maryam felt was directed at her. "Do not avert your eyes. Whoever does not watch all the punishments will be beaten." At the front of the crowd, Maryam could see young boys watching in silence. Close to the condemned, she also saw the man who called himself her husband, crouched on his haunches and holding a stone.

MARYAM AND SADIYA WERE NOW EMIRAS, WOMEN WHO'D BEEN COERCED into marriage, a designation that brought benefits and a new social status but the horror of constant abuse. Separated from their classmates and for the most part, each other, the sisters were part of the growing number of hostages beaten down by Malam Ahmed's pressure. By the end of 2015, more than eighteen months since their abduction, perhaps two dozen of the Chibok students had been coerced into what the Malam called marriage. Many had been relocated to new camps elsewhere in the northeast. Their days were now spent learning about the sanctity of Boko Haram's mission and the evils of the life they'd left behind. They attended hours-long classes at a new Islamiyah school for fighters' wives, where veiled women and their children peeped through the doorway to catch a glimpse of the Chibok girls they had heard about. Their new dress code, a full-length black niqab that covered everything but the eyes, announced they had now married into the sect and were no longer deemed hostages.

They had disappeared into another world, where women wearing precious stones on henna-dyed fingers would gather to talk about the bravery of their husbands and the evils of Nigeria over hearty meals served by captives they called their slaves. Some of these women wore makeup and fine new clothes and talked about the perfume they would wear for the men when they came back from battle. They boasted that women marrying into the sect were lucky to be freed from *arnan*, paganism. But Maryam and Sadiya were still being watched. The Boko Haram wives viewed them with suspicion and would inform their husbands of any hints of deviation from their new creed. They could never fully be trusted, and it was no longer clear if the sisters could confide in anybody, including each other. "We weren't allowed to talk about home," Maryam said. "Or to discuss anything sensitive."

Some of their classmates who had not been coerced into marriage viewed life as an emira as the easy choice. Maryam and Sadiya had slept in fine houses in Gwoza and now slept under bed linen in grass-roofed homes. They were sheltered from the elements and protected from the insects that had once burrowed into their skin. They ate hearty plates of rice and sometimes beef or fish, delicacies they had rarely enjoyed even in Chibok.

But inside their homes, they were victims of other forms of predation, serving men who saw them as their property, a reward for the holy war. Mornings would mean washing clothes, wiping boots, cooking, and tidying the home. At night, they would lie in the dark and wait for the militants to come home, knowing they were expected to conceive a child.

The man who had chosen Maryam was Abu Walad, a fighter who helped guard the Chibok girls. Sadiya was married to a more senior commander, who already had three other wives, but Abu Walad had been unmarried when he "met" Maryam. He was known as an enforcer, a tough, unforgiving true believer, well over six feet tall, who had been quick to flog hostages who disobeyed him. He was rarely seen without his rifle, and he often disappeared for days. When he returned, he kept the gun under their pillow during the day and propped next to their bed at night. Maryam was terrified of sleeping next to a loaded gun, but at least it was on Abu Walad's side of the bed, so if it went off, he would be more likely to be shot. He never talked to Maryam about his "work." She knew only a couple of details about his life, that he had been born in Maiduguri and that his father was from across the border in Chad. Beyond these few mumbled exchanges, they had rarely spoken at all.

"We never had a real conversation," Maryam said.

One thing Abu Walad didn't conceal was his love of violence. As far as Maryam could see, his favorite pastime was watching videos of Boko Haram military operations and executions. For months after she arrived at his home, he would sit and flick through grainy clips on his phone. He ordered Maryam to stay home unless he gave her permission to leave.

Maryam was too afraid to think about escaping. Besides, where would she go? Malam Ahmed had said over and over that Chibok was now part of Boko Haram's caliphate, and she wasn't sure if her mother and father were even alive.

Back home, in the Gana Hostel, Maryam and Sadiya had been so close they'd shared a single mattress, but now they were rarely alone together. Sadiya was always accompanied by her co-wife when she came to visit Maryam, because the sisters' loyalty to their captors could not be taken for granted. The co-wives would instruct Maryam and Sadiya on the *true* practice of Islam, which bore little resemblance to religion of the same name that the sisters had observed faithfully since birth. Talking about home or the families they left behind was considered suspect. Maryam wanted to play *haptiri* with her sister, a game they used to play at home, taking turns to throw small stones into a hole. But the guards had said it was haram, that it was "the devil's game," and Sadiya feared what would happen if they were caught.

Instead, the sisters would sit with other wives for hours drawing henna tattoos on each other's hands or plaiting each other's hair. The women would pepper their sentences with praise for the fighters' cause. The more they repeated these exhaltations, the more Maryam wondered if Sadiya was starting to really mean them.

IN THEIR RARE MOMENTS ALONE, SADIYA WOULD SOMETIMES TELL SELF-deprecating stories that revealed flashes of her old self. When her husband remarked that Maryam was more beautiful than she, she told him to propose to her sister instead. That was the Sadiya Maryam had known, willing to laugh at herself. But the fighter appeared to take her seriously, responding regretfully that it was not permissible to marry two sisters.

Most of the time, though, Maryam worried that Sadiya was becoming too sympathetic to their captors. She was parroting more of their teachings, lacing her conversations with diatribes against Nigeria's godlessness or the evils of Western education. One day she announced that she was happy to be married, as it was "God's will." Her husband had brought her a ring, and she was hoping to have a child. Maryam nodded, unsure whether her sister was being truthful or toeing their captors' line. She began to doubt herself, and felt her own sympathies starting to shift.

As the months went by, both sisters became less sure where the other's loyalties lay. Sadiya had once confided to Maryam that she was having nightmares and struggling with stress, but now she was defending her captors' beliefs and said their murders would be forgiven in the afterlife. God would look kindly on "our fighters," she said.

"I was worried," Maryam later said. "I observed she was gradually changing to adopt the group's ideas."

ONE DAY WHEN ABU WALAD WAS AWAY, MARYAM FOUND HIS PHONE AND began to rifle through it. She knew that the storage disk was full of gory battle scenes and executions. They repelled her, but from boredom, she swiped through, clicking on the videos.

Most of the clips were the now familiar gruesome scenes she'd seen him watching, but one stood out. It was some sort of newscast from many months earlier. Curious, Maryam pressed play. The phone lit up with footage of thousands of people marching in cities in Nigeria and around the world. They were waving banners and shouting in unison: "What do we want? Bring back our girls!" The voice of a narrator explained in English that the crowds were campaigning for the release of the "Chibok girls." Maryam couldn't recognize the cities some of the protestors were marching through.

Watching those few minutes of footage under a thatched grass roof in an encampment with no Internet, Maryam was perhaps the first Chibok hostage to have any sense of the full extent of the global campaign calling for their release. Her classmates had understood that they were different from other Boko Haram captives, piecing together the chatter they overheard among the guards and snippets of radio broadcasts wafting around the camps. But none of them realized there was a worldwide campaign that had ricocheted through social media to Hollywood. None of them had a Twitter account, and very few of them had visited even Abuja, let alone the cities thousands of miles away where thousands of people were marching in their name. Maryam didn't know any of the American celebrities who were demanding her safe return.

She sat there in spellbound silence as the clip went dark.

She watched the video again, several times in a row, trying to connect her almost two years inside Boko Haram's caliphate with the images on the screen. It was bewildering, disorienting. When she'd seen it enough times, she said a prayer that she would soon find her way home. "I realized that the world really cared about us," she said.

She wanted to tell Sadiya, but there was no obvious time or place. The wives might overhear them talking and report her to Abu Walad. Moreover, Maryam didn't know how her sister would react to the news, so she kept it to herself.

Weeks later, it was Sadiya, during a visit to Maryam's home, who broached the subject. "Maryam," she said. "Did you see the video of the world's reaction to our abduction?" Maryam had hoped that her sister would have the same reaction as she, but Sadiya was racked with guilt for men on both sides of the conflict. "The whole world is concerned about our plight," Sadiya said. "Many soldiers died trying to rescue us, and our fighters also died in the conflict." She was in tears: "I hope we have God's mercy."

Maryam told her not to cry. "You should instead pray for our freedom," she said.

SOMETIME SHORTLY AFTER, SADIYA RETURNED TO VISIT MARYAM WITH A new secret to reveal. "Our father is alive," she said. "I heard him on the radio." Sadiya had been lying around her home, listening to a newscast from a portable radio her husband had inadvertently left lying around. The station was BBC Hausa, a channel the girls used to listen to at home, playing in the fields as they helped on the family farm.

A familiar voice leaped from the speaker. Their father was offering an emotional plea for his two girls to come back. It was harvest season and he missed their company, reaping his crops and returning to an emptier home. The dad they knew was old-fashioned, a tough disciplinarian who didn't normally talk about his emotions. Yet on the radio his voice was cracking. "I miss my hardworking girls," he had said.

After Sadiya's co-wife slipped away, she hurriedly recounted the interview to her sister. They managed some laughs through tears. "He had always said we were more obedient than our brothers," Maryam said. The broadcast confirmed that Malam Ahmed had been lying. Their father was alive, and if he was still in Chibok and being interviewed by BBC Hausa, it was not part of the caliphate.

When the girls had wiped their tears, Sadiya shared another detail from the interview. Their father had concluded by saying he would swap his three sons for his two daughters. "They work harder on the farm, help me in the home," he'd said. The sisters laughed again. Talk of Chibok, discouraged in their households, had somehow beamed its way in. "My dad was still alive," Maryam said. "The news stirred my desire to go back home."

MARRIED HOSTAGES WERE KEPT AWAY FROM THEIR INFIDEL CLASSMATES, but as the months wore on, Maryam found that they were sometimes being kept in encampments close enough that she could slip away to visit. One day she decided to walk the mile or so from her home to the latest holding site for the other students. She worried that they judged her and would curse her as a Boko Haram member. As she approached the Tamarind, she expected her friends to turn their backs. But as she lifted the niqab, some rushed to greet her.

Naomi Adamu hung back. "Maman Mu!" Maryam said, locking eyes. "Our mom."

"Oh! You haven't forgotten that name you called me," Naomi said, her body language cold.

As her classmates plaited Maryam's hair, they peppered her with questions about what her new life was like. What was she eating, and where did she sleep?

"Are you OK?" Naomi asked. "Do you know that your husband is very tough? He likes to beat Chibok girls."

Maryam said little, assuring them she was fine, though she was not. Her life was changing, and she wasn't sure how much she could tell her friends. Her husband seemed happier than he had been in the early months of their marriage. Maryam was finding a way of dissociating from their life together, a mechanism to cope with an imposition that only felt real when she said it aloud. She was pregnant.

24

The Imam

The headlights of a pickup truck lit a checkpoint made from rope. Masked fighters were standing on a dirt track that was almost enveloped by the forest, its clusters of thorn trees and grasses over a yard tall. The guards halted the driver with an outstretched arm and asked questions in Kanuri. They seemed to be on high alert.

The driver was Tasiu, a tall, athletic commander who spoke with a husky voice. He was Malam Ahmed's boss, just a couple of ranks below the man he called the Imam. The pickup was carrying the *arnan daji*, the "pagans in the forest" who had most defiantly refused to accept Boko Haram's creed.

The hostages were jammed into the truck bed, sitting in each other's laps or dangling their legs from the open tailgate. Naomi said little as the vehicle carved its way deeper into a forest landscape none of them recognized. Already they had passed several roadblocks and were given authorization to access thin dirt tracks that blended into the forest floor and were shrouded from the sky.

Their destination was a place the Nigerian Army and the world's top intelligence assets had spent half a decade trying, but failing, to find. Gaba Imam, the House of the Imam, was Shekau's hideout. "You are going to see our leader," Tasiu had shouted back at them, his hands clenching the wheel. "Since you refused to convert to our religion."

In hushed Kibaku, the girls speculated about the punishment Shekau would sentence them to. They feared they could be killed. Some were also curious. Tasiu had assured them not to worry, and they wanted to believe him. Either way, they had agreed they should look brave in front of the man who controlled their fate. "We told ourselves . . . that we should not be afraid if he tries to frighten us," Palmata Musa would later say.

The militants untangled the roped checkpoint and waved the truck through. The forest path on the other side became too thick for vehicles, so Tasiu ordered the girls down, all of them stooping under thorn branches that sheltered the pathway into the base. They arrived at a small clearing around a white plastic chair, and he commanded them to gather. Each of the girls sat on their knees in silence. Tasiu stood at attention, staring ahead. Several minutes passed as the girls scanned their surroundings in silence.

There was a small structure nearby, made of thatched reeds, with a tree in front that obscured the view as its inhabitant limped out. Four bodyguards in crisp uniforms walked alongside him, trailed by two more escorts each holding a tablet, filming the scene for the Imam's records. The light from the devices bathed his face.

Shekau liked to capture everything on video.

A scarf partly covered his face, but the girls recognized the Boko Haram leader immediately. They had seen his videos. He wore a beanie more suitable for winter than a sweltering Nigerian evening, a padded soldier's uniform, and military-issue jackboots. In one hand, he held a Kalashnikov, and in the other, two Qurans, which he stacked on his lap as he took a seat. Through the air wafted a strong and unmistakable scent: the warlord had doused himself in perfume.

Tasiu bowed. "Is this the real Shekau?" Naomi whispered to Hauwa Ishaya, who whispered back, "Yes, it is him!"

The Imam propped his rifle against his thigh and with eyes closed, began to intone a long prayer in classical Arabic that his captives couldn't understand. Then he pulled out an enormous *miswak*, brushed his teeth with it, and spat out a set of questions in Hausa, words the girls understood perfectly.

"Do you know who I am?"

"Yes."

"Are you the Chibok girls?

"Yes."

"Is it true you refuse to pray?"

Naomi studied his face. Hauwa Ishaya thought to herself, *We are going to be killed.*

THEIR JOURNEY TO THE IMAM HAD BEGUN AT MIDNIGHT THE PREVIOUS night, when Malam Ahmed had gathered a group of about a hundred Chibok girls under a tree and announced that they would again be separated. The hostages had by now been divided and reunited several times. This time, he called out names, and the chosen girls had to step forward. There was no explanation of what he was designating them for. The students listened, trying to deduce his intentions, as he ticked through a roll call.

Naomi's name was second on the list. As soon as Hauwa Ishaya heard "Naomi Adamu," she knew this was a register of the girls Malam Ahmed found the most insubordinate, the *arnan daji*, and that her own name would also be called.

When Malam Ahmed had finished the list, twenty-two girls had been split from their classmates without a pause to share good-byes. As the Malam ordered the chosen onto trucks, other militants fired insults and threats. They shot guns in the air and shouted the *takbir*: "Allahu akbar!"

NOW, SAT BEFORE THE IMAM, THE GIRLS LISTENED IN SILENCE AS HE launched into a lengthy extemporaneous monologue. The leader seemed more collected in person than the madman the girls had seen in his propaganda videos. He wanted them to know the true Shekau, he explained.

"I laugh when people call me insane," he said. "I am not a lunatic. I behave strangely just to infuriate Nigerians."

Naomi had a sense this man was intelligent. She had heard him switch through several languages, and he asked himself aloud if he should address the hostages in English. Shekau was also wounded, she could see. His leg dangled limply from his bench, and it was obviously giving him discomfort. She too, was in pain, and at one point he looked at her and could see she was holding her lower back. "Don't cry," he said.

Instead, they should listen. The Imam had much to say to the young women who had made him famous around the world. It had been more than a year since their disappearance captivated a fleeting global audience. But he was still working through what he should do with them. "What do you wish I would do for you?" he asked.

"We responded by saying that he can do whatever he so wishes with us," Palmata said. "He then told us that our parents have been insulting him."

There were so many lies spoken about him, Shekau complained, and as he listed them, he showed flashes of the screaming warlord they had seen on the screen. He had not forced any of the Chibok girls to marry, he declared. He said his enemy was not Nigeria's masses, but the elites.

He began to fulminate against Christian practices he saw as heresy. He could tell the difference between faithful Christians and unbelievers, he said. Some women went to the altar with their shoulders exposed, insulting God.

"There is nothing in the Bible I do not know," he said. "Some people go to church to dance. They did not go there to worship the true God."

"It is the true religion I am practicing," he said. "Chibok girls were sleeping with men at home before the abduction."

The hostages were glued to the spot, watching him. Their host now seemed to be saying whatever landed in his mind. He was neither smiling nor scowling, but spoke with an eyes-wide intensity, his head slightly cocked.

Eventually, he arrived at the question that had been troubling them for more than a year. Why had they been kidnapped?

"I didn't kidnap you to spoil your lives," he said. "You are the ones who spoiled your own lives.

"I didn't kidnap you to get married," he continued. "Even for those who were married it was not by force."

Rather, he wanted a simple trade. Nigeria had arrested so many of his compatriots and shoved them into jails. Did they know how Nigeria treated its prisoners? They should understand that he was caring for them better than Nigeria handled his men in return. In fact, one of his own wives had been arrested.

He spoke for so long that it took a moment for the girls to realize he was posing a question when he asked them something: "Are you ready to go back home?" The hostages, still processing what they'd heard, stumbled over a response: "We replied in unison by telling him that we all wanted to go home," Palmata said

"You should go away for a thorough reflection," he continued. "Decide whether you are going to convert to Islam and stay with me or if you actually wish to go home."

He asked if they had anything to say, but his audience was working through their confusion. Hauwa had assumed he would order their death. *Is this not the boss of those who have been killing people?* she wondered. Palmata was also in shock.

Naomi wanted to chime in and complain about the way they had been treated, but Hauwa pinched her before she could. Naomi concluded that Hauwa was right. *He already knows everything that's happening to us*, she thought.

They would have two weeks to make a choice, Shekau said. "If any of you decide to become Muslims, you will be married and will no longer be treated as slaves," he said. "Those who get married will be fed, and two rams may be slaughtered for them."

He concluded with a threat. "Those who make the wrong choice will burn in hellfire."

Before he left, he turned to Tasiu with an order: These girls should not be kept with the others. The meeting had to stay secret. If the rest heard what he had promised, they would also not practice religion.

The women nodded their heads, and he lifted the Qurans off his lap, limping away with the rifle. His entourage followed. None of them had said a word. For the first time in months, Naomi could imagine the prospect of freedom. "Our feelings changed," she said. "That's when we thought we could maybe get out."

THE NEXT DAY, THE TWENTY-TWO YOUNG WOMEN ATE THE BIGGEST MEAL they'd enjoyed in a year. Tasiu had driven them to what appeared to be a storage depot some distance away. This would be their home while they reached their decision, he informed them.

He had good news. They were free to eat any of the food and cook with any of the utensils.

Stacked in front of them were bulging burlap sacks of uncooked rice. Heavy yellow jerricans sloshed with palm oil. There were packets of spaghetti, macaroni, and crunchy deep-fried *chinchin* biscuits. All day and night, Naomi and the others watched as trucks arrived with fresh deliveries or to load up and drive off with rations needed elsewhere in the caliphate. The depot housed large aluminum cooking pots, and on the first day, they prepared vats full of jollof rice, stirring tomatoes and spices into the grain. A militant approached them as they were eating. "Eat whatever you want," he said. "But be prepared to give Shekau a favorable response. Otherwise we will kill all of you here in this bush."

Ever since leaving Shekau's base, the women had been debating the upside-down position in which they found themselves. For

Shekau to put the decision onto their shoulders was unusual. Several were convinced the offer was a trick. Hauwa, normally one of the more upbeat voices in the group, was scared and reasoned that agreeing to convert and marry would mean they would never go home. But saying no could prompt a death sentence. Naomi applied the same logic in reverse: the group had no choice but to push on and hope for the best.

As the debate wore on, the women were plagued by second thoughts about their convictions. Nobody had asked them to make any significant decisions since they'd been kidnapped. Two of them had begun to report terrible nightmares. One had seen their group abandoned in the wilderness after telling Shekau they wanted to go home. Another dreamed that after refusing to convert, they had learned they would all be beheaded.

Shekau had told them they would "burn in hellfire" if they made the wrong choice. The women debated and agreed on a plan. Everybody should fast to reach one answer together. For several days, they prayed and skipped meals. Some went hungry for two days; others, for three. Fasting had become a collective ritual, and they were getting better at it.

"I became tough," Naomi said. "I told them they should not be scared and that nothing was beyond the power of God."

The communal fasting brought resolve. After a week, the women decided they didn't need any more time. They told Tasiu they were ready. Sometime later, he returned to hear their decision.

He was holding a piece of paper and two pens, one red, one black. "The red is for Christians," he said, "and black is for Muslims." Anybody who wanted to stay with the group, get married, and go to heaven should write their name in black. The rest should choose red.

The first student took the pen. She wrote in red. Then the next student, one of the women who'd been suffering nightmares, took her turn; she too wrote in red. Saratu Ayuba went third, followed by her cousin, who draped an apostrophe in Na'omi, just as she had on her schoolwork. When it was finished, the paper was signed with twenty-two names in red. Tasiu scanned the page.

"So, all of you want to go home and meet your parents, which is better than our religion?" he asked. Someone in the crowd replied. "Yes, that is what we want."

Very well, Tasiu replied. He would take their answer to Shekau.

After he left, the hostages discussed the selection they'd made. They still weren't sure if it was a ruse and they'd be killed for their choice. Maybe it was the first step toward their release?

Whatever it was, it was done.

"If it was going home, we had signed," said Palmata. "If it is death, we had already signed. We said we did not care either way."

25

The Escapade

A few weeks after declaring their decision to go home, Naomi, Hauwa, and Saratu were crawling through waist-high grass on a secret mission to find their friends. It was rainy season in Sambisa, and the blades of grass had turned bright green and were thick with bugs. Only yards away, some of the *rijal* were strolling past, unaware that three Chibok students had thrown themselves onto the ground and were lying on their bellies so the men wouldn't spot them. The journey was risky, but Naomi felt it might be the last chance to see some of her classmates. It was taking what felt like hours.

Weeks of silence had ticked by after the *arnan daji* had written their names in red, but they had heard nothing about Shekau's response. Still, the twenty-two young women agreed that their choice had been correct. Sequestered at the food depot, they had eaten oily meals of spaghetti or jollof rice and snacked on chin-chin, restoring flesh to their bones. There had been no more nightmares. The steady stream of freshly butchered beef, cooked over firewood, had left them optimistic that the militants might keep their word. "God will see to it," Naomi had said.

"If you go home, we want you to look healthy," Tasiu said one day. "We don't want you to go home looking skinny."

Naomi had been lying on a tarpaulin mat before another meal

one day when two of her friends came back from collecting kindling, whispering excitedly of a new discovery. On their walk, they had heard the distant sound of voices they thought they recognized. Palmata Musa was sure it was the classmates they had been separated from, living nearby, in thatched houses. It sounded like dozens of them could be there.

Now, the women were sneaking along the forest floor to find them. More than a month had passed since they had seen any of their fellow hostages. Shekau himself had banned all twenty-two from any contact with the others. The students had deliberated on what they could do. Disobeying the Imam could risk everything, and yet Naomi and her friends had concocted a plan.

The rains had brought out the seasonal cloud of grasshoppers, long-bodied brown and yellow leaf-eaters that would bounce between the towering blades of grass. Back home, some locals would trap, box, and deep-fry them. Dipped into hot chili powder, they were a crunchy delicacy. The bugs could become an excuse to slip away on a foraging run. At dawn one morning, Naomi, Hauwa, and Saratu had asked their guard's permission for an outing that would last until noon. "We would like to go catch grasshoppers," they said, with a promise not to go far.

The guards had become lackadaisical about keeping watch over the young women, and they granted permission. The three of them slipped away, heading toward the far-off clearing where Palmata had heard the voices. In the heat and humidity, the group stopped three times to rest, particularly for Naomi, whose back and kidney pain had been getting worse. When they heard men's voices, they stayed low and waited until the sound faded into the distance. Eventually, they approached a cluster of grass houses perched under a large tamarind, and in the distance, they could hear girls' voices. There were more men there too, seated under nearby trees, staring back nonchalantly in their direction. They crawled closer, painstakingly sifting through the overgrown tangle of tall grass, taking care not to make a sound.

Parting the blades with their hands, they could see dozens of shrouded figures sitting in clusters behind a small fence. They instantly recognized their faces, although the expressions looked drawn and their body movements languid. The three women wanted to grab their attention without alerting any guard stationed close by. Naomi cupped a hand over her mouth and made a hooting sound, like an owl: *Woo-woo . . . woo-woo.*

The schoolgirls had used the call at previous camps when they were held in nearby homes. The noise blended with the ambient sounds of the forest.

Woo-woo . . . woo-woo.

The first to register the sound and look to see where it was coming from was Mami Ibrahim. As she waded into the grass, Naomi was struck by how emaciated she looked, her frame almost skeletal. She looked like a head with no body attached. Mami's eyes widened as she saw her friend. "I never thought I would see you again," she said to Naomi as she ducked down into the grass, then embraced her, Hauwa, and Saratu. All four of them began crying. Through her tears, Mami apologized for her appearance. She was caked in dirt. "We have no cleaning materials," she said. They were not only malnourished, but had run out of soap.

The women could feel Mami's bones through her shawl. Naomi was shocked at their condition. "How come you are starving," she asked, "while we are sleeping on foodstuffs?"

WHILE NAOMI AND THE OTHER MOST DEFIANT HOSTAGES HAD BEEN SPIR-ited away to face the Imam, the classmates left behind were facing a new kind of torment. Malam Ahmed had been depriving them of enough food. With his most insubordinate hostages removed, the chief guardian had been lavishing rations on those women who had been pushed into marriage and denying those who remained. These girls, about one hundred in all, were languishing in a gray zone. Fear had forced them to accept Boko Haram's faith, and they prayed and fasted daily according to their captors' instructions. Sometimes, they sat alone, leafing through a Quran. But they were still refusing to marry.

"If you are not married, we will not bring you good food," Malam Ahmed would repeat to them often. "Marrying is God's will. If you get married, we will give you whatever you want."

At times, the Malam mingled with the girls, telling them he loved them and encouraging them to smile. He said he would try to secure meals for them. But if they wanted to eat better, the best food was reserved for the wives.

The strategy was working. The hostages were surviving on a diet of foraged roots and baobab-leaf kuka soups, which were becoming thinner and less sustaining. The hunger weakened their bodies and their resolve. Some women had been falling sick, then stepping forward to agree to marriage, sometimes alone, sometimes in groups of two or three. Their fellow hostages would bid them good-bye with tips to try to avoid pregnancy. "Mix salt with hot water" to wash yourself when he falls asleep, said one. Then the resigned girls would walk to Malam Ahmed's grass-roofed home, which was close by, where he could keep an eye on them. He would stand up to lavish praise on the soon-to-be brides. "Good!" he would say with a smile. "That is why I like you." Eventually, a man would come on a motorbike to deliver the woman to the fighter chosen for her. Often it was the last time her classmates would ever see her.

Aside from the Malam, there were more than ten guards observing them closely, and the girls were too scared to openly discourage their friends from accepting the offer. Few had the energy to protest. They feared the beatings or lashings if they did.

That was why none of the hostages had spoken out on the day that Lydia John addressed a few of her classmates and announced that she too was giving in. "I'm tired of this, everything that I'm suffering," she had said. "I want to get married." For weeks, the eighteen-year-old had been sick and feverish. She had been shivering and talking to herself in strange tongues. At first, the women seated around her had assumed she was joking. They grew worried as her ramblings lasted the whole day and she refused food, even throwing a meal on the ground. It continued for days, and her classmates feared she'd been possessed by spirits.

They had tried to soothe her. They had washed her body with buckets of water, then asked for help. Guards had burned charcoal and veiled her head, then pushed it over the smoke to banish the evil spirits from her body. "Inhale the fumes," they had said. The episodes had come and gone. Lydia had remained exhausted and disturbed. Finally, one afternoon, she insisted on walking into Malam Ahmed's house, telling her classmates she was suffering and had endured enough. They were afraid to stop her.

Lydia had developed a persistent pain on one side of her body and limped as she made the short walk to Malam Ahmed's house. "I want to get married."

"That is why I like you," he said.

NAOMI, HAUWA, AND SARATU SNUCK BACK THROUGH THE GRASS TO THE food depot where the other holdouts were waiting for them. There, the twenty-two young women huddled together and outlined a plan they hoped would relieve their friends' suffering. They were sitting on a food surplus while the others were being starved into submission. They had to do something.

Mami had told Naomi about what had happened to Lydia John, but if Naomi was grieving for her friend, who had helped her write the diaries, which were still tied in a bundle beneath her clothes, she didn't show it. She said little as she listened to the story. "She shouldn't have joined them," Naomi said. In Lydia's absence, she had stopped keeping a diary.

What was urgent, the women felt, was to bring their other classmates food. Palmata called their plan the escapade. "We will take the food to them whatever the consequences," she said. "Even if I will be killed, I will take food to them." They all agreed. "We are saving their lives."

The first delivery would be a heaping supply of jollof rice. They boiled an entire burlap sack of Red Cross rice with spices and tomato paste, then piled the meal into jerricans that the women had sliced open to make into containers. Gray clouds were forming, so they coated the jollof in plastic wrap and waited for the rain. Their guards would retreat into houses once the sky unleashed. They wouldn't expect the hostages to sneak off during a thunderstorm.

As the downpour began, four of them set out, carrying the food on their heads as they walked through the driving rain. Several times they dropped to the ground and crawled through the grass when they spotted militants. As they reached the edge of their classmates' camp, they cupped hands over their mouths, and repeated the owl call.

Woo-woo, woo-woo.

It was Hauwa Musa, hungry and lying on the ground, who this time heard the sound and told a guard she needed to ease herself. Slipping into the grass, she found the couriers and embraced them. "Oh, God has the power to make things easy," she said. Then she grabbed rice with her fists. Nine other girls followed her down and did the same. Years later, she recalled it as the first time she had tasted salt in over a year and said, "The food they brought saved us."

The success of the first mission gave Naomi's group a road map for regular deliveries—a breadline. Along with more jollof, they took chin-chin, comforting the women with a snack they knew from home. They brought jugs of fresh water and containers of grilled meat and cooked beans, spiced with Maggi stock cubes left lying around the depot. They managed to catch a few of the grasshoppers they'd been pretending to chase, then deep-fried them to bring on the next trip.

Deliveries were precisely timed. They embarked when supervision was lax, when guards were praying, or in the evening. They would deliver more food in the rain, when few men would be walking around. They would meet in a prearranged location in the tall grass somewhere between their camps, a place where the starving teenagers could feast in secret. "I felt when they ate, it's like *we* ate," said Palmata. "We could not allow them to live in hunger."

The deliveries didn't stay secret for long, however. When a guard caught one of the women carrying a bag of sorghum to the hungry camp, word reached Tasiu that Naomi and her friends had been straying from the area around the food depot. He gathered them to issue a warning. "Whoever is caught moving around will be killed!"

"We kept going anyway," Naomi said. Instead of retreating, they became more brazen. Sometimes her friends guided schoolmates all the way over from the hungry camp, sneaking through the grass to the depot to help with the cooking. If a guard came around, they would wrap their visitor in a blanket to hide them.

The secret food source was not only nourishing the starving captives but also shifting the balance of power between them and their guards. The steady stream of women walking dejectedly into Malam Ahmed's house dried up. They were becoming stronger, sleeping better, their minds clearer.

They were also making sense of the strange proposal Shekau had put to Naomi's group. How could it be that he would just let them return home? And all they had to do was reject his religion? Shekau had warned Naomi's group not to tell the others about the offer, but they had immediately shared the news with their sisters, passing word that there was a possibility to go free. "Don't ever get married," Palmata would say. "Let's see the end of this."

As the others recuperated from hunger, they were also recovering their resolve. Many began to renounce the faith that they had adopted only out of fear, and as a group they ceased to participate in Malam Ahmed's daily prayers. Several now spoke back to their guards. "We only prayed out of fear, because we were threatened with death," one of the women said.

Word of the rebellion soon reached Tasiu.

Naomi was sitting with her classmates one afternoon, when his vehicle sped up and stopped and he stepped out. He was bursting with rage. "Was there a meeting between you and the other girls?" he asked.

Definitely not, the hostages assured him.

"We never had a meeting," Naomi told him. When he left, she laughed harder than she had for some time. "We lied!"

26

The Deadline

"My name is Naomi Adamu," she began in a slow, stone-faced mono-tone. Naomi was standing against a cream-colored wall, her face wrapped in a dark gray veil, repeating scripted lines into a cam-corder held by a militant. He was watching her image on a flip-out viewfinder. "I am a student at Chibok Government Secondary School for Girls."

Nearby, Saratu, Hauwa, Palmata, and the other holdouts sat under a tree, taking turns to record the same message, offering proof that they were alive in a video they hoped could help send them home. "Your government is saying you have been killed," Tasiu had explained to them before they began filming. "They are saying there is no way the Chibok girls could have survived."

Tasiu appeared eager to be done with the nearly two dozen *arnan daji* in his care. After finally finding out about their secret food supply, he had erupted in fury, sending the more pliable cap-tives away on trucks to a holding site near Gaba Imam. There, in Shekau's secret forest encampment, they would be closely moni-tored, away from negative influences. Gathering Naomi and her group, he had issued another threat. He would kill them all, he said, if their insubordination continued.

The death threats, followed by more death threats, had come

to sound hollow to Naomi and her friends, cut off from the other hostages for months since meeting Shekau. They were being instructed to film videos they assumed were puzzle pieces clicking into some sort of negotiation back in Nigeria. They could see they were eating better than many of Boko Haram's own *rijal*, a sign of their value. At one point a tailor had visited with a measuring tape, took their dimensions, and left. Days later, the seamstress returned and then, without explanation, handed each of them custom-sewn matching wrappers and dresses.

The group of twenty-two had taken off their hijabs and folded them. They wanted to believe the new clothes were a prop in preparation for their release, an image for the television cameras to show they had not been mistreated. But they could also be their costume for another, more morbid type of video. It was a relief to pull on clean, colored wrappers for the first time in a year and half, but Naomi and her friends reminded one another not to let their guards down. These men could change their mind in the blink of an eye.

Shortly after, a convoy of seven Toyota pickups and one Jeep had arrived, packed with armed fighters wearing spotless fatigues. Tasiu announced that he would lead the operation and ordered the hostages into the vehicles. Hauwa and Naomi clambered into one of the truck beds, pulling themselves over the sides, both convinced they were on their way home.

But when Naomi settled herself into the truck, she saw something that stopped her cold. There, laid out in the bed by the feet of the fighters, were shovels, industrial tape, and strips of a white cloth that she immediately recognized. They were burial shrouds that Boko Haram used to wrap their dead. It dawned on her that they weren't going back to Chibok. They were going to be executed for refusing to convert.

Hauwa had seen the shrouds too and stared into Naomi's eyes, the friends sharing the same realization. The men in the front seat and in the truck bed were all holding assault rifles, and at first the hostages didn't feel free to speak to each other. Then, as the convoy began to snake its way through the forest, they began to whisper final good-byes. In Kibaku, they took turns apologizing to their friends for anything they'd done wrong. As the engine revved to bump the truck over uneven ground, they held hands and said their last rites. "Everybody now forgave each other, whatever it is, because we were thinking we were going to be killed," Hauwa said. She added a silent prayer: "May God give me paradise." At least, she thought, they would die together.

Naomi, meanwhile, was thinking darker thoughts. The trucks were headed out of the wilderness, she noticed. The forest was thinning and dirt trails widening into well-trodden tracks that suggested they could be headed toward a town. She worried that Boko Haram had summoned their parents on the pretext of a release, but was instead planning to murder them too. "I was thinking they are going to take us home and kill our parents right before our very eyes."

AROUND THE SAME TIME THE HOLDOUTS WERE SAYING THEIR FINAL prayers, a battleship-gray C-130 Hercules military transport plane was readying itself for takeoff on a runway in Abuja, the roar of its propellers echoing off the capital's rock-strewn, grass-capped hills. Nearby the private Bombardier and Gulfstream jets of governors and businessmen shimmered in the heat haze, readying to take off to foreign cities.

Inside the C-130 cabin sat three handcuffed Boko Haram members, freshly extracted from prison and still disoriented by the preceding hours. One was a senior qaid who had led some of the group's largest attacks before his arrest in 2011, including a jailbreak that freed more than seven hundred prisoners. Another had been Shekau's "accountant," the only man trusted by the Imam to handle the cash deliveries for big operations. With them was the blogger who hoped to trade this group for the first set of Chibok girls to go free in one and a half years. Ahmad Salkida, better known to Twitter as @ContactSalkida, had returned to Abuja on a secret assignment.

President Buhari wanted the girls back by his hundredth day in office, in early September, and the freelance journalist still seemed to represent the quickest channel for contacting Boko Haram. But coaxing him back into Nigeria had required trust on both sides. In the suburbs of Dubai, the exile had come to feel at home. His wife moved easily through the Arab metropolis and spoke its language. She was happier now that he had left behind a calling that had brought detentions, death threats, and a degree of infamy among his countrymen. The children were in good schools. Salkida had finished doing night shifts at the grocery store, and he was sleeping better.

Back in Nigeria, there were still plenty of people in government and the general public who remained skeptical about the motivations and allegiances of the man behind this well-sourced Twitter handle. He was a member of the insurgency, trying to play both sides, accusers claimed. Trolls on Twitter hounded him. The unproven allegations of double-dealing had shadowed Salkida even in exile.

The only way to defend his reputation was to air his feelings and motivations in the open. On April 15, 2015, the one-year anniversary of the kidnapping, he had given a long interview to a BBC reporter, who at one point had asked, Are you a member of the insurgency? He knew that Boko Haram and the incoming government were both listening.

He said no. When the interview ended, his cell phone rang. It was the DSS, the Department of State Services, Nigeria's main spy agency. Would he like to come in for a meeting?

On July 17, the president had authorized a new round of negotiations. Salkida began liaising with contacts close to Shekau. One of the journalist's main sources in the insurgency was a man with a husky voice he knew as Abu Zinnira. Naomi knew him as Tasiu. Often the man most closely watching her group would call

Salkida on a Thuraya satellite phone to discuss whether a deal
could be possible.

A breakthrough came when a video landed in Salkida's lap.

It was ten Chibok girls, lined up one after another against
a cream-colored wall. "My name is Naomi Adamu," one of them
had said, speaking slowly with a vacant expression. She gave
little away about her state of mind, but she looked relatively well
fed. "I am a student at Chibok Government Secondary School for
Girls."

Within two weeks, Salkida had secured an agreement. With
the DSS's approval, he had squeezed his way through jam-packed
prisons, combing through cells to identify a group of known Boko
Haram members in custody that he could deliver in exchange for
the girls' freedom. He expected Boko Haram to deliver more than
twenty girls. It would be a confidence-building step, the president
was told.

By August 1, the president had assented. The girls' freedom
was a moral imperative. Three days later, the C-130 flying Salkida and the prisoners to Maiduguri touched down on a tarmac
swarming with soldiers and police. A convoy carried them to a
three-bedroom apartment inside the military's sprawling Maimalari Barracks, at the northern end of town. The base had mas-
sively expanded in the years since Salkida had moved to Dubai.
The long, gray cement wall surrounding the compound had moved
farther north, swallowing up farmland belonging to neighboring
villages. It was now the forward operating base for an expanding war, the headquarters of Operation Lafiya Dole, Peace by All
Means, wherein Nigerian commanders would sit alongside the
US intelligence officers deployed by President Obama. On one
side of the base was a graveyard filling up with bodies marked
by plastic placards that listed each soldier's name, rank, number,
and religion.

The Boko Haram prisoners arriving into the apartment, accompanied by Salkida and their government escorts, were thrilled. To

men who had spent years in forest camps and Nigeria's toughest jails, the accommodation was almost unfathomably luxurious, with twenty-four-hour electricity and oversize couches. Their handcuffs were removed, and one of them asked for Salkida's phone to dial his wife. He broke down as she told him she would have waited thirty years for him in prison. Salkida had takeout menus for the local restaurants, and deliverymen brought meals on request, including bananas and watermelon, food the commanders hadn't tasted in years. After satisfying his appetite, one of the detainees turned to Salkida in gratitude. This was the best day of his life, he said.

WHILE THE MILITANTS WERE SETTLING INTO THEIR APARTMENT, A CONVOY of pickups carrying Naomi and the other *arnan daji* was snaking over the dust-strewn flatland outside Maiduguri. It was nighttime when the trucks halted at a small set of ramshackle houses woven of dry grasses and reeds. The women clambered down, past the shovels and burial shrouds, for what they thought would be their final few steps. In the distance, they could see what appeared to be the glow of a bustling city, a halo of fluorescent light they hadn't seen since the kidnapping. Solar-panel streetlamps hung over a distant highway that hummed with the steady passing of vehicles. Naomi thought it could be a stretch of road that led into Maiduguri, the regional capital. Her aunts lived there, and she'd spent nearly a year in the city as a teenager to receive medical treatment.

Tasiu ordered them into one of the nearby houses. They filed in and sat on a makeshift carpet of stitched-together cement bags. Inside, it was almost pitch-black, but they could see a mosquito net hanging from the ceiling. They waited for their captors to follow them in.

When the fighters finally appeared, they were not carrying guns but bulging sacks of rice and beans. They handed the women the food and explained that they would stay at this location for a few days as final preparations were being made for their release. Naomi was overcome with relief and looked at Hauwa, who was quietly weeping. The women built a fire and began to cook.

The next morning, they woke up to fresh bread and plastic bags filled with roasted peanuts. Tasiu arrived to address them. "If all things go well, we will release you to the Nigerian government," he said. "But if the Nigerian soldiers try to play smart and shoot us? It means sorry, you people will not be released."

IN THE APARTMENT NOT FAR AWAY, SALKIDA'S PHONE BEGAN TO *PING* with voicemails. Senior members of Boko Haram were sending him recordings on WhatsApp to check if their comrades were alive. The prisoners inside the apartment were fellow fighters who had been assumed dead.

Over WhatsApp, the insurgents began exchanging voice memos between detention and the front lines. One prisoner said he had been chained to a prison wall. Another said he felt as if he were being resurrected. The men had not been heard from in years, and in their absence, their families had mourned their deaths. The recordings flying back and forth were emotional. One message arrived from Aliyu, the murderous commander who had once threatened to decapitate the Chibok hostages and had accused them of lesbianism. Salkida knew his backstory. Aliyu had been born a Christian in Nigeria's southeast; he was a convert who felt he needed to be the most extreme man in Boko Haram to fit in. Now he was sobbing through emotional messages on the tinny speaker of Salkida's phone. "I married your widow," he told a prisoner. The prisoner thanked God, and began hugging the other detainees in the apartment. He was overjoyed that someone had cared for his wife. Watching the men embrace, Salkida, too, began to cry.

The deal could come together fast, Salkida hoped. The men had been brought to Maiduguri along with a ransom of €500,000 to sweeten the offer. The cash was in Boko Haram's preferred currency, which traded easily in the former French colonies surrounding Nigeria. This was meant to be the secondary aspect of the deal. Shekau had said he didn't want any money for the release. What was important was that the imprisoned fighters would finally walk free. But as the hours ticked by, more voice messages arrived, and the mood shifted. A furious debate emerged over the value of the hostages. It became clear that different individuals in Boko Haram's leadership had different price tags in mind.

Some of the Boko Haram negotiators wanted more money, an amount closer to the tens of millions of euros the Islamic State had demanded for its European hostages. The schoolgirls, known around the globe, were worth a fortune, Tasiu argued. Weren't these women worth just as much as Americans or Europeans? This was the value the Internet had given them. But Buhari's administration balked. They had never agreed to or even discussed these demands, the government said.

Salkida begged for more time, but as hours turned to days, the deal began to flounder. Both the state and the insurgency were losing patience. The prisoners inside the apartment were growing nervous. Finally, the military's hospitality abruptly ended, and security forces arrived to escort the fighters back into detention. Shekau was said to be furious, and his underlings were livid.

Salkida was also enraged and embittered. He refused to fly back to the capital on the C-130, slipping out of Maiduguri and back to Dubai, at each step worried that Boko Haram and the security services would both be looking for him. Shortly after reuniting with his family, he penned a cryptic tweet: "I'm glad for what I tried to accomplish & for what I've done silently. There is no contribution more rewarding than one that goes unnoticed."

In time, Salkida would question whether Shekau ever intended to release the Chibok students. Through the distorted prism of social media, the hostages had become the source of the Imam's fame and power, the two things he had always craved. A poor boy who'd raised himself begging for change in the traffic of Maiduguri, it was #BringBackOurGirls that had somehow given Shekau the recognition he felt he deserved. "He had fallen in love with them," Salkida later said. "The Chibok girls made him world famous. And it was the attention he had always wanted."

NAOMI AND HER CLASSMATES HAD BEGUN COOKING FRESH MUTTON from another food delivery when Tasiu showed up covered in dust. It had been several days since they had arrived at what they believed to be their final prison. Palmata Musa had counted two rams and two goats the insurgents had purchased for the students to eat, not including the already butchered meat the men were preparing on their own, a kind of farewell feast.

But Tasiu was flustered and in a hurry. "You should enter the vehicles for us to move," he said. The women didn't ask why. Abandoning the meat crackling on the fire, they raced to the pickups parked nearby and helped each other clamber into the truck beds. The moment had finally come. "We thought we were going home," said Naomi.

The convoy of trucks sped off, raising a cloud of dust over the rutted earth. As the car shook over the dirt road, the guards seated near them gripped their rifles tightly. Naomi looked across the landscape, as flat as a pancake, and after a few minutes, she realized they didn't seem to be headed into the city. They were headed back into the countryside.

She whispered to Hauwa across the backseat of the pickup. "This road does not look like we're going home," she said. "It looks like we're going back into Sambisa."

"Maybe we have just not followed this road before," Hauwa replied, hopefully. Perhaps the convoy was entering Maiduguri through some back route, to avoid detection.

Hours of dust-dry wilderness swept by as Hauwa scanned the horizon and concluded with dismay that Naomi was right. They rode in silence for hours, the vegetation thickening around them, until the convoy deposited them at a familiar site. They were back in one of the forest camps where they had been held months earlier.

The hostages surrounded Tasiu and pleaded with him, shouting over one other and demanding an explanation. Had they done something wrong?

"Was it because of the meat we ate?" one asked.

"Was it because of the new clothes?"

"Your people have rejected you," Tasiu said, shrugging them off. "You should just take it as your fate."

The women protested, shouted in frustration, and asked more questions, at first to the guards, then among themselves. "Most of us were crying," Hauwa said. The young women argued over their predicament late into the night, searching for meaning or logic, until their need to understand was overpowered by exhaustion. They were helpless, still prisoners with no control. Perhaps they would never go home. Hauwa suggested they pray to find answers. "Maybe there is something God wants us to do," she said. "Maybe there is some reason God has left us behind here in this place."

27

The Windows Are Closed

ABUJA, NIGERIA
NOVEMBER 2015, NINETEEN MONTHS INTO CAPTIVITY

The work now felt like a drug, a roller coaster of adrenaline rushes that was ultimately leading in circles. At odd hours, the phones of the Dialogue Facilitation Team *pinged* with messages from the insurgency, demanding to know what the government thought of their latest proposal. Zannah was continuously shuttling back and forth between Abuja and Maiduguri, the roadblocked city held under curfew. Pascal, the Swiss diplomat, could be summoned at any time, day or night, by Nigeria's intelligence agencies for a readout of their work or to quiz them on their sources. The team had learned to speak in code, careful to avoid words or names that could trigger the antenna of digital spies. Each new contact would bring hopes of inching toward an agreement, and when the trail inevitably went cold again, it was followed by a crushing low.

In late 2015, almost two years had crawled by since the kidnapping, and after the failure of Ahmad Salkida's attempts, the Swiss team was running what seemed to be the sole remaining effort to free the Chibok students. They had spent untold hours gradually expanding their contacts inside the insurgency, building a regular flow of correspondence with some of the terrorists they'd once studied on a whiteboard in the embassy. Members of the team had been mounting motorbikes to travel to the edges of the Sambisa Forest to deliver messages to contacts, hoping not to hear the whistle of an incoming air strike or the pop of gunfire from soldiers or insurgents. Fulan, the fundamentalist turned terror analyst, had become practically a resident of the Swiss embassy by now; he would vanish from Abuja, crossing unseen over a front line to exchange communications. Zannah's friend Tijani had become the team's chief courier, making calls to Boko Haram contacts, then sometimes venturing out to meet them, slipping into Cameroon on a rowboat moving up the Benue River. "Try to come back," his teammates would nervously joke, mindful of the risks he was taking. Tahir, whose encyclopedic knowledge of the sect's hierarchy had won him the nickname "The Elder" from his teammates, was secretly meeting his own contacts to cross-check the information flow. Safehouses in northeastern cities under military patrol had become locations for face-to-face meetings with insurgents who could take information to senior commanders. The team had permission from the presidency to pursue this dialogue, but the bosses of the military and security agencies had their own ideas of who should talk to Boko Haram, and the risks were constant. At any moment, a soldier could have kicked down the door and thrown team members and their sources into jail to join the ranks of Nigeria's disappeared.

To crack their way into the insurgency, the Dialogue Facilitation Team had started from the fringes and methodically worked their way toward the core. At first, the terrorist group had been hard to penetrate. But the team had followed a key lesson that Zannah's instructors had repeated in his classes in Switzerland's Lake Thun: "Meet the diaspora." Every armed group, they explained, had an orbit of ex-members, scholars, or businessmen sympathetic to its aims, people who could teach the basics, like how to address its leaders, and connect you to people closer to the center.

Zannah had spent 2015 flying around Africa and the Middle

East with other team members to meet as many of these Boko Haram exiles as he could find. The Barrister had visited followers, sympathizers, and former companions of Shekau in Chad, Niger, Benin, and Ghana, semi-retired insurgents trying to melt into civilian life. But the most important location was farther afield, the Sudanese capital of Khartoum. The desert city at the confluence of the Blue and White Niles had once offered refuge to Osama bin Laden and Carlos the Jackal but had more recently extended its invitation to a clutch of Boko Haram's most senior members who had refused to follow Shekau into armed insurrection. These men were out of the fight, but they retained serious theological influence over the militants in Sambisa. Dressed in flowing white *jalabiya* robes and thumbing prayer beads, they had met Zannah and Pascal in the Swiss embassy in Khartoum, a gardened oasis that visitors entered by walking under a floral archway flanked by palm trees. The men had talked for four full days over cups of sweet tea, building trust and drawing a map of how they could reach the Imam, aware that their conversations would percolate up to Shekau. When a meeting was going well, Zannah and the exiles would nod and say, "Toh, toh," meaning, "Alright, alright," in Nigeria's Hausa language. Soon, the Barrister noticed another man at the table repeating the same phrase: "Toh, toh," said Pascal.

The team was operating on a philosophical concept they called "the windows." For all the progress they were making, there could never be a deal until both sides were truly ready to negotiate. Then "the windows would open." Until then, they needed to meet as many people as they could, to build credibility with both sides.

Still, the more people the team met, the more they understood the myriad reasons why the windows remained closed. Boko Haram didn't trust the government to deliver on any of its promises. Worse, Boko Haram's leaders didn't entirely trust each other. The team's contacts described disputes between emirs over what to do with the hostages that boiled over into the Shura Council, the organization's ruling body. Shekau and his subordinates disagreed about whether to trade the women for money or for prisoners, and the friction between them was growing. Some of the militants wanted to demand billions of dollars for the schoolgirls, arguing that they had become the world's most famous hostages.

Complicating matters was the fact that several of the militants the Dialogue Team had painstakingly cultivated were dead, executed on the Imam's orders for dipping their fingers into Boko Haram's cash supply. "They touched the money," a contact had explained to Zannah. The Barrister was always on the hunt for new contacts to replace the ones who'd been killed or jailed or who'd simply stopped responding to his messages.

That is why, near the end of 2015, the team started talking with another fugitive militant on Nigeria's watchlist. The army knew him as the Nyanya bomber, and he was accused of dispatching explosives that had killed eighty-eight people at Abuja's Nyanya bus station on the same day as the Chibok kidnapping. The member of the team tasked with reaching him was Tahir, the elderly friend of Zannah's who had once been a disciple of Boko Haram's founder, Mohammad Yusuf.

Tahir was kneeling in the middle of prayer in a central mosque in Abuja, when he realized that the soldiers who had entered the building were looking for him. Nigeria's military intelligence had been monitoring his phone calls. A mutual friend dialed Zannah in a panic and told him that Tahir had been frog-marched out of the mosque's vaulted entryway. The Barrister immediately understood. For a terrorism suspect, disappearing into the custody of the military or the security agencies could mean torture or death. Zannah and Tijani boarded the next flight to Abuja, then headed straight for the Swiss embassy, but it was too late. Tahir was beyond the reach of Swiss diplomacy. In two years, the team had failed to free even a single Chibok student. Instead, they had now cost one of their members his freedom and possibly his life.

AT THE START, THE DIALOGUE FACILITATION TEAM HAD WORRIED THAT talking to Boko Haram would be the hard part, but by late 2015, problems were piling up on the other side of the ledger.

Nigeria had two separate spy agencies, the Department of State Services and the Office of the National Security Adviser, with overlapping mandates, and their rivalry had deteriorated into a bitter turf war over strategy and access to the billions of dollars in contracts the war brought. The Swiss team was caught in the middle. The men who ran the twin agencies loathed, and were constantly trying to undermine, one another.

Pascal would book separate face time with each, imploring them to bury their differences and accept that one agency should run the Chibok file. One of the agencies had launched an internal task force to determine whom the Swiss were speaking to. The feuds made it nearly impossible to deliver a message up to President Buhari himself, and it wasn't clear that the commander-in-chief was being fully briefed. People who managed to meet the president were walking away with the same impression, that Buhari had too little information on what was happening in the north. At times, he doubted that the Chibok girls were still alive.

A huge amount of work was needed just to keep the two sides talking. The team members were not just passing messages, but also doing a lot of "message management," translating and interpreting for the other side to avoid any missteps that could suck momentum from the talks. Short, blunt messages from Boko Haram would be repackaged to ensure they didn't offend. Government messages would be edited to include the correct salutations to show appropriate respect for the Boko Haram leadership. When there was no response from the government, the Swiss team would assure Boko Haram that the delay was because the message had gone up to the president. It was like keeping a body on life support.

Each member of the team was under tremendous pressure, coordinating hundreds of messages with the insurgency and counting on Pascal to shuttle between the intelligence agencies and extract answers from the Nigerian state. In times of crisis, he was sending his own audio messages over WhatsApp and Telegram to Boko Haram commanders to insist that the talks could be trusted. His teammates began to call him the Glue.

And the stress was starting to show. "Breathe," Fulan, the analyst, would tell him. "Do you feel better?" Pascal would usually say no.

WHEN HE WASN'T MEDIATING BETWEEN THE GOVERNMENT AND BOKO HA-ram, Zannah was still trying to run his orphanage in downtown Maiduguri. Driving into his hometown from the airport, he would see a city utterly transformed from the place he had known before the conflict. Vast tent cities stretching for miles had sprouted on boulevards leading into the city center, the squalid new residence for some of the two million people that had been forced from their homes throughout the northeast. Most of the new arrivals were children. Public buildings and abandoned factories had become squats for the displaced fleeing the war. Schools and even the cement blocks around Maiduguri's old flour mills, once the pride of the town's industry, were now housing thousands of people with no running water and very little food. Inside, hundreds of young children were near starvation, their bellies swollen. It was an image of wartime hunger Nigeria had not seen since the 1960s Biafran War.

On metal benches in front of the city's Federal Neuro-Psychiatric Hospital, crowds of women in flowing veils, and men in tunics and kufi caps squeezed next to each other, sat waiting for a two-minute session with the last psychiatrists left in the city.

There were only nine of them, overworked doctors who dealt with fifty-two thousand active patients, recording their post-traumatic stress symptoms into blue files that spilled on top of each other. Billboards from the election held months earlier still hung eerily over the city: WE WILL DEFEAT BOKO HARAM. THE POWER TO CHANGE NIGERIA IS IN YOUR HANDS.

Thousands of the people streaming into Maiduguri were former Boko Haram captives; women, girls, and boys who'd managed to run for freedom during the chaos of the war rumbling through the countryside. Many had simply been abandoned by the sect as it fled different waves of military offensives. They brought stories of forced marriage, children born of rape, children lost during airstrikes. But tellingly, not one of the women who'd gone free was a Chibok girl. The schoolgirls were too famous for Boko Haram to let them escape.

This sea of human suffering had washed into every corner of Zannah's city. He had spent nearly two years trying to free just one small fraction of that boundless toll of victims. Even though he told himself the Chibok hostages were the key, at times he wondered if he was the right man to unlock a conflict that defied easy conclusions. Thousands of people had died in the two years he'd been working with Pascal, and there were few concrete achievements to show for their efforts. The violence in his hometown seemed senseless and unending.

Monday Market, the commercial plaza where Shekau himself had once sold perfume bottles from a wooden box, had been bombed so many times that its security guards had lost count of how many blasts they'd lived through. The sect had begun dispatching children wearing suicide vests, usually small girls, into local markets. Some of them were as young as ten years old. Then there were the boys, a generation of them who seemed to have been lost. Sometimes, Zannah would scrutinize the latest videos of men riding stolen army vehicles or firing truck-mounted machine guns and recognize the faces of the angry men from the child beggars he had once seen sleeping on his neighborhood's streets.

One morning, he was hunched over, studying at his desk at the orphanage, trying to work above the chatter and laughter of pupils preparing for class, when a loud thud shook the buildings. It was an explosion, and very close. Screaming students sprinted into classrooms to shelter under desks. The street outside the campus was covered in dust and debris.

Moments later, a group of students ran to Zannah in alarm. One of his boys had been walking to school down that street, they said: an eight-year-old named Ammar whose father had been a Boko Haram member. Children throughout the orphanage began to panic as the news rippled across the schoolyard. *Ammar is dead.* The campus echoed with the cries of youngsters who'd lost family members on both sides of the conflict.

Then, in the midst of the melee, the iron gate swung open, and Ammar appeared, smiling, in his freshly laundered school uniform. The ecstatic cheers of hundreds of children filled the air as he was thronged by his friends. On a whim, Ammar had chosen to walk along a different street, he said. The school canceled the day's classes on the spot and declared a feast. Zannah watched the orphans of Boko Haram and its victims sharing plates of bright red jollof rice. He captured the moment with his phone. Even if their team failed, perhaps there was still hope.

"The new generation could make peace," he said. "They will grow up and make a settlement."

PART IV

A Breakthrough

28

"I'm Not Going Anywhere"

ASO ROCK PRESIDENTIAL VILLA
JANUARY 2016, TWENTY-ONE MONTHS INTO CAPTIVITY

Kolo Adamu had made the perilous journey from her home to the heart of the capital to join the activist Oby Ezekwesili and more than a hundred Chibok parents at the fortified gates of the Aso Rock Presidential Villa. Rifkatu Ayuba had traveled alongside her in a convoy of parents demanding to see President Buhari to protest that more than six hundred days had passed since their daughters had been kidnapped. Some of the mothers had needed to pawn their belongings to fund the trip, which cost more than their monthly earnings selling vegetables or homemade soap. They had ignored the orders of soldiers in Chibok who had told the women they had "no permission" to leave the town. Now the mothers had made it to Abuja, marching to Aso Rock in the baking midmorning sun. They were there to demand an update on the government's alleged rescue plan.

Kolo had expected that the president would tell her where the schoolgirls were, their condition, and then outline a strategy to bring them home. She had high expectations that the meeting would herald a swift resolution. "We thought they would be brought back home safely on a particular date," she said.

It was January 14, 2016, and Muhammadu Buhari's deadline to recover the girls within a hundred days had long expired. News about the kidnapping had virtually disappeared, and the global attention on the Twitter campaign had long evaporated. The protests opposite the Hilton that once made headlines had dwindled to just a handful of dedicated activists. The scene was beginning to feel more like a wake. The Chibok parents were feeling helpless and abandoned. Many of them were struggling with the physical and psychological manifestations of their grief and would tearfully call Oby in the middle of the night: "Ma, can you tell me what happened to our girls?"

"We wouldn't stop, we wouldn't stop," Oby later said, in tears. "How many times did I give that assurance?" To preserve the campaign's momentum, she had written repeatedly to Buhari's office, and when the presidency failed to respond, she would tweet about it. And tweet about it. Until finally they granted her an audience.

Now she was marching down the smooth boulevard leading to the presidential villa to take them up on the invite, trailed by a crowd of Chibok families brandishing megaphones and banners printed with the faces of their kidnapped daughters and the words #CryingToBeRescued. Dozens of activists wrapped red plastic around their mouths printed with the words We will not be silenced. Several held a list of the names of the hundreds missing.

Still, there was a problem. A covered pickup truck full of police officers refused to let the marchers through. The villa had locked the gates, a protocol official announced; therefore everybody should return home. A surge of voices rose from the crowd, parents erupting in anger.

Oby was furious, but had expected the authorities might try to subvert her plans. The Buhari administration, elected after courting the #BringBackOurGirls campaign, had come to bristle at the daily tweetstorms emphasizing the military's failure to recover any hostages.

The woman known as Madame Due Process refused to retreat and led the parents in the first of a series of fist-waving chants; "A fight for the Chibok girls is a fight for the soul of Nigeria." The parents, dressed in their finest *atampa* dresses, sat on the asphalt, creating a visually striking cordon and an image that would captivate social media. Oby and other activists began to hammer out posts on Twitter, a cascade of real-time dispatches reporting how grieving mothers after traveling more than five hundred miles were being blocked from entering the seat of their government. Two of the parents, exhausted from the journey, fainted in the heat. Other protestors rushed to revive them.

Oby kept tweeting and retweeting pictures, video, and the sentiments of parents, fomenting out rage inside Nigeria and beyond. Soon, photojournalists and reporters arrived at the gate to begin capturing the protest and interviewing the families. Nigerian Twitter erupted with sarcastic memes lambasting Buhari for his indifference to the tragedy that helped him get elected. The military man was being mocked.

For the president, the embarrassment was a serious matter. Finally, a message reached the protocol officer stationed at the entrance: Open the gates. As he complied, a convoy of white air-conditioned buses rolled out to collect the families, and carry them to the commander-in-chief.

TWO HOURS OF ANXIOUS WAITING LATER, KOLO WATCHED A GOVERNMENT spokesman walk into the marble-clad banquet hall in the heart of the presidential villa where she and the other parents had been sitting. The official was flanked by cabinet ministers and security chiefs. He spoke into a microphone to deliver his announcement: "I regret to inform you that the president will not be able to see you."

A murmur rippled through the crowd, as parents looked at each other in bewilderment. The delegation, the spokesman assured them, would faithfully attend to their plea. Oby rose to her feet and requested a microphone, trying to restrain her anger.

"We appreciate the president sending his delegation," she said. "But we haven't come here to meet the delegation. . . . The president told us he would do everything to bring the Chibok girls back. And not declare victory without bringing them home. We are here to see the president."

The government officials took turns expressing their sympathies to try to mollify the families. Officials were engaged on the issue, they said. There had been great progress on the war front, and new batches of other hostages were being freed often. As the crowd became more restless, the women's affairs minister, Aisha al-Hassan, the only woman in Buhari's cabinet, spoke to calm them. "We understand how you feel," she said.

A voice came from the back of the room: "You understand how I feel?"

It was Esther Yakubu, the mother of Dorcas, fifteen years old on the night of her kidnapping. In high school, Dorcas had been a popular student and careful note-taker who would hand-copy the love letters she received from an admiring boy. "You are the remote control of my life," her crush had written. Now her mom was demanding to see the president.

"No, you don't understand how I feel," Esther said. "My baby is with terrorists and your daughters are home with you. So I'm not going anywhere."

The moment brought shouts and cries of agreement from the roomful of angry parents. Some parents burst into tears, others stood up from their chairs gesticulating wildly as they spoke over each other. Nearly two years of grief and exasperation was curdling into rage. Oby live-tweeted the scene, her posts shared by thousands across the world. The national security adviser appeared to have understood the potential political damage from the unfolding action and ducked out of the room. About an hour later, another male voice bellowed into a microphone: "Ladies and gentlemen, please rise for the president."

The Chibok parents would later remember the unpleasant body language as he strode into the room, lip curled. "He came in with a terrible stiffness," Oby would recall. "The president was visibly unhappy," said Kolo. President Buhari began listing the military's recent gains. Your daughters were abducted under a previous government, he reminded the room.

"I heard some of you are angry that I told the truth," he snapped. "We have no intelligence."

"Why not?" Oby shot back.

The problem, Buhari said, is that the government had not figured out how to open a credible negotiating channel. He was unaware of any mediators that could reliably connect him to the men who held their daughters. Kolo said nothing. After a few minutes, he put down the microphone and left. The parents shuffled out of the room in a dispirited silence. They had come to the capital with the expectation that the president would offer a reason to hope. The journey home would be long and dangerous. The extravagant expense of the bus ride appeared to have been for naught.

That evening, Kolo and Rifkatu shared a hotel room, venting frustrations from their twin beds. The entire trip was a waste of time, Rifkatu argued, and Kolo nodded. The fact that the president could not even tell them the likely location of their daughters was terrifying, they agreed. Neither of the two wanted to say it, but they felt it meant their daughters might be dead.

The two said a prayer before they switched off the light. "We commit our daughters' safe return to God," Kolo said, and she lay herself across the itchy sheets for an uneasy rest. Oby, meanwhile, was thumbing out tweets.

"Our President @MBuhari MISSED OPPORTUNITY of emotionally connecting with GRIEVING PARENTS," she wrote. "HOPE MAKETH NOT ASHAMED. OUR #ChibokGirls ARE NOT BACK. WE STAND ON. WE DEMAND ON!"

29

The Fire

The pilot had the coordinates scrawled on a paper sheathed in plastic across his thigh, numbers directing him through thick gray clouds to a site in the Sambisa. His target was the forest head-quarters of the man who had eluded Nigeria's military for almost a decade. The Chibok hostages knew it as Gaba Imam, the home of Abubakar Shekau. From above, the hideout was nearly invisi-ble, hidden among a cluster of camps camouflaged under a dense canopy of trees. But by August 2016, Nigerian and American sur-veillance planes were closing in on him. In the cockpit, the pilot guided the joystick toward the target, watching the needles on the flight instruments and waiting to push the trigger.

In the preceding months, Nigeria had received a lot more in-formation about what, and who, was hiding on the forest floor. At the start of the war, its military had only a few broken-down drones with missing cameras, but now they had a spy plane they called the Beechcraft, one of the company's long white twin-engine turboprops equipped with a bulbous node of sensors and cameras on its belly that would regularly spot structures worth examin-ing. Noting the coordinates, the Nigerians would ask the US to take much higher-resolution photos and video recordings. The Americans were studying the area with drones flown by the Night

Stalkers, a special-operations force that had flown surveillance for the raid that killed Osama bin Laden, now based at a fenced-off airstrip across Nigeria's eastern border with Cameroon. The US drones fed their footage to the CIA station in Abuja, whose chief would regularly meet with President Buhari's office to share information. The wellspring of intelligence was transformative, a "game changer" one senior American official boasted. It had helped Nigeria dramatically expand its air war in a campaign that had weakened Boko Haram but also cost civilian lives.

The pilot was flying an Alpha Jet, a light and unsophisticated fighter plane, one of four that the Nigerians had purchased from an American company in Quincy, Illinois, with American government approval, just as Buhari was beginning his presidency. The Alpha Jets had arrived without any weaponry, because US laws against providing arms to governments with records of human-rights abuse restricted American businesses from selling Nigeria arms. But there had been little doubt what the Alpha Jets would be used for. The Nigerian Air Force had simply paid a local car-parts manufacturer $13,000 to cheaply weld a rocket launcher onto the sides of these jets, saving hundreds of thousands of dollars. Some of the world's most sophisticated drones were now feeding targets to a fleet of outdated, imprecise warplanes with rockets crudely rigged to the wings.

On that cloudy Sunday morning, the Alpha Jet pilot knew it would take only a half hour to reach his target, coordinates identified by a complex multinational chain of intelligence. He was undertaking a risky shot. Boko Haram had hidden heavy Soviet Dushka antiaircraft guns in the bushes around the camp, and the range of the Alpha Jet's rocket launcher was short. As he came close enough to fire, there would be a tense few seconds when the insurgents would be able to shoot back. A year earlier, a pilot had been shot down, captured, and beheaded with an ax on YouTube by Aliyu, the Boko Haram commander who had once threatened to do the same to the Chibok girls. Air force officers had become skittish, firing rockets, then quickly climbing to the safety of higher altitudes while rocking the wings to make them harder to hit. The pilot checked his instruments for a final time, unaware that he was about to fire a rocket into a group of young women who had just woken up.

SUNDAY MORNINGS ALWAYS FELT CLOUDIER, HANNATU STEPHEN THOUGHT. The teenager was lying by the open doorway of a corrugated zinc–roofed hut, staring into an overcast sky. She could hear the breathing of a friend sleeping next to her and the chatter of the young women outside. A half dozen schoolmates were already awake, braiding each other's hair with sculpted wooden forks they'd carved themselves. Metal bowls, ceramic plates, and yellow jerricans full of water were scattered on dust-caked raffia mats, ready to be used when food arrived. Several of the students were flipping through the Quran. There was still an hour before breakfast, and the hostages hadn't yet prepared a fire.

At home, Hannatu would have spent a morning like this getting ready for a service at the Church of the Brethren. Her boyfriend, Ibrahim, a Muslim from central Chibok, would sometimes drop by the house as she and her seven siblings put on their Sunday best. Ibrahim was going to be her husband, she'd decided. They'd talked it over and agreed. Hannatu didn't mind school, especially literature and the logic of English grammar, and she had loved the required reading of "Chike and Chukwu," a morality tale of two brothers who helped each other graduate and advance professionally. But she wasn't planning to attend college. Sometimes when the teacher asked the class to recite from the board, she would hum along and move her lips vaguely because she hadn't done the homework. She was more interested in enjoying herself, helping her mother make spicy okra soup, and hatching plans with Ibrahim. Her school circle knew her as a joker, her tiny frame releasing a bellowing laugh that sometimes made her double over and slap her thigh. Her friend Ladi Simon called her the smiling type.

In captivity, Hannatu had been beaten on the hands for failing Quranic recitations and across the back for venturing too far to fetch baobab leaves for kuka soup. She knew what it was like to be whipped by an electrical cable. The routine punishments had made her harder, and she noticed she wasn't as quick as others to cry in painful moments. Her group was now being kept near Gaba Imam, the home of Shekau, closer to the leader's supervision and away from Naomi's troublesome crew. The forest was thicker here, tree branches overlapping to create stretches of canopy that shaded them from the sun and made them feel safer from the jets patrolling the sky. They had now been captives for two years and four months. The camp was awash with rumors of a feud between the Imam and his top commanders, and the leader wanted to keep them close.

Their lodgings had become larger, squat tin-roofed structures that could shelter more than a dozen hostages each, nestled under the branches and ringed by fences woven of cornstalks. That barrier was the hostages' prison wall, and to cross it required permission. The long barrel of machine guns meant to shoot low planes from the sky poked up from nearby bushes, the gunners crouched under the branches of trees. The *rijal* stationed around the base were more serious in their posture than those the hostages were used to seeing. The militants were also nervous, constantly studying the sky for the next jet to screech past. For two days, a long white propeller plane had been circling the area, sparking worried conversations among the guards. They had nicknamed the turboprop Buhari and suspected it was flying surveillance sorties to mark targets for coming airstrikes. Sometimes the guards and their captives alike could hear the buzz of drones high above.

Hannatu knew it was Sunday because it was two days after Friday prayers, when crowds would head to the camp's mosque

for the week's largest communal gathering. Several of her classmates believed that Sundays, their holy day, always felt different.
They could sense it. Looking upward, Hannatu could see a thin
black dot dancing in the distance, just below the clouds. The others didn't seem to notice, chatting under a tree and stitching their
hairs into plaits. Hannatu craned her neck as the plane tilted
sideways far above them. "It was so small, like a bird," she remembers thinking.

"Before I knew it I just saw the fire."

The students don't recall ever hearing a warning sound. Just
a cloud of hot dust that cleared to reveal a wall of flames chasing
into the branches above. Their vision was blurred by fumes and
the fog of concussion. Ears began to ring with a high-pitched puls-
ing that muffled any other sound. Hannatu was dazed and almost
dreamlike as she squinted through parting ribbons of smoke.
She was flat on her back, her body limp, as the heat choked her
breathing, and she began to panic. Discombobulated thoughts
were clarifying into an internal alarm. Something in her body was
not working. Her mind was telling her to get up but her limbs
were not complying. As she tried to raise her knees to stand, she
realized she couldn't feel her right leg. She lifted her head and
saw that where her knee had been was now a tangled mess of
blood and shredded fabric. Her bones were shattered. The circuit
connecting her brain to her leg had broken. Her head fell down
onto the dust, and she passed out.

She woke up to screams of panicking teenagers and a militant
standing over her. "Raise your head! Raise your head!" He was
shouting in her face, trying to figure out if she was dead or dying.
All around, young women were running and screaming in distress. Behind her, in the hut, was a jumble of bodies, some caked
in blood. One woman was trying to stanch the bleeding from a
friend's torso. Another, who'd been braiding hair just minutes
earlier was dead, lying with her eyes and mouth open next to
a classmate who was staring into the ground and repeating the

words "Please hold me." Next to them, a yellow-colored jug of water stood perfectly upright, untouched. Militants were weaving their way through the crowd, inspecting the bodies on the ground. In pairs, they began to carry them away.

There would never be a funeral, just two mass graves, shallow and unmarked. The names of the dead were not recorded by the men who'd taken them hostage. Even some of the surviving students would remain unsure for a long time of who had died. Years later, what several would remember about the airstrike that killed at least ten of their classmates is that, in that moment, instead of helping the distraught and wounded teenagers, some of the militants reached for their phones and began snapping pictures and recording video.

"We were screaming and running around everywhere," one of the young women said. "Dead bodies, blood everywhere. And they are just videoing us with their phones. Then they just came and packed the dead bodies of our sisters in their trucks."

THE FOOTAGE THE MEN WERE TAKING WITH THEIR PHONES WAS FOR A YouTube video to provoke an audience on social media. They were capturing the moment when a global rescue mission to free the Chibok schoolgirls, inspired by millions of people clicking a screen and discovering their own collective power, had culminated in the worst possible outcome. The drones and surveillance aircraft first sent by the world to find the young women had guided the warplanes that fired on them as they sat on the forest floor. While the hostages lay dying on the ground, one of the men managed to capture a few seconds of an Alpha Jet quickly climbing, the pilot flying much too high to survey the consequences of the button he'd pressed.

THE EXPLOSION WASN'T THE SOUND THAT CAUGHT NAOMI'S EAR, BUT THE shouting that followed immediately after. The thunder and rumble of airstrikes had become so common that they barely registered.

Usually the jets missed their target, and rockets often were shot into tarpaulins or scooters placed in open ground as decoys. But the commotion after this strike sounded different. From their latest holding area, a set of thatched grass–roof homes a few minutes away by foot, Naomi and her friends heard screams and panicked wails in a familiar language. Somebody was shouting in Kibaku.

"It's our girls screaming," Naomi said to her classmates. Just hours earlier, they had been discussing a dream in which an airstrike had killed a large group of Boko Haram fighters kneeling in prayer. Now they were imagining what had happened to their sisters. Plumes of smoke were billowing from the treetops, twisting into the branches. "Whatever it is, we will get to the root of it," Naomi said.

But when they tried to leave their fenced-off enclosure, the guards stopped them. For hours they sat or paced, anxiously listening as the screams faded into the mournful afternoon sounds of the forest. Finally, a man arrived with a bowl of food and some water for them., He was wearing a pained look across his face. "Kai!" he said, shaking his head, "You girls have to take heart . . . many of your girls are dead."

Naomi and the group sat stunned as they processed the words. "It was the worst day, the one we all remember," she said. They were in disbelief, frantically running through the names of friends who may not have made it. Every one of them wept, their moods lurching between grief and unrestrained anger. They talked about the beatings and hunger their friends had endured and asked God to punish their captors. "We were shouting and raining insults on Boko Haram," Naomi said. "If they had not brought us to this forest, they would still be alive."

She felt their classmates had made the right decision not to get married and that God would reward them for their choice. "They suffered because they refused to get married. And now a bomb killed them," she said. "We could not believe it." The women prayed that their friends would be welcomed into the next life and for their own lives to be spared. "We couldn't do anything," she said. "We realized when it was our turn to die, it would come."

MEDICS HOVERED OVER HANNATU, PRICKING HER ARM WITH A NEEDLE threaded into an intravenous drip. She was lucky to be alive, but the risk of septic shock was rising. If her right leg wasn't treated, the gangrene would spread into her bloodstream, the men warned, and would shut down her organs then stop her heart. Her injury required surgery, but the only instrument available was a pair of scissors. The clinic was a thatched-roof hut with little in the way of supplies. The drip was hanging on a stick of firewood. Hannatu's bed was a thin plastic mat laid on the dirt. The sting of an anesthetic pierced her skin, and before the painkiller took effect, she looked up at the faces analyzing her condition and felt that she knew the man holding the scissors. It was her family doctor, a Christian and a familiar face from Chibok. He, too, had been kidnapped and pressed into medical work for the insurgency, he told her. The procedure took more than an hour.

The surgeon finished the amputation by wrapping off the leg with dressings and medical tape. Hannatu lay there feeling parched and alone, until three of her classmates came with water. She drank and fell back to sleep. In the days that followed, friends would visit with meals or to sit and keep her company. They were often silent. When Hannatu was ready, they shared the news of who had been killed. More than thirty Chibok girls had been injured, she learned. Many had shrapnel wounds, and the metal would remain embedded in their flesh because it was too difficult for the clinic to remove. One of the classmates would die of such injuries three months later. Another had gone mute.

From her mattress on the floor of the clinic, Hannatu could tell that Gaba Imam was in chaos. Hundreds of the *rijal* were leaving Shekau's hideout; others were relocating weaponry to different sections of the secluded camp. Hostages had been moved elsewhere as well. The fighters were no longer just watching the skies but were nervously eyeing each other. Who would leave next? At one point, a man in thick, military-issue boots walked into the clinic, flanked by a group of armed guards in combat fatigues, and approached Hannatu's bed. He was wearing a white robe with a bulletproof vest and a balaclava and spoke in a voice she recognized.

"How are you feeling?" he asked. He wore a machine gun hung over his shoulder. In his hand was a large *miswak* chewing stick and 7,000 naira, about $40, a gift he offered his injured hostage. "Do you recognize me?" he asked.

"Yes," Hannatu told Abubakar Shekau. "You are the Imam."

30

"Never Trust
a Breakthrough"

ABUJA

AUGUST 2016, TWO YEARS AND FOUR MONTHS INTO CAPTIVITY

Zannah's phone buzzed with a number he didn't recognize, but when he answered it, the voice sounded familiar. The man on the call spoke slowly, his voice crackling over the patchy line.

"God has tested us," he said. "Many of the girls have gone."

"How many girls?" Zannah replied.

"We are facing our *ibtila*," the voice responded.

The call went dead.

This was urgent. Zannah messaged Pascal to relay the news, then bought a plane ticket to Abuja. They would need to cross-check it with other sources inside the insurgency. The Dialogue Facilitation Team knew the military's air campaign was routinely claiming Boko Haram casualties, and civilians too, but the Chibok girls had been spared until now. Zannah was thinking of his own daughters and cursing that it was taking so long to bring the warring parties together. He reflected that the militant's use of the word *ibtila* was striking. In Arabic it meant a "trial," a crisis or hardship interpreted as a test of religious faith.

When Zannah landed, he raced to Pascal's house to discuss how it could affect the process. Both men were in shock, and as they drank tea around the diplomat's table, they wondered aloud whether they could have moved faster or done things differently. Boko Haram had not sent video verification or revealed the names of the dead, but Zannah's contact had never lied about something like this. "God will punish us for not getting them out before the strike," Zannah said.

In the Swiss embassy over the following days, the team could see that news of the airstrike corroborated a stack of intelligence foretelling Boko Haram's gradual rupture. Abubakar Shekau appeared to be losing his hold over his army and the war. Nigeria's military was advancing, and Shekau's commanders were frustrated because they had been pushed into smaller and more remote corners of the countryside, where food was becoming scarce. Grievances had been brewing over the Imam's maniacal leadership style, they had learned. Shekau and his rivals had been circulating audio files as propaganda messages, wherein they accused each other of deviating from God's law. In retreat, the Imam saw enemies everywhere, and traveled with more bodyguards, always dressed in a bulletproof vest. He had come to suspect that followers plotting his assassination had planted a tracking device on him.

Commanders were turning their guns on one another, and thousands of people began to flee Shekau's territory to breakaway camps far north of Sambisa. They named their exodus the Hijra, after the migration of the Prophet and his followers from Mecca to Medina. Four thousand miles away, in the Syrian city of Raqqa, the Islamic State's leaders were receiving missives outlining the drama playing out in the caliphate's Nigerian province. It was the global outrage to the Chibok kidnapping that had first convinced them to ink an alliance with Shekau in 2014. But by 2016, Shekau was proving too fanatical for even the Islamic State. He had kidnapped thousands of Muslim girls, something which the caliphate's top clerics had outlawed. Its leader, Abu Bakr al-Baghdadi, came to a decision: Shekau would need to be sidelined.

He stopped answering Shekau's handwritten letters, which were being smuggled over thousands of miles to reach him.

In the embassy, the Swiss team was analyzing tapes, encrypted texts, and voice messages that spelled out the schism in Boko Haram. One source of division was the girls, and some of Shekau's followers felt he'd mishandled his prized asset. Why hadn't he traded them for money months ago? It seemed they were witnessing a profound shift. The militants had spent more than two years trying to divide the Chibok hostages by forcing them into marriage. Now the young women were dividing Boko Haram.

FOR SOME TIME, ZANNAH AND THE TEAM SEEMED TO BE THE ONLY PEOple who had heard that something tragic may have happened in the forest. Days passed with no reports and no verification. The airstrike might have remained a rumor if not for a tweet sent from the suburbs of Dubai on August 14, posted by the exiled blogger Ahmad Salkida pointing his followers to Boko Haram's new eleven-minute video, uploaded onto YouTube. "Just In: Link to Video on #Chibokgirls." On the screen stood Tasiu, carrying a Kalashnikov, his face covered by a camouflage-colored scarf. He was pacing in front of dozens of the Chibok girls and speaking into a small lapel microphone, his gravelly voice restating Boko Haram's demand for their prisoners to be released.

Then he held the microphone up to a student dressed in a gray veil and asked her a question. "How long have you been held captive?"

"We have been captive for two years," she said, staring at the floor. "There is no kind of suffering we haven't seen."

"What do you want to tell the government?"

"I want to tell my parents and the government they should release the Boko Haram members into custody of the security services. . . . Military jets have killed some of the girls."

The video ended with thirty seconds of graphic footage of young women's corpses, most of them face down in the dirt, some covered by shawls, one laid out in a zebra-print dress, mouth open, as if she'd fallen asleep. It ended with a quick shot of a plane fluttering away.

The announcement that Chibok hostages had been killed in an airstrike barely made the news. Only a smattering of people commented. The Nigerian Air Force publicly denied that the airstrike ever happened, although senior officers later conceded that the deaths were collateral damage in a war that needed to be won. Some wondered if the footage was doctored by Boko Haram. US officials said the Nigerians who received their intelligence often didn't disclose if or how they acted on it. The incident wasn't reported to senior Obama administration officials in Washington, and was seemingly never included in the president's daily briefings. The White House Council on Women and Girls, which had asked for updates, was not told. On Twitter, where the conversation had long moved on, not a single celebrity expressed any sympathy for schoolgirls killed in an air campaign that had been fueled by the drones once encouraged into service by their tweets. Hours after he posted the video, Nigeria's army declared Ahmad Salkida wanted.

Oby Ezekwesili had begun to fire off a series of pressing tweets directed to the president's account, @MBuhari. "How can any leader watch the recent video released by TERRORISTS that have 218 of Daughters of Nigeria & NOT DO EVERYTHING TO RESCUE THEM?" Only a few dozen people liked or weighed in on her calls for action, and major news organizations largely skipped the story.

For Zannah, the footage, however hard to watch, contained an opening. Mediations rarely proceeded in a straight line. In a complex world of unforeseen consequences, a setback could allow one more small step forward. And the eleven-minute video Salkida had just posted divulged two important revelations. It corroborated his phone call from Sambisa. Moreover, it suggested that

Shekau was eager to negotiate before more died, perhaps to quell a mutiny by followers. With the Imam in a position of weakness, it seemed a window could be opening. In Zannah's literature, there was a term for the moment when both sides are in enough distress to have reason to make a deal: "the ripeness of conflict."

Around the same time, Zannah received an even clearer signal from Sambisa. It was an audio recording sent on Abubakar Shekau's behalf. The Imam had heard that Zannah's team was trying to speak with him and wanted to confirm it was the same Barrister he had once known years ago in Maiduguri.

"If the Zannah I know is the Zannah I am talking to, let him tell me when and where we last met," said the voice on the audio message. It was a test. Boko Haram was doing their own verification.

Zannah sent back a recording of his answer. The last time he had seen the Imam was outside Maiduguri in 2008, at a bus station. Shekau had been chewing two sticks of sugar cane, on his way to preach. The Barrister had just dropped his daughter off at college in a nearby city. The two men had exchanged halting greetings, then moved on.

He waited for an answer. A few days later, another message landed on the phone: "The leader accepts you." The next contact was a phone call from a man Zannah knew by his rank, emir, one rung under Shekau. "I have very good news for you," the man had said. "Shekau said we should talk to you." Zannah snuck away from the orphanage to board another flight to Abuja. He remembered a mantra he had heard during the Swiss Alps sessions: "Never trust a breakthrough."

THE WINDOW FOR A DEAL SEEMED TO BE OPENING. NOW THE CHALLENGE was to get the government to walk through it. The Dialogue Facilitation Team had spent two years "building trust" on both sides of the conflict. They had risked their lives to talk to hundreds of people from the fringes to the center of the sect: qaids and *rijal*, former militants and their wives, and the truck drivers and deliverymen who serviced the insurgency's territory. And in Abuja, Pascal had worked tirelessly to win the confidence of government security bosses, patiently explaining the mediation effort over mugs of Nigerian Star lager and spiced beef kebabs at a beer garden, or on the oversize pleather furniture of government offices.

But just as there was conflict at the top of Boko Haram, there was another fight raging between Nigeria's spy agencies. And the more the Swiss team learned about Boko Haram, the more they seemed to become caught in the crossfire. The Office of the National Security Adviser and the Directorate of State Security Services were still trying to undercut each other. They were using phone-hacking tools and presumably trying to intercept their conversations. And one of the security services still had a team member, Tahir, locked up somewhere in custody. Aggravating matters, President Buhari's health was deteriorating, his face now gaunt and drawn from an illness the government said was only an ear infection. He had begun taking medical trips to London for treatment, and in his absence, messages went unanswered and weeks rolled by.

The Swiss embassy meanwhile had begun drawing up plans to evacuate Pascal from Nigeria after Fulan had warned him that he was being followed. His growing profile was attracting enemies. "You're not bulletproof," he had said. The embassy beefed up the security at Pascal's home. But after a long discussion, the Team agreed it was better for him to stay. The moment seemed to be ripening for an agreement. Instead of hiding from the spy agencies, the embassy invited an intelligence officer from the DSS to come sit in on their sessions, and offered to send him to Switzerland for training. There was no trace of the man's name, Iliasu Alhassan, on the Internet, but this covert-operations specialist with rimless spectacles had his ear on the right doors in Abuja. He could speak to the kind of deal the government might consider.

To seal it, they needed Nigeria's ailing president to give them his blessing. Buhari was withdrawing from public life and shielded by a maze of bureaucracy, but there was one way to reach him. The commander in chief would soon head to New York, where world leaders were gathering for the United Nations General Assembly. He would definitely want to see the Swiss. Of all the countries in the world, they were the ones who had Nigeria's money.

31

Meeting in Manhattan

THE UNITED NATIONS

SEPTEMBER 16, 2016, TWO AND A HALF YEARS INTO CAPTIVITY

Nigeria's president was dressed in a starched black kaftan, a white round-necked shirt protruding like a priest's clerical collar, as he made his way to the first of a series of high-level meetings on Manhattan's East Side. The presidents of three countries were scheduled to see him privately, each of their governments intricately involved with his war on Boko Haram.

It was the seventy-first annual United Nations General Assembly, and world leaders were descending on New York, their motorcades creeping through cross streets choked with traffic under sticky, 80-degree weather. The chyrons spinning along the sides of Rockefeller Center spelled out a toxic US election campaign lurching toward a nervous climax. At the UN, Barack Obama would deliver his final address as president. New British prime minister Theresa May arrived, touting her pledge to unite her polarized electorate around a swift and harmonious exit from the European Union. Vladimir Putin decided the gathering wasn't worth attending, opting to stay in the Kremlin.

Muhammadu Buhari's first big meeting was with Obama, a short but cordial one that focused on the latest developments on the Boko Haram insurgency. "I wish you well in your retirement," Buhari said. François Hollande, also attending his last UN meeting as president, told Buhari that France would boost humanitarian assistance to the areas most affected by the conflict and increase investment in its agricultural sector. "We will invest in Nigeria; we believe in her," he said.

In the earlier years of the Boko Haram conflict, bilateral talks with the American and French presidents, who commanded by far the largest military and financial assets in West Africa, would have ranked as the most important. France had decades' worth of intelligence sources in its former West African colonies, and the US had stationed hundreds of troops along Nigeria's borders. But on Chibok, the Old World powers had failed. Africa's most important nation was addressing its highest-profile problem in a meeting with a head of state virtually no one had heard of—Johann Schneider-Ammann, a bespectacled former mechanical engineer, the president of Switzerland.

The first and stated reason for the meeting was not the Chibok issue, but another missing prize Nigeria was trying to bring home. More than $1 billion of Nigerian money had vanished into Swiss banks, government revenue that Buhari wanted to retrieve. Two decades earlier, a Nigerian military dictator, Sani Abacha, had died of a heart attack, reportedly overdosing on Viagra after a long night in his palace with three sex workers. His death ushered in Nigeria's return to democracy, but it also left the question of up to $5 billion that had gone missing during his five years of repressive rule, money his successors suspected he'd partly stashed in Swiss banks. A few months before the UN meeting, Switzerland had finally agreed to return $321 million of the looted funds, the first such payment in thirteen years.

But the most important, off-the-record agenda item was the fate of the Chibok girls. The Swiss president offered his country's services to resolve the kidnapping. His embassy had a team that was putting together the outlines of a deal, but needed the approval of the president. Abubakar Shekau was ready to settle, and yet the delay was in Abuja. Every time the team had asked their contacts in Aso Rock to pass along a proposal for presidential consideration, no answer had come back. Meanwhile, one of their members had been jailed, and the others were demoralized, wondering whether to just call it quits.

"Do you want to do this?" a member of the Swiss delegation asked. "Do you want to do this?"

FOR FOUR DAYS, PASCAL WAITED, SITTING IN THE SWISS EMBASSY IN Abuja until an answer arrived on the 007, his encrypted fax machine. The offer had been approved by President Buhari. To organize the deal, they would have to move quickly, and in the Swiss embassy, the Dialogue Facilitation Team scrambled to piece together a proposal that could be acceptable to both parties, each now seemingly serious about a deal. It was all systems go, operating in a window that the slightest misunderstanding could slam shut.

The ultimate agreement would be a three-phase process. Phase one would be confidence building. The insurgency would release a small batch of Chibok students in return for the first tranche of a €3 million ransom, Nigerian money to be hand-delivered by an officer from the DSS. The young women would be hidden in a government safehouse, and the government would say nothing, even to the parents. If that first release went smoothly—and remained a secret—the sect would promptly free the remaining captives who hadn't been coerced into marriage, and Boko Haram prisoners would be freed in exchange, along with the second tranche of the ransom. Then the world could know.

That would unlock phase two, de-escalation. The military would stop air strikes, and Boko Haram would stop deploying suicide bombers. Finally they would move to phase three, disarmament, demobilization, and reintegration, or DDR. The Boko Haram members ready to leave the struggle would be welcomed to a safe location for years of counseling that would culminate in their return to society.

The team reasoned that they would be freeing scores of Nigerian lives at a price just under what the French government had paid in 2013 to free a family of seven kidnapped by Boko Haram. They knew that the Imam, facing airstrikes from above and an insurrection from below, needed the money. Whether he would obstruct the next two phases would have to wait. First, they would need an agency willing to drive into hostile territory and collect the young women.

A CALL CAME IN TO THE GLOBAL HEADQUARTERS OF THE RED CROSS, A Belle Époque mansion perched on a hill in Geneva, overlooking the vast columned palace that once housed the League of Nations. The humanitarian agency's unique assets were urgently needed for a potential hostage release thousands of miles away. On an upper floor of the mansion, the operations team would regularly discuss such requests around a long table in a conference room decorated with four Nobel Peace Prizes. The call was from a field officer in the Maiduguri Subdivision, who had detailed a complicated operation that could pose serious risks to the agency's personnel and civilians along the route. Success would require meticulous planning. Every detail would need to be kept in the utmost secrecy.

When most people think of the Red Cross, they picture bags of rice delivered by aid workers to refugee camps, but that is only the public-facing part of what its fifteen thousand employees do. The agency's more sensitive caseload takes place away from the cameras, when Red Cross field officers comb through dingy jail cells to confirm which prisoners are alive and which remain unaccounted for, or when they cross battle lines to retrieve foreigners kidnapped for ransom or civilians trapped on the wrong side of a shelling. When a person goes missing in conflict, the Red Cross is often the call of first and last resort. In its basement, the headquarters keeps an archive detailing the names of civilians who disappeared during the wars of the twentieth century, from children kidnapped in the Spanish Civil War to fighter pilots lost over

Vietnam. The folders would stretch for three and a half miles if laid end to end.

Officially, the Red Cross is neutral, not beholden to any nation. But there is one government with which the Red Cross remains umbilically linked. Every member of the agency's governing council, the Assembly, is a Swiss national. Its president, Peter Maurer, was a founding director of Switzerland's Human Security Division, the agency that deployed Pascal to Nigeria and funded Zannah's work. Now a formal request had arrived in Geneva's Red Cross headquarters to provide the logistics for a complex operation facilitated by the team in the Swiss embassy. They needed cars, drivers, a translator, and assurances from every armed group along the route that each side would honor a cease-fire and that nobody would shoot at a convoy on its way to an unmarked point somewhere in Boko Haram's territory. If everything went according to plan, they would return with twenty hostages, who would need immediate medical care. Every step would have to be done without the media or the less predictable social media getting wind of the operation. A single unexpected tweet or Instagram shot from a curious bystander could botch the operation and endanger lives.

THOUSANDS OF MILES FROM GENEVA, NAOMI HAD BEEN LYING ON THE forest floor, studying Hauwa's face and wondering if hers looked the same. Weeks of hunger and vanishing rations had gradually exhausted her friend and had started to eat away at her muscle tissue. Hauwa's full cheeks had grown sunken, her hair had become stringy, and her skin was starting to peel and crack. She couldn't stand without help. At one point, she was so desperate for food that, as Naomi watched, she killed an ant to eat the crumb of food it was carrying.

Boko Haram was splitting apart and running out of food. When Naomi and Hauwa scanned the other faces around them, they saw the same hunger. Friends they had known as youthful teenagers

Abubakar Shekau, Boko Haram's extremist militant leader, became Africa's most wanted man, with a $7 million reward for his capture. *(Courtesy of Boko Haram video)*

Hundreds of inmates escape from the infamous Giwa Barracks military prison after a jailbreak by Boko Haram in March 2014. Many immediately joined the insurgency. *(Courtesy of Boko Haram video)*

A "100 Most Wanted" Boko Haram fugitives poster released by Nigeria's army with Abubakar Shekau in the center. *(Courtesy of Jonathan Togorvnik)*

US drones searched an area of remote, rugged topography the size of West Virginia but were unable to track the girls as they vanished into Nigeria's northeast. *(Courtesy of Glenna Gordon)*

Nigeria's military units patrolled the fringes of the Sambisa Forest for more than three years, fruitlessly chasing tips of the girls' whereabouts.
(Courtesy of Glenna Gordon)

Muhammadu Buhari, a former military dictator, won the presidency with a pledge to restore order, vanquish corruption, and return the Chibok girls.
(Courtesy of Muhammadu Buhari official Facebook account)

Aso Rock, the sprawling complex of modernist buildings and manicured lawns that houses the Nigerian president's office and official residence.
(Courtesy of State House Nigeria)

Zannah Mustapha, a retired barrister, founded an orphanage for children made homeless by the war.
(Courtesy of Glenna Gordon)

At Zannah's school, the Future Prowess Foundation, the children of Boko Haram and its victims alike studied and lived together.
(Courtesy of Glenna Gordon)

Ahmad Salkida, a journalist and blogger, became an authority on the insurgency. He kept a running count of the dozens of times security forces detained him for his work.
(Courtesy of Glenna Gordon)

The serene Lake Thun in the Alps, where Zannah trained with Switzerland's Human Security Division.
(Courtesy of Esin Efe)

Boko Haram paraded in a vast convoy to celebrate the capture of the city of Gwoza and Ababukar Shekau's declaration of an "Islamic caliphate." (Courtesy of Boko Haram video)

فرحة المجاهدين بإقامة دولة الموحدين

The war wreaked devastation across Nigeria's northeast, leaving 2.7 million homeless, many of them fleeing torched and abandoned villages. (Courtesy of Glenna Gordon)

Bodies, shrouded for burial, after a 2017 suicide bombing by Boko Haram in Maiduguri. (Courtesy of Glenna Gordon)

Civilians could only travel the roads safely with a military escort. (Courtesy of Glenna Gordon)

A screenshot of an unreleased video shows Naomi Adamu, several rows back in a purple headscarf. Her insubordination was rewarded with beatings and a nickname: "Chief Infidel." *(Courtesy of Boko Haram video)*

A diary kept by Naomi and Lydia John in captivity, then smuggled out to freedom. *(Courtesy of Mohammed Bukar)*

Naomi wrote letters to her family inside elaborately sketched hearts. "I want you to help me in prayer." *(Courtesy of Mohammed Bukar)*

Throughout her captivity, Naomi managed to surreptitiously hold on to her trusty Nokia 3200, a durable phone with a long-lasting battery held together with a rubber band. *(Courtesy of Mohammed Bukar)*

Trapped between hope and grief, the Chibok mothers tried to keep their daughters' memories alive. Yana Galang laid her daughter Rifkatu's clothes on her bed, in case she came home.
(Courtesy of Glenna Gordon)

Kolo Adamu holds a photo of Naomi in March 2017, when her daughter had been missing for almost three years.
(Courtesy of Mennonite Central Committee)

In the summer of 2016, a masked Boko Haram commander known to the hostages as Tasiu, appeared on a video to state that the insurgents were interested in a deal.
(Courtesy of Boko Haram video)

Maryam Ali is escorted toward a military helicopter at the army base in the town of Pulka after she escaped with her ten-month-old son, Ali, in November 2016.
(Courtesy of Maryam Ali)

Red Cross Land Cruisers en route to the rendezvous point to rescue eighty-two hostages in May 2017. *(Courtesy of International Committee of the Red Cross)*

Eighty-two Chibok hostages—some on crutches, others faint from exhaustion—wait at the exchange point, May 2017. *(Courtesy of Zannah Mustapha)*

Armed Boko Haram fighters greet their comrades, freed from Nigerian prisons as part of an exchange. *(Courtesy of Zannah Mustapha)*

Zannah Mustapha inside the C-130 transport plane with the freed Chibok hostages, flying back to Abuja to the president's residence. *(Courtesy of Zannah Mustapha)*

The liberated Chibok hostages pose with a gaunt-looking President Buhari. Zannah, dressed in black (*back left*), looks on. Naomi, in grey (*second left row*), stares ahead. (*Courtesy of Getty Images*)

Almost seven years after their abduction, 112 of the hostages remain missing. (*Courtesy of Joe Parkinson*)

All that remains of Chibok Government Secondary School for Girls is a sign, with the word "Girls" painted over. (*Courtesy of Glenna Gordon*)

Naomi and Kolo Adamu, June 2020. (*Courtesy of Mohammed Bukar*)

were now skeletal, with large dark rings framing hollowed eyes. Some were hallucinating about where they might find a meal or water. The weakest had already passed that phase, their stomachs no longer demanding sustenance. When they were handed a bowl of rations by stronger classmates, they struggled to lift it to their mouths.

For days, the hostages sat listless or lay down flat, trying to conserve energy, staring up at jet fighters circling in the skies above. On the ground, a familiar voice was blasting through speakers strapped to the back of a pickup. "The lack of food is the will of God," said Abubakar Shekau. The Nigerian military had cut off his supply lines and captured their food stocks, he said. It was a recorded sermon for his followers, his explanation for their suffering.

After more than two years in captivity, the Chibok students were beginning to starve. They had experienced hunger before, but the new shortages were more severe. Still grieving for classmates killed by the air strike, the hostages had seen their diet gradually deteriorate to the point where they were lucky to eat once a day. Sometimes they ate bark or the sinewy grass called *yakuwa*, which their mothers used to weave mats and baskets. There were small seeds in between the blades, which they plucked out. Some drank hot water, believing the heat would give them energy. A thought began to trouble Naomi. If her friends passed away, they would be too weak to bury them.

The men were refusing to share jugs of water—except for washing before prayer. But one day, a guard walked over with a phone and said he needed the teenagers' help. There was a Swiss man on the line, and he wanted to talk to them. Rebecca Mallum, a classmate of Naomi's and one of the best at speaking English, stepped forward. But when she took the handset, the voice on the other end sounded like a Nigerian.

ON THE SECOND FLOOR OF THE SWISS EMBASSY IN ABUJA, PASCAL WAS huddled over a smartphone, other team members listening in.

"Hello . . ."

It was a girl's voice. She was speaking in English.

"My name is Rebecca."

She sounded weak but also animated. This was the first time they'd heard the voice of one of the hostages they'd been seeking for two and half years.

"I am one of the Chibok girls."

32

Praise God

Ladi Simon woke with a start as a flashlight hovered over her face. "You, what's your name?" the figure in front of her demanded. Ladi barely found the strength to answer, her cracked lips woven shut by dehydration. She mouthed her name and the guard scribbled it down on a scrap of paper. "Move there," he said, shining the flashlight against an adjacent tree trunk.

Anyone back in Chibok who had known Ladi as a trendsetter who'd research fashion tips to share with her dorm-mates would have struggled to recognize her under the October moonlight. The teenager was no longer just hungry, she was starving. Her face had been chiseled away by months of malnutrition. Dust-coated clothes billowed over her skeletal frame. She hadn't eaten a thing for several days and could no longer stand without help.

More than one hundred hostages who had refused to get married had been gathered under the tamarind trees, but Ladi was one of the weakest. That, she assumed, was why her name had been called. The insurgents were hastily picking the frailest young women from the crowd, almost two dozen in all, and corralling them under another nearby tree.

Ladi was too exhausted to ask why the guards were separating them again, this time in the middle of the night. A militant commander addressed the group. "We are going to try and get a deal with the government." He continued with a warning: "After the first group is set free, if they tell lies that it was the military who rescued them, then the rest of you are not going home."

The group needed to move immediately, he said. There was no time to stall or collect belongings. In the confusion and rush, one of the women complained she'd forgotten her shoes, a pair of red flip-flops still under a nearby tree. She would have to walk barefoot, the commander said. Ladi had also left behind the small bag that contained her possessions: a hair tie, a toothbrush, and a school uniform. The militants were too pressed for time to allow good-byes, and Ladi only managed to say a few parting words to her friend Deborah Samuel, who had helped lift her to her feet. "If I don't come back, you know I have made it out," Ladi said.

Naomi watched as the teenagers proceeded on foot, guided by a handful of fighters carrying flashlights. As the lamp beams slipped away from view, she felt a growing anger. *Her* group was meant to go first, and now she was watching others on their way out of Sambisa.

Ladi and the chosen girls walked through the darkness until they saw the red taillights of a large truck and, near that, at least ten pickups. Each was full of men clasping machine guns, rifles trained outward as they watched the young women clamber into the vehicles. The convoy set off, bumping in nauseating lurches down a dirt path, cutting between waves of reeds and thorn trees. The men were peering out across the horizon and conferring, gesturing to a far-off point.

Then came the crackle of gunfire, and a rumble of shelling echoed in the distance. The driver and another militant were chattering nervously in Kanuri. Ladi could understand enough of their conversation to know they were talking about the army. The trucks edged forward, then halted at a burned-out shell of a building, and a militant ordered the hostages down. "Anyone who needs to urinate, do it in that building," he said. Ladi stepped through the grass growing in tufts around the ruins and saw that the alphabet had been written on the walls. The building had once been a school.

The militants were talking rapidly into cell phones, and they seemed stressed. They instructed the hostages to gather for a head count. The roll call added up to twenty-two names, and an argument ensued. They had changed their mind. They would be exchanging only twenty-one girls. A commander pointed at one of the students and ordered her sent back. Ladi watched her face fall, reality sinking in, before a truck carried the lone hostage away into the darkness. She returned to the camp in tears. A few months later, despondent and suffering from starvation, the teenager would relent into a coerced marriage to a member of Boko Haram.

THE AGREEMENT CALLED FOR NO AIRCRAFT IN THE SKY OVER SAMBISA except one, a helicopter with its lights switched off, carrying the Dialogue Facilitation Team. In the cabin, over the thrum of the rotor blades, Zannah and Pascal ran through the crucial stages of the operation. Zannah could see the diplomat was anxious, chatty even, but he assured him all would go fine. Shekau had given his word.

The lawyer folded his hands on the crisply ironed kaftan he'd picked out, with an embroidered kufi cap and Calvin Klein spectacles. In his top pocket was a piece of paper, a list of twenty young women set to go free. He hadn't found the time to thoroughly study their names. Shortly before dinner the night before, he'd slipped from home to take part in an extraction so secret that he said nothing about it to his wife. He waved good-bye to his daughter Aisha and drove his gray Toyota straight to Maiduguri's air force base to board the helicopter heading southeast.

Strapped in next to him, Pascal was running through his own copy of the list and the nerve-racking ways the mission could go wrong. Tijani, the team's chief courier, was clutching a satellite phone, fully charged. The specifics had come together over four sleepless days of furious phone arguments between the Dialogue Facilitation Team and Boko Haram. On the night of October 7, a militant commander had called Tijani to say Shekau would free five captives as the first release, but when that proposal had reached Buhari, the president had demanded twenty, pushing the militants to fire back with an enraged counteroffer: "Just one girl." To double down, they had sent a voice message the next day, threatening to decapitate a group of young women, who could be heard screaming in the background. The team had chosen not to tell the government, for fear the deal would collapse. Instead, they had improvised, pleading with their contacts in the Sambisa Forest and as far away as Sudan, Egypt, and Niger, sympathizers who prevailed on the insurgents to calm down and relent. But there would be a price.

Sitting in the helicopter was a DSS officer carrying a dark-colored bag stuffed with euros in high-denomination notes. The team's destination was a landing pad near the Cameroon border, where a convoy of Land Cruisers would carry them to a rendezvous point set in the countryside where the insurgents felt comfortable. They wouldn't know the exact location until minutes before.

They would be traveling under cover of darkness when nobody could spot them, traveling a vulnerable route at a dangerous hour. The roads there were well known for improvised bombs and land mines stolen from army barracks. Lookouts would be charting their progress, and Shekau himself would be following their movements in real time, they assumed. To escort the hostages, he would send his most battle-hardened fighters.

The team also worried that the Nigerian security forces could sabotage the mission. Pascal and other team members had spent days petitioning the military and intelligence agencies to withdraw from the area and respect the brief cease-fire necessary to complete the release. If any military or security forces were spotted, the deal could collapse in a violent confrontation. There was also the risk that a rival faction of Boko Haram would foil it first.

The helicopter touched down on the dirt, and the men separated. Pascal would have to stay near the helicopter, monitoring remotely and hoping for the best. Zannah and Tijani pushed ahead, at the front of a convoy of white Land Cruisers emblazoned with a red cross. The Barrister was carrying nothing on him except for a cell phone, the list, and the jangling keys to the car he'd left in the airport parking lot.

The first vehicle carried the money. The rest were Red Cross vehicles driven by employees from the agency's Abuja and Maiduguri offices, ferrying the team to the exchange point. One of the drivers was a man from Chibok who they hoped would be a reassuring presence. He spoke Kibaku, and his name was Hurray Peter. Their progress was being continuously tracked from Geneva in real time by a Red Cross operations officer at headquarters, sitting near the first Nobel Peace Prize ever awarded. The early hours had passed without incident, but then the officer's phone rang with an update of immediate concern. Somebody was firing off gunshots.

A half dozen soldiers, pacing in flip-flops at a roadblock, had somehow not received the order to retreat. When they saw the headlights of the Red Cross vehicles, they had panicked, shooting their rifles. The boom of a mortar echoed from some distant point. A contact from Boko Haram called Tijani's satellite phone to tell the team the deal was off, that this was an ambush. In the darkness, it was unclear whether the insurgent fighters' next move would be to attack. The team begged their contacts for time, then radioed the base commander of a nearby town to pull back his troops, shouting into the handset, "Get them to understand!" Monitoring from the nearby base, Pascal feverishly dialed senior insurgent commanders to reassure them that there was a misunderstanding and no breach of the agreement. The deal was still on.

Several tense hours crawled by until the sun began to rise, the

Barrister wearing a look of practiced nonchalance. He couldn't show fear, he told himself. He was protected by the prayers of his orphans. Finally, the troops left, and the chain of white Red Cross Land Cruisers continued to thread their way north as instructed, headlamps bobbing over a dirt path, each driver steering into the tire treads of the car in front to mitigate the risks of roadside bombs.

It wasn't long before Zannah could see the shapes of men moving around them, close by. The red brake lights of vehicles flashed through acacia trees as his convoy stopped at an unremarkable patch of dirt. Zannah cranked open his door, then stepped out into the dust. They were surrounded. In the surreal glow of headlights, cell phones, and the first hints of sunrise, he could see more fighters than he could attempt to count. Most of them were wrapped in turbans with their faces laid bare. Zannah greeted the men with his gap-toothed smile and glanced around. But there were no girls.

"HAVE YOU SEEN ANY MEN IN UNIFORM?" ONE MILITANT ASKED ANOTHER. Ladi had watched them draw their weapons as the sound of explosions thundered somewhere in the distance. Now they were discussing the vehicles that had just pulled up.

"No," the other man shot back. "They're just white vehicles. No military."

In the inky dawnlight, she could make out men clambering from four or five cars and greeting uniform-clad fighters. In the foliage beyond, dozens of shadows were standing in groups, pointing weapons. Within minutes the hostages were given an instruction: line up in two rows and start walking in silence. Boko Haram were in front and behind them. As they approached the vehicles, Ladi saw several men wearing white bibs, each decorated with a red cross. They were the first men she'd seen in two and a half years who weren't members of Boko Haram.

A feeling she hadn't allowed herself to entertain through hundreds of days in captivity rushed over her: "I thought, *Wow, maybe this is actually real?*" As the group neared the white vehicles, Ladi traded glances with her classmates. She could see a tall man in an ash-colored kaftan wearing thick-rimmed spectacles. He had a piece of paper in his hand.

ZANNAH TOOK A BREATH AS THE FIGURES APPROACHED, STEPPING slowly through the grass in two rows. Their bodies were weighed down by dark-colored shrouds that hung from emaciated frames. Several limped, and some stared at the ground, disoriented and shuffling in evident discomfort. Their faces were gaunt, eyes hollow, and his heart skipped as he tried to hide his horror.

For more than two years he and the team had been working to release a group of young women he'd never met, who had unwittingly become global symbols, and now they were walking toward him, veiled and almost spectral figures. The smiling class photos of high-school seniors he'd studied in the embassy and the orphanage had been reduced to ghosts. They looked like victims of a famine. One was carrying a baby boy, conceived before her abduction, on her hip.

As they drew closer, Zannah smiled but tried not to stare at them. Their condition was dreadful, but at least they were alive, he told himself. He counted them as they filed into a single line on the road opposite. Then he counted them again. The deal had called for twenty hostages to be released, but there were twenty-one standing on the other side of the road.

"This girl is a gift," said one of the militants, gesturing to a veiled figure in the group. Rebecca Mallum had been known in school as a bubbly and gifted senior, fluent in English, and a chorister in Chibok's Glorious Singers. It was her voice that had crackled through the telephone into the Swiss embassy days earlier. Her attendance was symbolic. It was to show the Dialogue Facilitation Team that Boko Haram wanted to take the process forward.

Zannah nodded in solemn appreciation, then glanced at the list. He realized he couldn't properly pronounce the Kibaku names. He beckoned Rebecca forward, then handed her the piece of paper. The teenager began to call off the names of the students she'd spent two years in captivity with.

"Ladi Simon . . ."

One by one, the hostages crossed the road, joining Zannah on the other side. "I walked across and I thought, *Thank God*," Ladi said. The men shook hands, and the militants began to filter away, returning into the forest with barely a word. "Now we are parting from you," one said to the twenty-one women taking their final steps as captives. There was no explanation or note of sympathy for their suffering. Their captivity was one more casualty of the war.

The Barrister lifted his arms up, like a priest offering a blessing. "You are free today," he said. Then he fished through the billowy fabric of his pockets, lifted his cell phone, and proceeded to do what came naturally. He snapped pictures, material for posterity, or possibly his Facebook page, until a rifle-clasping militant ordered him to stop, shouting, "Delete the photos!" Pictures were strictly against protocols.

Ladi and her friends were helping each other climb into the Red Cross vehicles, discovering care packages left for each of them: meat pies and juice boxes of Five Alive to lift their blood sugar levels. The driver of Ladi's car turned to his passengers and greeted them in the comforting tone of a language that had been banned in the forest, Kibaku.

Hurray Peter asked if they were ready to go. They would be driven to the city of Banki, taken in a helicopter and flown to the capital for a meeting with the president. Ladi had never been on a plane, or to Abuja.

The day was bright enough by now to see the dirt road as the convoy sped along it, racing back to the airbase where Pascal was nervously waiting. As the sunlight reached across the road, Zannah thought to himself, *I've achieved something in life*. In the back of his car, several teenagers were singing a hymn from Chibok, "Praise God!"

The notes were stupefying high and louder than the rumble of Land Cruisers bumping over the dirt road. He could hear the singing carrying across the air from the other vehicles too. It was the first time the young women had been able to fully lift their voices and speak their hearts. Ladi was overcome as she heard herself in a chorus with her classmates.

The Barrister lifted his hands to the side of his face. "They are disturbing my ears," he complained with a chuckle. As the convoy pulled into the base, the passengers descended and saw a small group of men on the tarmac, staring back at them. One of them, a white man with a pouf of red hair, was dressed in dark trousers and an untucked light blue shirt. He stood there in silence as they spilled out of the cars, the women clapping and singing in a high-pitched chorus as they assembled on the dirt. The man was watching them, his expression fixed as if he were looking for the words. "Now you are free," Pascal said.

33

The One You've Been Looking For

Maryam was confronting the final, most heart-wrenching item on a to-do list she'd been secretly working on for days. Lying awake on a mat outside of her thatched-roof home, her thoughts had ticked through the things she would need to decrease the chance of failure. Escaping Shekau's camp would require meticulous planning and total discretion.

That's why she had waited until the last possible moment to tell her sister. Maryam didn't know what Sadiya would do with the information. It was a revelation that could get either, or possibly both of them, killed. *If I reveal it to her, and she doesn't want to go, she might report me*, she thought.

It was November, the early weeks of the long dry season, and the camp where they both lived with the men who called themselves their husbands had become cloaked in chaos and paranoia. Fighters continued to leave at night in an exodus from Shekau's dwindling authority that was stretching into its fourth month. In retribution, the Imam had ordered more surveillance and a string of executions, some of which Maryam and Sadiya had been instructed to watch.

Food was now scarce, even for the commanders and their wives. The beef intestines and ram's meat of previous months had faded into memory, and fighters' families were now stalked by hunger. Maryam and Sadiya had begun to mix the bitter *aduwa* fruit with water, and eat *doyan daji*, a "forest tuber" so acidic it needed to be dried for three days before it was edible. These days, when Maryam's husband Abu Walad spoke, he only complained, mainly of hunger.

In the disorder, word had reached the sisters that twenty-one of their classmates had been released. Maryam had long consoled herself that marrying into the sect would at least offer more material comforts during a life sentence with no prospect of release. But news that her classmates would be returning to their homes and families revealed that exiting the caliphate was possible. Perhaps trying to escape was worth the risk. "I realized that some of the fighters themselves were leaving . . . I revived my hope," she said.

Maryam had spent hours rehearsing what she would say to Sadiya, but when she finally tracked her down, the moment was unexpected. A vehicle full of women was passing by, and a voice cried out her name. As the driver slowed down, Maryam saw her sister's eyes through the opening of her veil. Delivering her message in front of Boko Haram members was impossible, and even trying to speak in code was too dangerous. She wondered if she could communicate her escape plans with just a widening of her eyes or by sucking her teeth. Maybe Sadiya would understand. But just as the driver seemed to be pulling to a stop, he accelerated again, throwing dust in the air. Maryam ran after the car, but it was useless. She ran a few steps then stopped to catch her breath and adjust her precious cargo, the baby boy she was carrying on her back.

MARYAM HAD SPENT EIGHTEEN HOURS IN LABOR AT THE BEGINNING OF 2016, a rag in her mouth and a hand in the palm of another classmate from Chibok. The two had repeated a prayer: *Hasbunallahu*

wa ni'mal wakil, "Allah is sufficient for us, an excellent guardian."
She knew that many women in the camps had died in childbirth.
"I recited it like a song while I cried," she said.

Maryam had wanted to name the child Khairat if it was a
girl. The word, meaning "blessings," or "good deeds," had a melody
and a meaning she liked. After cutting the umbilical cord, she'd
decided to deceive Abu Walad, who came in eager to know if she'd
given birth to a boy. "No," she said.

The father's face had hardened when he learned his firstborn
was a girl. He turned away and sulked out of the room. A few
minutes later, word reached him that the newborn was a boy, and
Maryam heard him fire a rifle three times in celebration. He ran
in, lifted the child skyward and chanted the nickname he'd prepared for his son, "Soldier Shekau! Soldier Shekau!"

Maryam was overwhelmed from the physical pain and the
emotion of the birth. She watched in exhaustion as a woman
rubbed crushed charcoal to treat her umbilical wound. A visitor
brought a tamarind pod for the newborn, to rub on his dry skin.
Now it is real, she thought. *This is going to be my life. This will be
my destiny.*

The naming ceremony took place days later in a mosque.
Maryam was not allowed to attend. As the militants shared goat
meat, a few people came by her home, one by one, to congratu-
late her. She didn't feel there was much to celebrate. Whatever
the future held for her and her son, she was now bound to his
father, and Boko Haram, forever, caring for a baby she loved by a
man who abused her. It wasn't until the ceremony ended that she
learned what Abu Walad had named her child: Ali, after his older
brother, a militant jailed in Lagos.

"I was depressed on the day," she said. "First my marriage
to the person I called my husband is illegal because it was done
without the consent of my parents and secondly, it was just me. . . .
I know how a normal naming ceremony would have looked if I
wasn't in captivity."

Ali's future became clear as Maryam observed how the other children conceived by the insurgents or kidnapped from other villages were being treated. Food was becoming scarce, and when she deposited the trash bin outside her home, packs of malnourished boys would run up and fight over any edible remnants. Military training was mandatory for boys as young as five. One day, as she was contemplating what kind of mother she would be, a woman told her to walk to the edge of the camp. Lying under a tamarind tree with his palms facing skyward was the body of an infant boy, no older than four, dressed in polyester pants, his mouth open and eyes permanently closed. Maryam muffled a scream. This was somebody's child. Her thoughts burned with an anger that she'd been holding in for months.

"The image stuck in my head," she said. "I knew then I would leave . . . Even if I lost my life, I never wanted my son to grow up in this Boko Haram camp.

"But I couldn't tell my husband . . . I didn't know what his response would be."

ALI HAD BEGUN TO CRY AFTER HIS PARENTS HAD WALKED THE FOREST paths for more than five hours. They had begun their escape shortly after dawn, traveling with a single small bag to avoid arousing the suspicions of fighters tasked with hunting defectors. The baby boy had slept soundly on Maryam's back, behind her long gray veil, as they breezed past several Boko Haram checkpoints, claiming they had been summoned to visit other commanders.

The journey out of Sambisa and into Nigeria would take days, skirting along the foothills of the Mandara Mountains that lined the border with Cameroon. Abu Walad knew the roads. He had traveled them as a fighter to lead attacks on Nigerian troops. Still, the route was riddled with danger. They had only a small amount of food and water. As they neared government territory, they could be killed by Boko Haram lookouts or vigilante patrols.

As they trudged under a baking summer sun, Maryam watched

the taciturn young man who held her life entirely in his hands. When he first told her he wanted to leave the forest, she had said nothing, fearing it was a test to expose her disloyalty. She barely knew him, and his allegiance seemed more fixed to the Imam than to her. Yet this was the caliphate, and Maryam's captor—her husband—was also her only way out. She would have to nudge him along, without letting him think escaping was her idea. "I didn't show him any encouragement initially," she said. "I didn't want him to think I too was planning."

Now, the three travelers would need strength and luck to survive.

On their second day, they came across a group of herdsmen in the wilderness, who offered them food and shelter, then warned them to take a different route. Boko Haram were patrolling the area and executing anyone found wandering without cattle.

They took another path, walking in silence through an abandoned village where the only sound was dogs barking. Their meager rations went to Ali. Maryam became irrational with exhaustion. There was nobody around and yet her footsteps sounded so loud that she worried the echo would reveal their location. *Maybe they should turn back*, she thought.

As night fell, the travelers could see the small town of Pulka by the lights from its tin-roof homes, clustered next to a rocky outcrop. Ali was sleeping on Maryam's back when the first gunshot popped by her ear. Next came a whistle and a crash of explosives. They'd been spotted by soldiers, who were skittish and shooting rounds into the darkness.

Maryam ducked to the ground, and Abu Walad followed. Remove your hijab, he said, to avoid suspicion. Flashlights were beaming toward them, sweeping the ground to locate the intruders. Maryam counted five shots. Ali didn't stir.

They were crouched against the ground.

"You lead the way," Abu Walad said.

"How come?" she shot back.

"You are a woman. They will not hurt you."

Maryam was taken aback that this seasoned warrior was asking her to carry his child toward the soldiers, but she also knew that he would be arrested and vanish into a prison. Whatever else he'd done, he was also Ali's father, and had helped her to escape with their child.

Maryam began to crawl toward the source of gunshots then took a breath to steady her voice.

"*Oga!*" she shouted. Sir! "It is me."

She heard the crackle of a radio.

"Who's there?"

"The one you've been looking for. One of the Chibok girls."

34

"Now There Is Trust"

Pascal was scrutinizing a new video from Abubakar Shekau and trying to understand if it might affect the process. The Imam was standing in a forest clearing dressed in military fatigues with a black-and-white checkered scarf and his gun, a short-barreled Kalashnikov similar to the one used by Osama bin Laden, swinging across his waist. Behind him stood masked men brandishing rifles next to armored cars and long-nosed artillery. It was a New Year's message, and the Imam had some words for his enemies.

"I am here, well and alive. And our battle is just beginning!" he said, manically waving a blue plastic folder of notes. "Kill the infidels, detonate bombs! Kill, slaughter, and abduct!" He began to tick through a list of people to insult and threaten: President Buhari and a string of Western nations that he said were advising Nigeria's government, like the US and Israel. The Swiss diplomat noted with relief that the Imam made no mention of Switzerland and no mention of the Chibok girls, since months earlier he had issued a vague threat in his direction, shouting, "Geneva people, all your plans according to Allah are not up to a cobweb!"

It was the beginning of 2017, and months had passed since twenty-one of the hostages had been freed, each day painful for parents who were asking questions that were utterly reasonable, heartbreaking, yet unanswerable. Which of the missing would come home next? And when?

But the main reason more students hadn't yet gone free was an image that a bystander had thoughtlessly posted on social media. The deal to free the twenty-one had been confidence-building, a test of whether the two sides could trust each other, and its terms had called for complete secrecy. But just after the team had taken off in a helicopter packed with the young women, they checked their phones and saw that a soldier had snapped a photo of them hours earlier, a tweet that was being triumphantly shared by thousands of people unaware that, with a touch of a screen, they were prolonging the other women's captivity. Boko Haram was still furious over the lapse and refusing to release any more.

The team knew that more food was starting to reach the starving insurgency through Red Cross deliveries, which bought them some time. But they were struggling to stay ahead of Boko Haram's internal war. Their sources suggested that some of the hostages were now trapped with breakaway factions and beyond their reach. Shekau was executing some of his most senior commanders, including some involved in the talks and the care of the teenagers. He had beheaded Aliyu, the Emirul Jesh who had once threatened to behead the Chibok girls. Tasiu, the tall commander with the husky voice who had driven Naomi to see Shekau, then escorted her group for their abortive release, was also executed. "I killed Tasiu. Hear me well," Shekau said in a fifty-minute recording. "Tell me, what is the punishment . . . for the people that plot against their leader?"

The other challenge was the unwanted attention to and scrutiny of a team that had tried to stay in the shadows, eschewing recognition. The success of the first release had attracted publicity. Zannah's face was in the news, and the Barrister was warming to the limelight, a concern for his partners. One day, to the team's shock, they checked their phones and saw he'd violated their code of discretion: he'd tweeted for the first time, posting an article

titled "Zannah . . . Modern Day Hero." The clock, meanwhile, was
ticking. The team still lacked an exact count of the number of teenagers who had been killed. Then there was the matter of the president, Muhammadu Buhari, who looked sicker with each public
appearance—gaunt and pale, withdrawing from public life. Their
teammate, Tahir, had now been in jail for more than a year, with
little word on his condition or fate. Pascal counseled patience, but
the ups and downs of the operation were frustrating, and he was
speaking from a privileged position. Eventually, his assignment
would end, and he would move to another diplomatic post. Team
members like Tijani and Fulan would live with this portfolio for
the rest of their lives.

NAOMI WAS LYING FLAT ON A PIECE OF TARPAULIN, IN ANOTHER MAKE-
shift clinic somewhere in the Sambisa Forest, struggling to retain
hope. She had now been a prisoner for almost three years. The
relentless grind of life in the forest had aggravated her chronic
kidney pain to a stage where she often struggled to speak. She
spent weeks on end lying next to Hannatu Stephen, who had lost
her leg. Ali Doctor was no longer caring for them. He had been
killed in an airstrike outside his home while telling a nurse to
take cover. News of his death had moved Naomi to tears. "Don't
cry," Malam Ahmed had said. "It was the will of God."

Months had passed since the twenty-one hostages had been
released with a promise that other unmarried captives would soon
follow. The militants returned from that exchange with one hostage and told the group they refused to exchange her because the
government hadn't brought enough money. The absence of any
follow-up had left Naomi and the remaining hostages feeling dev-
astated. She was happy her friends had made it home but was
bitter she wasn't among them.

The only consolation was that the supply of food had improved. Yellow jugs of oil and big white sacks of rice had arrived,
emblazoned with the French name of a city she had never heard

of and couldn't picture: Genève. The remaining hostages had begun to put on weight, their sunken faces starting to fill out again.

They had tried to get Malam Ahmed to explain why their release had been delayed, but the old man had become furious with their questions. Perhaps it had been canceled, again? "Your parents don't want you," one militant snapped. That was why it was taking so long. "Maybe the government could say that," Naomi replied, "but not our parents."

Another time, Malam Ahmed tried to restart Quranic classes, and one of the teenagers replied, "Please, just come and kill us."

"So you want to be infidels?"

"Yes."

"OK. You will be executed in the morning."

Naomi was spending long, empty weeks in the clinic, often alone with her thoughts of home and the injustices of captivity. On some days, sprawled on her back, she would have to muster all her energy just to withstand the pain. She had always tried to be strong for her classmates, but she was starting to wonder again whether she would die in the forest.

One afternoon, an order rang out through the camp. Everybody needed to leave, immediately. Staring into the clear sky, the guards had spotted the silver glimmer of a plane drawing near and their position was exposed. Militants were sprinting for cover under trees or scrambling into Toyota pickups and cranking the starters. Trucks began peeling off in various directions with men clinging to the sides.

Naomi barely made it from the clinic into a vehicle, climbing in with six other hostages, who helped hoist her and Hannatu inside. Their pickup was the last in a four-truck convoy, accelerating down a thin dirt road. Looking through the windshield she noticed that the truck bed of a pickup in front was loaded with stacks of leather-bound books.

The plane was heading closer, a thunderous whoosh filling the sky as it circled overhead. The trucks braked and passengers clambered down, racing out from the road to shelter under a crop of trees. The men around them reached for their rifles, raised them to the sky, and began aiming, until a voice shouted an order.

"Don't shoot."

It was Abubakar Shekau, descending from the driver's seat of a truck while fixing the back knot of his turban. Naomi watched him from her hiding spot beneath the branches. He had put on weight since she'd last seen him, his face fuller but still etched with the scars of a life on the run.

The Imam gestured for the men to lower their guns and Naomi realized why. *If the plane located us, it would bomb us all into pieces*, she thought. Shekau was followed by two women in niqabs made from expensive-looking fabric, who were gesticulating at him, feverishly waving hands laced with gold rings. "We should let them bomb us and we can go to paradise!" they said. Naomi braced for the worst. Shekau spat back at them, "You should not seek death on your own. It will come when Allah chooses." He walked to the back of his pickup and took one of the leather-bound volumes to a lone tree where he leaned against the trunk and started to read. Naomi could see that it was a copy of the Quran that he was leafing through, revisiting the scripture he'd memorized as a child. He did not even flinch as a screaming rocket exploded near a Toyota in which another group of Chibok girls had been seated moments earlier.

For three years, she'd wanted to challenge this man and force him to justify what he had done to them. She felt entitled to answers. Still, beneath her anger, she started to understand the method behind his madness. Shekau couldn't show fear or let his followers succumb to theirs, so he had to bolster their courage with extreme displays of uncompromising principle. In the looking-glass world of his forest prison, she'd learned what he knew, how to endure and encourage others to persevere. These were the strengths she had summoned to help her and her friends survive.

She would later reflect on how life in his captivity had transformed her: "I felt I'd become a leader." The jet whooshed off, and in the silence, Shekau ordered the women onward.

WHEN IT CAME TOGETHER, THE SECOND RELEASE WAS MORE PRACTICED than the first.

A caravan of Red Cross Land Cruisers, longer this time, cut north through a dirt path under the clarity of daylight. It was May 6, 2017, the end of the dry season and a scorching 100 degrees, just before seasonal rains would make the journey an impossible slog. This time Zannah had rehearsed the names, reading them before the moment arrived.

Number 62 was easy to pronounce.

Naomi Adamu.

Her friends were on the list as well.

20. Palmata Musa.

35. Hauwa Ishaya.

67. Hannatu Stephen.

Second from last, 81. Saratu Ayuba.

Every single one of her *arnan daji* were going home. So were dozens of other captives, each a few hours from the end of her ordeal.

On the morning of the exchange, a drizzling rain had given way to an overcast sky that seemed to blend the scrubland's palette of greens and browns. Zannah, his team, and the Red Cross officers had descended from their vehicles with their own human cargo, five Boko Haram prisoners fished from maximum-security jails. One of the Boko Haram militants was filming the young women, lined up in two rows on the other side of the road, dressed in dark mud-stained veils. Zannah asked the militants if he could take some pictures and when they acceded, he began snapping away with his phone.

In those images, Naomi can be seen wearing a gray hijab.

Unseen, tucked around her waist and swaddled around her thigh, was the pouch she'd stitched from fabric, containing the contraband she'd carried on her person for more than three years: a cell phone, a folded blue-and-white checkered school uniform, and three notebooks full of diaries and letters home, written so that she could give them to her mom to read. Lydia John's parents would want to hold what their daughter had written, too, messages hidden inside Naomi's clothes.

A militant had turned to her before she reached the road and said something that surprised her: "We are sorry," to which another woman shot back, "No problem." Naomi stared at the man, saying nothing. The lawyer from Maiduguri and the Red Cross were waiting on the other side. Before the swap could take place, the militants demanded a final gesture from the Barrister, who'd raised their fighter's children in his orphanage but had never made his loyalties quite clear.

"Hold the flag," one demanded, handing him Boko Haram's white banner inscribed with black Arabic lettering.

"Hold the flag."

Zannah demurred. A militant asked him. "Don't you trust us?"

"We trust you," Zannah replied, "but we don't have to hold the flag."

"No. You have to hold it. If not, there is no show of trust between us."

"Why should I hold the flag? Am I a member?"

A militant walked over, thrust the flag into Zannah's hand, and for a few short seconds, the lawyer smiled awkwardly as militants snapped pictures of a man who had always kept his loyalties open to interpretation.

"Now there is trust."

Naomi had watched as one of the men handed a black bag to a militant. It was light, and the fighter handed it to another comrade with ease. *It must be money*, Naomi thought. Others would later describe seeing the same exchange.

In the middle of the road, the five prisoners, dressed in dark-colored kaftans given to them by the government, were trading clasped hands and hugs with comrades they hadn't seen in years. The final arrangement had called for Nigerian jails to release these terrorism suspects, whom Nigerian newspapers would later refer to as "senior commanders." Zannah knew they were much smaller fish who wouldn't threaten Shekau's authority. After all the grinding ups and downs of the negotiation, Shekau was more interested in the monetary aspect of the deal.

Naomi still couldn't believe it was real. At the sound of her name, she stepped across the road, her flip-flops slapping against the dirt, then stood awkwardly next to a Red Cross vehicle. When the roll call was finished, Zannah signaled for them to enter.

The young women climbed into the Land Cruisers, whose engines revved as the air conditioning bathed them in a cool breeze. The cars rumbled off, and as the schoolmates cracked open juice boxes, the men who'd held them hostage for three years became small figures on the horizon. The journey had barely begun when the passengers broke into a song from Chibok, loud enough that the entire convoy could hear and join in. Their voices arched and lingered over the *a* in *happy*, reaching for a note at the top of the melody.

Today is a happy day!
Everybody shake your body, thank God!
Today is a happy day.

35

Naomi and Kolo

CHIBOK

MAY 6, 2017, THREE YEARS AND

TWENTY-TWO DAYS INTO CAPTIVITY

Kolo Adamu was sitting on her porch, the hours of a Saturday evening ticking by. May was the hottest time of the year in Chibok, when the humidity built up in preparation for months of rains that left the town coated in muddy slush. From her wooden bench below an overhanging tin roof, she could hear the neighborhood children chattering and playing between stick fences. She heard another sound at the gate and looked up. Rifkatu Ayuba was hurrying into the courtyard.

Nine months earlier, when the first twenty-one hostages had been released, the two mothers had suffered a shared heartache. As other families flicked through their phones and tuned into radios to absorb the list of names, a relative in Abuja had called to say their children were not on it. The news had thrown Kolo into another well of depression. "I was downcasted," she said. Rifkatu and other neighbors had visited her to try to lift her spirits. "They advised me not to lose hope, and I continued to pray."

Now Rifkatu was rushing toward Kolo wearing an anxious look on her face. She had heard reports on the radio that many more hostages had been freed and would soon be flown to the capital to meet the president. A picture taken by a curious bystander of dozens of young women being escorted into a military helicopter was ricocheting around social media. Kolo and Rifkatu embraced, full of hope but also dread that their daughters would again be absent from the list.

That night, Kolo didn't sleep. Rifkatu didn't sleep. "I was just waiting for the dawn of a new day to grab the possible good news of my daughter's return," Kolo said.

Sunday morning came, but Kolo decided to skip church, afraid somebody in the service might have heard whose daughters had come out and whose hadn't. She wasn't ready to know or prepared to think of what it would mean if Naomi remained behind.

By evening, she didn't have a choice. The presidency planned to announce the names tactfully, in a controlled and respectful manner, but then somebody snapped a picture of a crinkled, four-page list left lying around, with surnames capitalized. The image went viral on Twitter, and the names began to filter out through the neighborhood. Kolo braced for the worst.

When she heard the good news, the first thing she did was to put her phone down on the dirt floor of her front porch, the speaker facing up, and press play on a gospel song that she and her daughter loved to listen to together.

"It is a must for trials to come," blasted Mama Agnes. *"But if we stand firm to the end, / We will receive the crown, / And be like the angels."*

The melodies from the tinny speaker filled the dirt yard as neighbors rushed to embrace their friend. Rifkatu ran through the gate with tremendous news of her own daughter. Saratu was on the list as well. "My heart was filled with joy," Kolo said.

NAOMI WAS SETTLING INTO A GOLD-RIMMED CHAIR, UNDER THE BEADED crystals of a giant chandelier, her brand-new sandals resting on the plush carpet of the Aso Rock Presidential Villa, six seats away

from her country's president. Wearing a champagne-colored kaftan and matching cap, Muhammadu Buhari stood with a microphone to address the eighty-two young women gathered before
him. Naomi noticed how frail he looked, sitting in a large arm-
chair propped up by a pillow embroidered with the presidential
seal. "Welcome, our dear girls, back to freedom," he said, speaking
slowly. "On behalf of all Nigerians, I will like to share my joy with
you."

In the preceding hours, the Chibok students had been whisked
through a series of military outposts, where each was given a brief
interview and a medical exam. They'd changed through cycles of
clothes, discarding veils for wax-print dresses, polo shirts, skirts,
and headwraps.

At one point, during a physical, an aid worker found the
things that Boko Haram never had: a carefully folded diary, then
another, and a third, all tied to Naomi's body under her Boko Haram clothes, along with her school uniform and a phone, its battery held in place with a three-year-old rubber band. She was the
only Chibok hostage who had held on to her uniform. Soldiers
gathered around, dumbstruck at the amount of material she'd
smuggled out of the forest. They called for the governor of the
state, who was filtering through the crowd, handing each of the
young women 10,000 naira, or $30, in pocket money. When he saw
what Naomi had kept hidden, he exclaimed in delight. Reaching
into his wad of cash, he showered her in a rain of three times
more money than he gave the others. Other men began spraying
her in banknotes as well, encouraging her to put her uniform on.
A television news crew captured the moment she returned in the
blue-and-white dress, flinched for a moment, then smiled.

A military helicopter had whisked Naomi and her classmates
high over the vast landscape of abandoned villages and forest encampments where she'd been kept, her first time in the sky. At
Abuja's international airport, a brand-new white bus escorted by
police cars drove her down Bill Clinton Drive, the flora-manicured

boulevard that leads toward the city center. The women stared out through tinted windows into the orderly calm of a capital city where virtually none of them had ever been.

Protocol officers spun them past the security gates and the languidly strolling long-legged birds outside the presidential estate, through its arched arabesque breezeways and into a banquet room with red padded chairs set out for each of them. Their president stood before them at attention, a soldier's posture, eyes trained up, under the thick green-and-white fabric of a Nigerian flag.

Naomi was tired and dazed as she took in her new surroundings, her eyes unaccustomed to the interior lighting and her body unused to the climate-controlled air. Like Hauwa Ishaya, she wore a fixed expression that suggested more bewilderment than relief. Many of the young women held their hands to their faces. Saratu stared into the plush carpet as a series of government officials took turns lifting the microphone to give speeches and praises to God.

Pascal had decided not to attend but instead to watch the TV coverage from the embassy office where he had spent hundreds of hours working with his team. This was a moment for Nigerians. Zannah wouldn't have missed it for the world. Sporting a pressed black kaftan and a polished silver watch, he sat in a corner, staring in awe at the rows of women and processing the stories some had told him on the flight over. He admired them for standing their ground. They reminded him of Shadrach, Meshach, and Abednego, the biblical characters who were thrown into a fire but escaped unburned.

"They survived," he would later tell the press, "because they stuck together."

KOLO SAT SQUISHED INTO THE BUCKET SEAT OF A SMALL BUS SPEEDING over the thinly paved country road toward the capital. It had been two weeks since her daughter's release, and the presidency had

finally invited the Chibok families to visit the government safehouse where their children were being held. After what felt like
an eternity, the moment had come for their reunion.

In a small bag, Kolo had packed clothes and her best jewelry, heavy necklaces of ceramic beads, then boarded one of three
special-purpose buses, curtains draped across their windows,
jammed with dozens of other parents eager to reach Abuja. They
were hours into the journey, weaving between overloaded trucks,
when she felt a bump. Then a jolt. A loose wheel shot from under
her vehicle as it spun out of control and careened off the road.

Naomi, meanwhile, was waiting, wearing a yellow floral-print
atampa dress and seated on a plastic chair as caterers passed
around boxes of Five Alive juice. The preceding days had been disorienting. Government officials had begun to debrief the women,
their memories being a treasure trove of intelligence, and therapists had questioned them to see if any had absorbed their captor's worldview. One by one, each of them was asked a question: Is
the Earth flat? No, each one replied.

The women had learned about a social media website called
Twitter, and the names of the celebrities who had called for their
release. Naomi had never heard of America's first lady, Michelle
Obama. Medical checkups and tests for signs of rape or sexual
abuse continued for days. At night, the women lay in bunks and
discussed what they would say to the parents whose daughters
remained unaccounted for. Would they tell the tragic news to the
parents of those who had died? They chose to say nothing for now,
a group decision. It would come in time. Naomi would make sure
Lydia John's parents would see her diary.

The safehouse was a recently built two-story villa with neo-
classical columns and a tiled courtyard filled with trees where
they were gathered to watch the gates creak open and the buses
slowly pull in, the passengers' faces pressed to the windows. Parents began to emerge from the folding doors to a chorus of screams
from their children, who began rushing toward the vehicles.

Naomi scanned the crowd for her mom, but Kolo wasn't there. Her confusion quickly turned to fear. There had been an accident on the road, an official announced, and Naomi didn't register the reassurances that followed. She didn't trust them. All she heard were her own thoughts arriving at the dark assumption that something terrible had happened to her mom just as they were about to be reunited.

Hauwa Ishaya was leaning into her mother's arms, crying into a yard of fabric, and shaking her head. Other parents lifted their children in the air, some fainted. Maryam Ali held her son on her hip, hugging old friends. One daughter ran so fast to her mother that she knocked an NPR reporter's smartphone from her hands. An overjoyed father was speaking into a journalist's microphone, addressing Boko Haram. "We forgive them for what they have done to us," he said. "To err is human, to forgive is divine."

Naomi was scrambling through the crowd of enraptured parents, hyperventilating in a lonely haze of fear for her mom. In the crowd, she found her cousin Saratu beaming in a bright green dress, her arms around her mother, Rifkatu, whose head was buried in her daughter's side, her body quaking as she sobbed. The three embraced.

Naomi was still in tears when the gate opened again. Another bus pulled in, a new tire in place, and a crowd of latecomer parents stepped out into the sun. Kolo felt Naomi's face with her hand, almost to see if she was real, and asked, "So I will see you again?"

Epilogue

The young women fell silent as their classmates walked to the stage, some limping from injuries they'd sustained in the forest, and took turns to spell out the same few words. "A . . . M . . . E . . . R . . ." the students said, stuttering nervously in English, their third language. "I . . . C . . . A . . . N!" Smiling at the applause, the students cleared their throats and began to tick through the next word: "U-N-I-V-E-R-S-I-T-Y."

It was freshman year at a new academic department established to help transform the Chibok girls from Boko Haram's captives into a standing repudiation of everything the insurgency fought for. One of the last remarks the militants had made to their prisoners was a threat: "If you return to school, we'll kidnap you again." The women were determined to defy that order, as well.

After months in a government safehouse, they'd boarded a commercial jet to a Nigerian college town hosting one of their country's top schools. As the plane taxied for takeoff, the flight attendants had tried without success to persuade the passengers to sit in their ticketed seats and buckle their seat belts.

"These are not world-class travelers," their new private security officer, a former US Marine Corps drill instructor, explained to the exhausted crew. "These are the Chibok girls."

Home was now an air-conditioned three-story dormitory block behind the high walls of the heavily guarded American University of Nigeria campus in the riverside city of Yola. Their classrooms were less than a day's drive from the Sambisa Forest where they'd been kept hostage, but college life was a strange, plush existence that felt like another world.

Along carefully mowed lawns studded with flower beds, the women strolled past the scions of Nigeria's metropolitan elite, rich kids who parked their SUVs and luxury sedans in front of painted curbs. Some of Nigeria's best-paid academics lectured in the classrooms, which were fluorescent-lit and decked with Smart Boards. Visiting professors from Europe and America ate suya by the pool, sharing beers with foreign correspondents on their way to the frontline of a war that felt far away.

Many of the instructors who dealt with the Chibok students were Americans, intrigued by a rehabilitation project without much precedent. Even psychologists who specialized in kidnap victims were unsure about the best way to simultaneously treat and educate such a large group of women after years of collective captivity and abuse. The motivational decor of the classrooms fused Western and Nigerian influences, and the tutors hoped it would make the women feel inspired and at home. THE BEST IS YET TO COME! read one banner.

Some of the young scholars flourished, responding to the US-style curriculum. Saratu Ayuba ranked among the highest performers, breezing through her subjects, which included algebra and computer science. Hauwa Ishaya's English improved rapidly.

Naomi's lagged behind. Even though she'd been a leader in captivity, old feelings of academic inadequacy returned to a student who had struggled in high school. She watched as her roommates rebounded from hardship and aced exams. Maryam Ali, it turned out, was very good at school.

It took time before the teachers could understand the full magnitude of the mental and interpersonal transition ahead. In the early weeks, the students filled empty bottles with water from the toilets, jealously storing a backup supply. Insomnia and nightmares kept the dorms abuzz with nervous energy. The harm-less rumble of a passing airplane could send students dashing for their lives under beds and desks. They spoke as a group, and when an instructor would ask one student how she was doing, the reply was often, "We are fine."

The university's goal became to turn them from a sisterhood into individuals, each educated to her own talents. This was pain-ful. The young women had survived by sticking together and now were being separated by academic ability. On a board, one group would be multiplying fractions, writing out, for example, $2/5 \times 2/5 = 4/25$. Another would sit uninterested in the corner, wearing a sullen, defiant look, texting, even though phone use in classrooms was technically banned under school rules. Convincing many of them to trust authority would take years.

A few of the students experienced the guilt of leaving young children at home, and on weekends, Maryam would call her fam-ily, speaking over the phone to the precocious and chatty toddler her mother was raising in Chibok. She kept a picture of Ali on her phone's home screen and in idle moments, she'd flick through pho-tos, the child growing older with each swipe.

The love she felt for him washed over the complicated emo-tions she felt about the crimes that brought him into the world. His father had vanished into military custody the day of her es-cape. After months of government-sponsored deradicalization, she no longer called him her husband. Instead, she asked that he be known as Baba Ali, Ali's dad.

Bearing a Boko Haram commander's child meant she would have to excel in class to escape the judgments of her community. She needed to be the best. This could be lonely, and the person she wanted to speak to about it was not there. What would Sadiya think about the difficulties of this new life of laptop-based curricu-lum and prep for interviews with *60 Minutes*? "I dream about her almost every night," Maryam said. "I have to believe she's alive."

As the semesters ticked by, things got easier. Some evenings, she and her classmates would set aside schoolbooks and be free again, college roommates playing mock karaoke with a hairbrush held as a microphone. They shared bowls of popcorn, watching Hollywood or Nigerian movies, cuddled up on a big comfy couch. They took yoga classes or joined the new debate club, where on a recent afternoon, they argued both sides of the question, Is social media a force for good?

Is it? . . .

In the grand sweep of an awful war, the freed Chibok students were the fortunate ones, their ordeal ending with full scholarships and a chance to become spokeswomen to the broader suffering outside their university's gates. All told, the conflict in Nigeria's northeast left some two million people homeless. The destitute filled tent neighborhoods and abandoned buildings just a few hundred yards from campus. In this context, what distinguished the Chibok classmates was not that their lives had been turned upside down, but that long after they came home, they were still valued.

BACK IN CHIBOK, THE TORCHED YELLOW BOARDING HOUSE THAT HAD once been their high school stood ghostlike in its abandoned lot for more than three years, until 2017, when a demolition crew finally erased what had become a monument to the town's pain. When the workers cleared the remains, they inadvertently left behind the twisted skeleton of a dormitory bunk bed, the sole remaining artifact. There was nothing else left of the only public high school accepting girls for miles around. The campus was swallowed by an overgrown patch of dry reeds, the same empty field that the Petre family of missionaries found in 1941 when they began teaching the alphabet under a nearby tree.

Although the kidnapping destroyed the innocence of Chibok's tight-knit society, it at least brought a sense of common cause. But the release tore the community in half. Shoving matches broke out at meetings of the Chibok Parents Association, which would end with the fathers of the still missing screaming at those whose daughters were free. Attendance dropped, and association leaders stopped inviting some of the most heart-stricken families. They were causing trouble.

The government routinely announced that it would restart efforts to free those high schoolers unaccounted for. But those talks made little progress, and today officials quietly doubt that there will be another major release. In the years after the exchange, several of the captive women took to YouTube to proclaim that they did not wish to return to their parents.

"We are the Chibok girls. We are the ones you are crying about . . . to come back," said one hostage in a twenty-one-minute video from January 2018, surrounded by a dozen others, all dressed in head-to-toe hijabs, with several bouncing infants on their knees. "By the grace of Allah, we are never coming back."

Of the 276 kidnapped schoolmates, 164 are now free. In the early hours and days after the initial attack, 57 fled; 4 escaped later. The Swiss team's mediation freed 103 in two batches. That number could have been 105, except that one hostage was forced to turn back by Boko Haram then agreed to marry and one backed out at the last moment. She appeared to be afraid of what would happen if she left her captors.

Officially, 112 remain missing. At least 40 have died, but Boko Haram has never counted. At least 2 teenagers coerced into marrying a fighter died in childbirth. Other reports of deaths from undisclosed illnesses and snakebites have reached the government.

Nearly all of the students would have been released, mediators broadly agree, if Nigeria's feuding intelligence and defense agencies hadn't called off so many of the early hostage swaps. There is also the troubling counterfactual that more might have been able to escape, as thousands of Boko Haram victims have, if social media had not elevated them into an invaluable bartering chip, guarded so closely for such a long time.

For the parents of those still missing, the agony is unending.

The Johns now live in a tin-roof home on the outskirts of Abuja, with their young child, the brother that Lydia has never met. They keep a five-by-ten-inch photo of their daughter, wearing a demure smile and an all-white outfit, but water drops have stained the picture, and the color is starting to fade.

The diary Lydia kept in captivity managed to reach them. When Rebecca John saw it, she wept. "I knew it was the handwriting of my daughter," she said. "I knew it was her. I wanted to see her here with me."

The Johns, and many in their position, feel that their country has simply moved on, and the world more so. The stream of well-wishers, charities, journalists, and others has virtually dried up.

But there is one member of Nigeria's elite who has not forgotten, who continues to advocate for the women still missing. Dozens of times daily, @ObyEzeks tweets, and on many Sundays, she returns to her patch of highway median to protest. In 2019, she ran a long-shot bid for president with a campaign promise to bring the rest of the women home.

THE WAR CONTINUES IN NIGERIA'S NORTHEAST, INTO ITS SECOND DE-cade. Within a few years, most Nigerians—half of its population is under eighteen—will be too young to remember when or why the conflict started. More than thirty thousand people have died.

Abubakar Shekau would only live for a few more years after the kidnapping of the Chibok girls, before he finally detonated a suicide vest during a battle in 2021. The abduction brought him the global infamy he craved until the end but could never quite repeat. In the months after the multimillion-euro ransom, he sent more than a hundred kidnapped girls, strapped to suicide bombs, into crowded places. It is difficult to either substantiate or dis-prove any causation between the cash windfall and the sevenfold spike in such mass murders that followed, aside from statements by government officials. "We had no choice," said one cabinet min-ister. "And if we had to pay the same price again, we would."

But officially Nigeria's government says there is no evidence that any cash was ever delivered. "If you have records of any such payments made, you can publish them," said a spokesman in response to written questions. "We have no such records here."

Shekau continued to kidnap foreign and Nigerian civilians for ransom. A month after releasing the Chibok students, he appeared on a video boasting that he'd seized ten policewomen. "We only abducted them to serve as slaves," he said, returning to a familiar script.

His rivals formed a breakaway faction along the Niger border, creating a de facto caliphate beyond the reach of Nigeria's military. In February 2018, they kidnapped 110 schoolgirls from a town called Dapchi, a clear attempt to replicate the fame and money that Chibok brought. Then they swiftly and unexpectedly returned all but one, the only Christian, in a process mediated by Pascal Holliger and the Dialogue Facilitation Team.

The Swiss diplomat left Nigeria after the carefully structured process hit more roadblocks. The release of the Chibok students was meant to unlock peace talks that would culminate in cessation of hostilities, but as the mantra goes, "Never trust a breakthrough." In the months after the release, the deal itself was criticized in Nigeria and Switzerland and gossiped about by other mediators.

Pascal had been run ragged. He could no longer maintain a low profile, and he faced conspiratorial rumors of Switzerland's ulterior motives in Nigeria. His colleagues in the Human Security Division would later say that the assignment took a significant personal toll on him and his family.

On the day he left Nigeria, Fulan, the former al Qaeda associate who had once scoffed at the thought of a white man injecting himself into African problems, wrote him a heartfelt good-bye. "It takes a lot to earn my respect," he said. "You earned it 100 times and more."

In Switzerland, the geopolitical shifts brought about by the global rise of nationalism has forced the country's peace industry to rethink its engagements. Terrorism has lost much of its once all-encompassing importance. Think tanks and universities in Geneva have planned a series of conferences in an effort to create new opportunities for Switzerland, such as mediating between the Latin American gangs whose fighting helped send thousands of refugees north toward America. Books about North Korea and China also lie scattered on department desks.

It is easy to dismiss Switzerland's grandiosity about brokering peace in countries continents away, but it's also possible to see in the Chibok story a parable for the breakdown of the Pax Americana. At the height of the trillion-dollar War on Terrorism, with a vast global surveillance program at its disposal, America proved unable to track a group of high schoolers hidden under trees. Neither Washington nor the onetime empires of Europe could extract their desired result from an enormous country whose two hundred million population is expected to grow to four hundred million by 2050.

Instead, freeing the classmates required strengths that don't come naturally to big, militarized countries. It meant forgoing moral judgment and ignoring terrorism watchlists while operating quietly and patiently in the shadows, long after public pressure moved on. It required, crucially, facilitating a ransom to a terror group that likely funded more violence. The complications of a post-American world are still only now coming into view, but if no single capital is powerful enough to write or enforce global rules, there will be endless such ethical dilemmas to solve. Myriad worldviews will collide, each confident of its own truth. And undoubtedly, Swiss diplomats will find themselves quietly playing peacemaker in the shadows.

"The Swiss do the Swiss thing, and they don't do it on the front pages of the newspaper," said one US official intimately involved with the Chibok brief. "It's not for us to tell them what to do or not to do. These are complex things and you're talking about human lives."

The freedom of the schoolmates also required the tireless efforts of unsung Nigerians who put their lives at risk, and in some cases, continue to do so as they try to bring their country peace. Those individuals, referred to throughout this book, and others not mentioned, may never receive their due credit. They include Tahir Umar, who came home after more than a year in prison, never charged with any crime.

There is also the public face of their success, Zannah Mustapha. In the months after the eighty-two went free, he traveled a victory lap through Europe, America, Turkey, and Brazil, meeting heads of state and accepting awards that he keeps on his office shelf. Funds flowed in to his orphanage, which now boasts two campuses and will become, he hopes, a school for peacemakers, teaching the best of African and Swiss mediation techniques: "I am half Swiss," the aging Barrister likes to say. "But only half."

His tour in the limelight felt like a betrayal to other members of the team, some of whom took greater risks and toiled in obscurity. Resentments broke the Agents of Peace apart. Mediators still living on modest means watched as the UN Refugee Agency gave Zannah its highest honor, the $150,000 Nansen Prize.

"Mr. Mustapha, you are an inspiration," said Angelina Jolie at a glitzy gala in Geneva in his honor, speaking to a crowd that rose to a standing ovation.

"I have a vision for peace in Nigeria," Zannah told the crowd. "We are on a journey to understand our differences and overcome our adversity."

Ahmad Salkida continues to try to both report and mediate, running an online news agency, HumAngle, while shuttling offline between the insurgency and a government that still doubts whether he can be trusted. "As long as I'm alive, they'll still call me," he says. Former president Goodluck Jonathan, having retired from politics, published a memoir in which he wondered whether the Chibok abduction was a conspiracy designed to topple him.

Muhammadu Buhari meanwhile won reelection in 2019, the

Chibok release one of the few signature achievements of his first term. Under his watch, the economy collapsed, as it did during his rule in the 1980s, and tens of thousands of young Nigerians took to the streets in protest until the military was deployed to remove them. His return to power, thirty years after his brief dictatorship, remains one of the stranger unforeseen consequences of the war and its best-known kidnapping.

The global Twitter storm lasted little more than a week, but its legacy continues to shape Nigeria and West Africa. American drones and foreign fighter jets still circle the conflict from airbases in Chad, Cameroon, and Niger, all built or expanded after #BringBackOurGirls. Many Americans are at best dimly aware that the hashtag they tweeted evolved into some eight hundred troops based in Niger, the Saharan nation to Nigeria's north, and another three hundred in Cameroon, to the east.

It could be decades before the final deployments of military hardware and manpower are brought home. Officials, senators, presidents, and generals still struggle to make sense of how a group of high-school seniors became the inflection point in which a localized insurgency became a multinational war.

The young women are also working to find meaning in the events.

A FEW DAYS BEFORE EASTER BREAK OF 2019, IN HER SINGLE-WINDOWED, yellow-painted dorm room, Naomi began piling clothes and possessions into her luggage, including an old friend, her battered Nokia cell phone. She was near the end of her second year at American University, and she had decided not to return.

Some of her classmates were flourishing in the competitive environment where academic rigor and strict routine were understood as pathways to healing. To keep their minds out of the forests, the curriculum was loaded with speech contests, supervised study sessions, basketball lessons, restrictions on social-media use . . . everything to keep them glued to the present. If all went well, several could one day mature into public figures with the platform and moral authority to speak for their community, like Malala.

That was not Naomi.

Speaking English was a struggle, and she found it difficult to express herself in written essays or debate classes. She talked back and challenged the authority of teachers who relegated her to the bottom tier. To one, she was "a bad egg."

In captivity, she had become accustomed to being Maman Mu, a leader, and now she was lagging on a campus chosen for her by government officials she'd never met. At first, she expected to slog through it, but the skills that got her through her forest ordeal were not always the attributes for university success. It was lonely, watching women who'd relied on her in captivity zip past her in classes she didn't fully understand. There was also so much trauma to come to terms with. Her kidneys still ached, and the local hospital proved poorly equipped for her needs. She decided to pursue her own path.

In the early months of reporting this book, the authors listened aghast as Naomi Adamu began to recount a story of inconceivable courage in the face of horrors that sounded fantastical in their depravity. Nevertheless, after more than three years of reporting her testimony and more than a hundred hours of interviews with twenty of the young women held in captivity, it became clear that her account often understated the schoolgirls' bravery.

Naomi and her friends had no reason to believe they would survive their ordeal and every expectation that each challenge to their captors' worldview would result in physical and mental punishment. They stuck to their principles all the same, staging a rebellion that led—although they had no reason to believe that it would—to their freedom.

"We stood our ground," Naomi told us. Several of her classmates echoed the same thought. "She gave us the confidence to resist," said Palmata Musa.

Holy week in 2019 began on a special date for the young women. Palm Sunday was on April 14, five years after the kidnapping. Back in Naomi's tin-roof church in Chibok, neighbors in sharp suits and billowing wax-print dresses would be dancing around the altar, lifting fronds in the air.

Naomi descended the concrete steps of her dorm carrying her bag through a covered walkway. The next chapter in her life would be without her sisters. She planned to attend nursing college in Maiduguri. She felt at home in the world of clinics and medical scrubs.

She lifted her bag onto the bus headed for Chibok. Kolo would be waiting.

Postscript

This book began with a slip of the tongue. A Nigerian cabinet minister was sitting at his desk, which was strewn with newspapers bearing graphic images of wreckage from a new spate of bombings by Boko Haram. It was late 2017. In the preceding months, insurgency attacks had doubled, prompting the question we had posed to him: Why? "Because we gave them millions of euros for the Chibok girls," the minister said with a nervous chuckle.

Millions of euros?

By then, the #BringBackOurGirls campaign had faded into distant memory in Nigeria's northern war, a celebrity-led aberration that burned at radioactive heat for just a few short days. We had begun covering the kidnapping for the *Wall Street Journal* just hours after the abduction, interviewing school staff, then family members, who pleaded for help over shaky cell-phone connections. We had raced to the northeast to cover a search that had fleetingly captivated the world, then just as quickly seemed to sputter into dust.

Now, three years later, the minister was drawing cause-and-effect between that long-forgotten frenzy of online activism and the deaths splayed across his daily newspaper. Was this the legacy of the world's passing interest in Nigeria? And what of Boko Haram's many captives still missing? Had the hashtag helped or hindered the efforts to free them? Was the West's adoption of #BringBackOurGirls good for the *girls*?

We began this project as a piece of accountability journalism, to interrogate the real-world impact of that celebrity-driven hashtag and find out why it had taken three years, at a heavy price, to free a group of captives that seemingly every important public figure on earth had championed. The article we published in December 2017, on the hostage talks, at 10,285 words, appears to be the longest in the newspaper's 130-year history. But as we plunged further into the secret world of hostage talks and drone surveillance, we confronted a leviathan of a story that was more complex, more clandestine, and more global than we could have ever imagined. Turning it into a book took us across the world as we tracked down the many people who had rushed in to play some part, for better or worse, in the girls' rescue: from northeast Nigeria to the White House, from the Swiss Alps to military outposts in the Sahara. We traveled from the glass towers of Dubai to the dusty backstreets of Khartoum to the leafy streets of St. John's Wood, where powerful Nigerians keep homes in London, and to a hotel ballroom in Casablanca where Africa's top intelligence officials gathered for a meeting overlooking the furious waves of the Atlantic.

It became clear that this was not just a story about a remote, tragedy-stricken town in Nigeria, but a parable, and perhaps a cautionary tale, about the flawed interconnected workings of our butterfly's-wings world. We wanted to make sense of a bizarre postmodern moment when a tweet by Mary J. Blige could propel the hurricane of online energy that launched drones over Lake Chad and delivered millions of euros to Boko Haram. How could any person claim to know the way forward in such a strange maze of morally incomprehensible choices?

We found an answer in a group of young women who had faced unimaginable hardship and somehow survived.

BRING BACK OUR GIRLS IS BASED ON SIX YEARS OF REPORTING ON FOUR continents and interviews with hundreds of people, from presidents to former child soldiers. It is not a history book, nor is it a complete account of the Boko Haram conflict, and we direct readers interested in further study to the many excellent texts mentioned in the notes. This is a work of narrative nonfiction. No details or dialogue have been invented or imagined. Any thoughts or reflections of people featured in the story have been described either to us or to others.

The story of the Chibok girls is not singular. We spoke at length to twenty of the survivors for many hours over a period of years, but this book conveys the experience of a handful. The young women spent three years together in the custody of Boko Haram, but every one of them had a different experience, and memories sometimes clashed. Collective recall is a slippery and sometimes subjective reality, especially when clouded by the haze of a traumatic ordeal, so wherever possible we have sought to corroborate individual accounts, and most of the scenes in captivity were described by multiple people or supported by diary entries. Interviewing Boko Haram victims requires ethical considerations that were familiar to us, and our work on this subject came near the end of a decade covering Nigeria and the war. Still, to deal with their unique circumstances, we sought the guidance of psychologists in the country's north and abroad, as well as their current and former care providers at the American University of Nigeria.

In time, others may step forward to tell their own story. We hope our contribution helps them do so, by sustaining public interest in their experience and by holding a spotlight on the young women who have yet to come home. It is crucial not to forget the larger suffering they represent. At the time of writing, some twenty-two thousand northeastern Nigerians have been reported missing to the Red Cross, the highest such toll anywhere on earth. We have offered guidance and professional support to those survivors who wish to commit their memories to print. One of their current university lecturers told us, "When Boko Haram kidnapped these women, they created a hundred Malalas." That may become true. Or, the young women may choose to pursue more private lives away from the sometimes heavy demands of fame and social media. In the meantime, proceeds from this book will fund educational needs in Chibok, especially for the survivors' families.

In our work for the *Wall Street Journal*, we follow a rule called "No surprises"; a standard meant to apply equally to Fortune 500 companies, heads of state, and ordinary people. It means that, before publication, all persons should be afforded a chance to know and respond to any major assertions of fact made about them. We extended this principle to the victims of the kidnapping, including the families of those who remain in captivity. Rather than erase those who have not come home from the public record, we have chosen to change their names, in consultation with their parents. We asked each of the young women who won their freedom that we interviewed if she wished to go by a pseudonym, but all except "Ladi Simon" preferred to be identified.

Likewise, some of the officials, negotiators, and mediators involved in the talks to free the Chibok girls remain engaged in the daunting struggle to bring peace to Nigeria's northeast. Several are unable to discuss the efforts they made. It wasn't possible to speak with Boko Haram, though we spoke with dozens of men and boys who once fought with what remains an enigmatic movement.

Nigeria, over the course of our research, became an increasingly difficult environment for journalists and their sources, and many of our interviews were conducted by "Woodward rules," named after the Watergate reporter. In short, information given to us was cross-verified and used, but the sources remain unidentified. Our reporting owes a special debt to the Nigerian journalists who contributed to the public record we drew on.

This book takes place at a twilight phase when both America and its tech giants could still consider themselves a great and global force for good. In navigating the often culturally dissonant space between the United States and Nigeria, there was another source that proved invaluable. Barely mentioned are a group of young women who, in the early hours of the kidnapping, jumped from pickup trucks and ran for their lives. Several were sent on scholarships to America, and a few spoke with us. One was Kauna Bitrus, whose adolescence—upended by the kidnapping, then by the social-media movement that followed—followed the course of the disruptions we were writing about. She met us in the booth of a Maine diner not far from the Bush family compound in Kennebunkport, and then, over a slice of pepperoni pizza, began to tell us about life in a pink tin-roofed boarding school once perched on the long dirt road that connects the world to a town called Chibok.

Acknowledgments

This book was made possible by the young women who agreed to share their detailed recollections. From our first meetings in 2017 with Naomi and Kolo Adamu, each of the Chibok Government Secondary School students we met welcomed us with generosity and patience, even when recalling the most traumatic events. Over the next three years, in hotels and restaurants in Yola, Abuja and Maiduguri, they helped us piece together the complex puzzle of their time in captivity.

Naomi Adamu not only shared her memories, but also showed us items that made her ordeal real in a way no interview could—a school uniform, a battered old Nokia, and hundreds of pages of secret diaries. Maryam Ali sat with us for countless hours and later took the time to write her reflections in a journal that will also one day, when she chooses to publish it, become an indispensable contribution to the literature on this subject. We are enormously grateful to them and to all of the Chibok class of 2014 who spoke to us, including Hauwa Ishayah, Palmata Musa, Hannatu Stephen, Ladi Simon, Saratu Ayuba, Mary Dauda, Hauwa Ntakai, Naomi Yaga, Rifkatu Solomon, Rakiya Abubakar, Yanke Shettima, Grace Dauda, Mwauda Dauda, Hadiza Yakubu, and Kauna Bitrus.

The high-school seniors who became the Chibok girls were a symbol of a conflict that has upended millions of lives, and we would also like to express our gratitude to the dozens of men and women, victims and protagonists, who agreed to answer the questions of foreign strangers so as to help others understand their experience. We hope our book does just that, making sure their actions are entered into the historical record.

The authors would also like to acknowledge the people who worked in the shadows for years to free the young women, those who have never sought the limelight and who declined to comment for this book: Tijani Ferobe, Tahir Umar, Iliasu Alhassan, Pascal Holliger, Abdulkabir Said, Ali Umar, Sadiq Mohammed, Muhammad Ibrahim, Dr. Hakeem Baba-Ahmed, Mohammed Umar Bolori, Dr. Cairo Ojougboh. There are other, anonymous members of the Dialogue Facilitation Team in Nigeria and abroad whose efforts may never be recognized.

Reporting in sometimes dangerous places is possible only with the help of others, and from its inception our research was aided and abetted by a peerless team of Nigerian journalists. Gbenga Akingbule, the *Wall Street Journal*'s indefatigable Nigeria correspondent, was utterly indispensable, accompanying us on countless trips throughout the northeast, deftly using his charm and smarts to push the reporting forward. Kabiru Anwar arranged and translated complex interviews and was our in-house encyclopedia on Hausa culture and politics. Julius Emmanuel made courageous trips to areas where Boko Haram still holds sway. This project would have been unthinkable without them.

At every stage, we have benefited from the generous feedback of respected colleagues, journalists, researchers, academics, and writers who have spent years studying and reporting on West Africa. Among them: Mausi Segun, Fatima Akilu, Chike Frankie Edozien, Gillian Parker, Alex Thurston, Matthew Page, and Bulama Bukarti. Adaobi Tricia Nwaubani, having interviewed more than two hundred of the Chibok parents, offered sage advice and perspective. Jacob Zenn reviewed chapters and responded to endless questions about the structure of Boko Haram and the shadowy figures who lead it. Joe Brock, Matthew Mpoke Bigg, Glenna Gordon, and Alex Perry read two separate drafts and responded with considered, detailed feedback on both. Their insights, honesty, and good humor helped us navigate sensitive subject matter and steer us away from cliché. Any errors, oversimplifications, or misjudgments that remain are entirely our own.

Nigeria's newspaper editors and reporters—professionals working under increasingly repressive conditions—were also unfailingly generous with their time and expertise. Manir Dan Ali, Azubuike Ishiekwene, and John Alechenu deserve special thanks for helping to steer us in our first years covering Africa's most important country. Ubale Musa, Isaac Linus Abrak, and Aminu Abubakar offered crucial support in the early stages of reporting. There is far more to Nigeria than the conflict in its northeast, and it is unfortunate that none of the events in this book take place in Lagos, Africa's economic and cultural capital and our base during some of the reporting. In the meantime, thank you to the hospitality of true Lagosians Jahman Anikulapo, Tilewa Adebajo, Patrick Fola Ayanbadejo, Will Connors, and Oo Nwoye.

Beyond Nigeria, friends and colleagues extended contacts and valuable advice. We're grateful to Rukmini Callimachi, Baba Ahmed, Julian Barnes, Andres Martinez, Chris Stein, Scott Stearns, Georgi Kantchev, and Ricci Shryock; also to Robert Christgau, the dean of music journalism, for helping us navigate the industry. Samantha Power, who correctly informed us that we knew nothing about the way Washington worked, connected us to a number of people who left us only slightly less confused. Achim Wennmann helped shine some light on the opaque mechanics of Swiss diplomacy. Joe Galvin at Storyful in New York helped us decipher and decode Twitter analytics to chart the minute-by-minute evolution of #BringBackOurGirls.

We would like to thank our editors at the *Wall Street Journal*, who encouraged us to pursue the story and counseled us to let the newspaper's motto, "Follow the money," inform our reporting. Special thanks are due to Editor-in-Chief Matt Murray and our other bosses—Matthew Rose, Sam Walker, Gordon Fairclough, Deborah Ball, and Thorold Barker—for giving us the space to conceive of this project and then execute it. Peter Wonacott and Rebecca Blumenstein were early champions of this story, and Bruce Orwall and Bill Spindle were unrelenting advocates for our journalism. In an age of twenty-four-hour news cycles racing after tweet-length commentary, the *Journal*'s commitment to patient, rigorous investigative reporting sets it apart.

Following the comet trail of the hashtag took us all over the world, but with so much of our reporting taking place in Nigeria, we are greatly indebted to Abuja's Nordic Villa, our home away from home, which hosted dozens of our most important interviews and allowed us to scatter endless notes across the floor as we tried to wrangle our mountains of material into a book. A special thanks is also due to the American University of Nigeria, in Yola, where we spent many months hosted and enlightened by staff members Dawn Dekle, Margie Ensign, Somiari Fubara, and Lionel Rawlins. And a lasting word of appreciation for our editor, Gail Winston, and our agent, Rebecca Gradinger, whose vision, enthusiasm, and good judgment helped propel this book into being.

One colleague advised us that writing a book of nonfiction is "like a series of marathons run back to back for years." We are grateful for the support of friends who helped us to stay sane, nursing both the authors and the manuscripts during the tougher parts of the process. Mackenzie Knowles-Coursin, Tom Wright, Anne Look Thiam, Brian Stefan, Michael Tate, Joseph Terranella, and Clem Hitchcock all listened valiantly over beers, coffee, or, in the age of COVID-19, FaceTime, offering thoughts on chapters, structure, and style. Will Hart and his design company, SubliminalVision, spent countless hours honing artistic concepts for the jacket. Daniel Levitt produced the map of Nigeria at the front of the book. To report for a widely read newspaper in another region is an immense privilege, and our time in West Africa would not have even begun without the extraordinary faculty of New York University's campus in Accra.

Most of all, we are grateful to our frankest critics and most tireless supporters, Hermione Nevill and Celeste Hinshaw, for their forbearance during a reporting project that became an overtime job but would never have become a book without their love and encouragement. We nearly missed the birth of two of our five young children in faraway hospitals while we were covering Nigeria. One day we will tell them of the promise and peril of an era when people sought to change the world using an arcane tool called a hashtag.

Glossary

007: Specially designed encrypted fax machines found only in Swiss embassies around the world.

Abuja: Capital of Nigeria since 1991, a planned city located in the center of the country.

aduwa: A bitter, yellow, single-seeded fruit also known as the desert date.

Al-Majiri: System of Islamic education practiced throughout northern Nigeria.

Alpha Jet: A Franco-German-designed light attack aircraft that is the core of the Nigerian Air Force.

Ansaru: A breakaway sect of Boko Haram loosely allied with al Qaeda.

arna: Pagans.

Aso Rock Presidential Villa: The workplace and official residence of the president of Nigeria.

atampa: Traditional fabric often made into long, fitted dresses worn by women throughout northern Nigeria.

Baba Go-Slow: Nickname given to President Muhammadu Buhari for his lackadaisical approach to policy making.

Babban Layi: Maiduguri's main boulevard, also known as Broadway.

Banki: Small town on Nigeria's border with Cameroon, close to the exchange point for the releases of the Chibok hostages and once home to the John family.

Bashin Gaba: Debt of Hatred, a northern Nigerian movie about star-crossed lovers.

bay'ah: An oath of allegiance.

BBC Hausa: The radio service listened to by millions of Hausa speakers throughout Nigeria, Niger, Ghana, and Chad.

Beechcraft: Military adaptation of one of the Beechcraft company's most popular civilian designs, turning it into a long white twin-engine surveillance plane equipped with a bulbous node of sensors and cameras on its belly.

Biafra: A self-declared republic that seceded from Nigeria in May 1967 but was defeated by Nigerian forces in 1970.

Book of Job: The Old Testament account of a man tested with endless suffering who refused to renounce his faith.

boulder in the road: An obstacle so fundamental that peace talks cannot proceed until that obstacle is removed.

C-130 Hercules: A large American-made transport plane in service since 1956.

caliphate: An Islamic state or empire.

Carlos the Jackal: Venezuelan Marxist convicted for a string of terrorist attacks during the Cold War.

Chief Infidel: Nickname given to Naomi Adamu by Malam Ahmed.

chin-chin: Crunchy biscuits made from deep-fried dough.

Church of the Brethren: A protestant denomination founded by American missionaries that has more than 150,000 members in Nigeria.

damina: The rainy season.

DDR: Disarmament, demobilization, and reintegration.

Def Jam Recordings: American hip-hop record label founded in 1984 that produced some of the world's most famous rappers.

Department of State Services: Nigerian domestic intelligence agency responsible for counterintelligence, internal security, counterterrorism, and surveillance.

"Dogorana": "Trust," a famous song by Mama Agnes.

doyan daji: An acidic tuber found in the forests of northern Nigeria.

École hôtelière de Lausanne: Swiss hospitality management school, consistently regarded as the world's best.

egusi: An oily stew thickened with ground seeds and popular in different variations throughout West Africa.

emira: A princess, ranking female member of Boko Haram.

Emirul Jesh: Head of Boko Haram military wing, ranked one rung below Abubakar Shekau.

Executive Outcomes: A mercenary firm, founded by a former apartheid-era soldier Eeben Barlow, which recruited men from both sides of southern Africa's liberation conflicts.

***Fim* magazine:** Kaduna-based magazine reporting on Hausa culture, movies, and gossip.

Fusion Center: A room in the National Security Agency building where Nigerian personnel worked alongside American and British colleagues to share intelligence on Boko Haram.

Gaba Imam: The "House of the Imam," Shekau's hideout.

gidan tabo: Thatched-roof homes constructed from air-dried clay-and-straw bricks, whose earthen walls require constant patching.

Giwa Barracks: A military detention center in Maiduguri that became a focal point of the war.

Global Hawk: "The workhorse," a drone with a bulbous head like a dolphin that could surveil a South Korea–size acreage daily.

Glorious Singers: The Chibok school choir, an exclusive club with only twelve members, chosen after an audition.

Gray Eagle: A highly sophisticated drone capable of flying uninterrupted for two days and reading a license plate from 15,000 feet.

halqa: A Boko Haram "cell" or department under the command of a qaid.

haptiri: A game commonly played by children in northern Nigeria, in which participants take turns trying to throw small stones into a hole. To Boko Haram, it was "the devil's game."

hijab: Islamic women's head covering.

hijra: An exodus, capitalized when referring to the migration of the Prophet and his followers from Mecca to Medina.

Hilton, Abuja Hilton: 667-room hotel in central Abuja where power and money mingle.

Hilux: Toyota pickup truck common across Nigeria and favored transportation of Boko Haram.

hisba: Islamic religious police deployed to enforce Sharia law.

Human Security Division: Unit of the Swiss foreign ministry that defines itself as responsible for "the promotion of peace and human rights"; formerly Political Affairs Division IV.

ibtila: A "trial," a hardship viewed as a test of religious faith.

#IceBucketChallenge: Global Internet campaign that came shortly after #BringbackOurGirls, to generate awareness for motor-neuron disease.

imam: Prayer leader, cleric.

Interdisciplinary Assistance Team: Group of thirty-eight officials deployed by the White House in early May to resolve the kidnapping, which included FBI hostage specialists, military intelligence analysts, and psychosocial therapists.

Islamiyah: Islamic school.

jasus: A spy.

jollof: One-pot dish of rice, spices, and tomatoes popular throughout West Africa.

kachau-kachau: Tamborinelike drums laced with strings of beads that are shaken to make a jingling sound.

kafir: An infidel, a non-Muslim.

Kanuri: Native language of much of Borno State, including Maiduguri.

katezi: A game of trust in which one person closes her eyes and falls backward into another person's waiting arms.

Kauwa Dima: Perfume Seller, nickname given to the young Abubakar Shekau.

kayan lefe: Clothing materials that a groom presents to his bride, well arranged in boxes or bags.

Kibaku: Language spoken by the approximately two hundred thousand Kibaku people living in and around the town of Chibok.

kufi: Close-fitting, delicately embroidered caps worn by men in northern Nigeria.

kuka soup: A hearty dish common in northeastern Nigeria made from kuka, powdered baobab leaves.

kunu: A light gruel made with rice and raw peanuts.

Lac Leman: Crescent-shaped Swiss lake with Geneva at its southern end.

Lake Thun: Swiss Alpine lake located on the northern periphery of Alps.

Land Cruiser: Toyota four-wheel drive vehicle.

League of Nations: Precursor to the United Nations, based in Geneva.

Maggi: Brand of bouillon cubes, ubiquitous throughout Nigeria.

Maghrib prayer: Sunset prayer.

Maiduguri: "The land of a thousand kings," regional capital of Borno state and birthplace of Boko Haram.

Maimalari Barracks: Sprawling military base situated on the northern outskirts of Maiduguri.

malam: Scholar, learned man.

Mandara Mountains: A 120-mile range straddling the Nigeria-Cameroon border.

mayafi: A woman's long shroudlike dress commonly worn in Nigeria's north.

miswak: A long chewing stick, used in the time of the Prophet.

Monday Market: Largest market in Maiduguri, open seven days a week.

naira: Nigeria's currency.

Nansen Prize: United Nations annual award given to recognize service to the cause of refugees.

nasheed: Islamic hymns, usually sung a capella, that reference to Muslim beliefs and history.

National Security Council: The group academics and intelligence officials who sketch out security strategy for the White House; commonly known as the NSC.

Niesen: Swiss Alpine mountain that overlooks Lake Thun.

Night Stalkers: An elite US Army special-operations force.

niqab: A Muslim woman's veil that covers the hair and face.

Nokia 3200: Entry-level Nokia cell phone first released in 2004, prized possession of Naomi Adamu.

NPR: National Public Radio.

Office of the National Security Adviser (NSA): Staff and chief adviser to the president on national security issues.

Operation Lafiya Dole: Operation "Peace by All Means," a 2015 military operation against Boko Haram, started after Buhari's election.

Park Oberhofen: A nineteenth-century hotel in the Swiss Alps that regularly hosted the Human Security Division's Peace Mediation training course.

pidgin: Any Creole language mixing English with a native tongue; West African pidgin is spoken as a lingua franca throughout Nigeria.

prayer warrior: A committed churchgoer who prays to banish evil spirits from the community.

PTSD: Post-traumatic stress disorder.

qaid: A Boko Haram commander.

Ramstein: US air base located in southwestern Germany, which serves as an important control center for America's drone wars.

Raqqa: Northeastern Syrian city that became the capital of Islamic State's caliphate.

rijal: Boko Haram foot soldiers.

Sabil-Huda: "Path of Guidance," a large Boko Haram camp in the Sambisa Forest.

sabon rai **hymns:** "Reborn soul" hymns, praise songs commonly sung in Chibok.

semo: Semolina, coarsely ground durum wheat middlings or the porridge made from them.

Shadrach, Meshach, and Abednego: Biblical characters who were thrown into a fire but escaped unburnt.

Shura Council: Twelve-man consultative body of the most senior Boko Haram leaders.

sitara: An Arabic word meaning "curtain" or "screen."

Snake: A video game on Nokia phones wherein the player maneuvers a line that grows in length until the line itself becomes the primary obstacle.

South South: The swampy coastal region of southern Nigeria that produces almost all of the country's oil and that is also the birthplace of Goodluck Jonathan.

"Stop Kony": A short documentary film produced by American filmmakers in an effort to bring fugitive Ugandan warlord Joseph Kony to justice.

sulhu: Reconciliation, cease-fire.

suya: A skewer of meat spiced and grilled.

takbir: A proclamation of the greatness of God, *"Allahu akbar."*

"Timbuktu": A Boko Haram military camp in the Sambisa Forest, the first holding station for the Chibok girls.

Toh, toh: Hausa phrase meaning "Alright, alright."

TweetDeck: A dashboard application that lets Twitter users view multiple timelines simultaneously.

Ukuba: A large Boko Haram camp.

Unity Fountain: Dilapidated white-tiled fountain built opposite the Abuja Hilton to symbolize the unity of Nigeria's thirty-six states.

verified: A Twitter account awarded the "blue check mark" to let users know it is authentic.

Xhosa: One of the official languages of South Africa; it uses a distinctive clicking sound.

Yerwa: The original name of Maiduguri, still used by locals.

zumunta mata: A female church choir.

Notes

Prologue

2   the skies above northeastern Nigeria were empty: The terms of the prisoner exchange, which dictated no flights in the area, were delineated in an interview with Zannah Mustapha (hereafter referred to as Z.M.) who was interviewed by the authors more than a dozen times between October 2017 and April 2020 in Abuja or Maiduguri, as well as by several other people involved with the exchange who spoke on condition of anonymity.

2   A soft rain streaked the windows: Z.M. in discussion with the authors, verified with historical weather reports from CustomWeather.com, https://www.timeanddate.com/weather/nigeria/maiduguri/historic?month=5&year=2017.

2   a Russian helicopter juddering through gray clouds: Z.M. provided photos of the helicopter he took in May 2017, which were verified by other people involved with the effort who spoke on condition of anonymity.

2   A Nigerian lawyer lifted a list and a pen from the chest pocket of his crisply ironed ash-colored kaftan: Z.M. provided photos of the exchange, verified by other team members.

3   Both had sworn to observe a total information blackout: Interviews with Z.M. and other team members.

3   The air power and personnel of seven foreign militaries: The governments of the United States, France, the United Kingdom, Chad, Niger, Canada, and Cameroon all confirmed in official releases that they sent troops, warplanes, special forces, or advisers into Nigeria to help with Boko Haram and its most famous captives. This tally doesn't include Israel, which offered intelligence and personnel; China, which offered satellite photos; Benin, which also sent a small detachment in a support role; or Switzerland.

3   a military outpost ringed by half-buried tires: Photographs of Banki base taken by Z.M. in May 2017.

4   rust-coated wheelbarrow: Ibid.

4   The area was notorious for land mines: Jean-Guy Lavoie, chief of operations, UNMAS, *Mission Report: UNMAS Explosive Hazard Mitigation Response in Cameroon*, United Nations Mine Action Service, April 2017: "Cameroonian military officials reported in 2015 that 'huge' numbers of landmines had been planted by Boko Haram along Cameroon's Nigerian border."

4   their bumpers mounted with fluttering Red Cross flags: Photographs courtesy of Z.M. and published on the International Committee of the Red Cross Twitter feed, @ICRC_Africa, May 7, 2017, https://twitter.com/icrc_africa/status/861138144758030336.

4   "The prayers of the orphans will protect you": Author interview with Z.M.

4   Fighters in fatigues, their heads wrapped in turbans: Photographs courtesy of Z.M.

4   Two of them were walking on crutches: Photographs courtesy of Z.M.

5   "We, the children of Israel, will not bow": Interviews with Naomi Adamu (here-
     after referred to as N.A.), whom the authors interviewed 11 times between 2017
     and 2020 in Maiduguri, Yola, Abuja, and Chibok, and with Hauwa Ishaya (here-
     after referred to as H.I.) in April 2020.

Chapter 1: "Gather Outside"

9   only four weeks from finishing senior year: The 2014 West African Examina-
     tions Council test schedule concluded the week of May 12 with health studies,
     religious, and drawing exams.
9   a three-hour civics exam: To retrieve a copy of the exam, reporters visited the
     WAEC headquarters in Abuja in October 2019 and sifted through thousands of
     old tests, which the testing agency responsible for all of Anglophone West Africa
     stored scattered on the metal shelves and floor of a dusty side office.
9   gathering in small circles in the prayer room: Interviews with N.A., H.I., and
     Palmata Musa (hereafter referred to as P.M.).
9   so hard it would make them cry: Ibid.
9   "Education for the Service of Man": photos of school uniforms, taken by Glen-
     na Gordon, "Abducted Nigerian School Girls," *LensCulture,* 2014, https://www
     .lensculture.com/articles/glenna-gordon-abducted-nigerian-school-girls.
9   just 4 percent of girls finished high school: "Education for the girl child in
     Northern Nigeria," *Africa Check*, July 2017, https://africacheck.org/wp-content
     /uploads/2017/03/July-info-graphic.pdf.
10   exams that were challenging for even the most high-achieving students: Only
     31 percent passed the math and English exams that year. "Mass Failure as
     WAEC releases May June Exam Results," *Vanguard Nigeria* (Lagos), August
     12, 2014.
10   At twenty-four, Naomi Adamu was one of the oldest: Interviews with N.A., H.I.,
     P.M., and Saratu Ayuba (hereafter referred to as S.A.).
10   "Should we run?": This scene, described by multiple survivors, was also corrob-
     orated by former CNN reporter Isha Sesay's *Beneath the Tamarind Tree* (New
     York: Dey Street, 2019).
11   Naomi threw herself flat on the cement floor: Interviews with N.A, H.I., P.M.
11   The school's security guard: Ibid.
11   Girls began bracing themselves to run, stuffing clothes into bags: Ibid.
11   Mary Dauda stared into a phone that had dropped a call to her brother: Inter-
     view with Mary Dauda (hereafter referred to as M.D.) in 2018.
11   Margret Yama managed to briefly speak to her brother, Samuel: Interview with
     Samuel Yama, May 7, 2014.
12   "There is nobody here," a male voice said: This scene and other subsequent
     dialogue, described by multiple survivors, was also corroborated in Amnesty
     International's May 9, 2014, report, "Nigerian Authorities Failed to Act on
     Warnings about Boko Haram Raid on School," https://www.amnesty.org/en
     /latest/news/2014/05/nigerian-authorities-failed-act-warnings-about-boko
     -haram-raid-school.
12   a stocky man with jutted front teeth wearing a red cap: description of Malam
     Abba provided by N.A. and Maryam Ali (hereafter referred to as M.A.), whom
     the authors spoke with eight times in Yola and Chibok between December 2017
     and June 2020.
12   Rhoda Emmanuel, a pastor's daughter, grabbed a blue Bible: Interviews with
     N.A.
12   Maryamu Bulama carried her own copy of the Bible: Ibid.
12   Naomi furiously stuffed a bag: Ibid.

12   3,600 naira (about $23) in pocket cash: The naira has since crashed relative to
      the dollar, and at the time of writing, this amounts to only about $9, but on the
      night of the abduction, the naira traded at 160 to the dollar, per the Dow Jones
      Factiva Exchange Table.

12   "I'm very sure it's Boko Haram": Interviews with N.A. and S.A.

Chapter 2: The Day of the Test

13   Built at the foot of a solitary hill studded with jagged gray boulders: Reporter
      visits to Chibok in May 2018, December 2019, and May 2020.

13   It is home to around seventy thousand people: An exact measure of the town's
      population is difficult. The 2006 Nigerian census put the entire Local Govern-
      ment area, which extends beyond the town, at sixty-six thousand.

13   At dawn, Chibok's families step out together: Interviews with Kolo Adamu,
      hereafter referred to as K.A., whom the authors met several times in Yola,
      Maiduguri, Abuja, and Chibok between 2017 and 2020, and N.A.

13   The grocery-stall shopkeeper . . . is nicknamed Dangote: Interviews with K.A.,
      N.A.

14   April is traditionally a month of celebration: April and December are harvest
      seasons, and weddings are scheduled in those months.

14   the sound of a sandal slapping in mud: "A Brief Walk into the Homeland of One
      of Africa's Bravest People," *Pulse Nigeria* (Lagos), November 16, 2017.

14   Just one-tenth of 1 percent of the country understands it: Ethnologue: Lan-
      guages of the World, https://www.ethnologue.com.

14   *Mimigai? . . . Yikalang*: Interview with Naomi Yaga (hereafter referred to as
      N.Y.), October 2019.

14   home to 206 million people: Nigeria's population is a subject of dispute. This is
      the UN Population Division's 2020 estimate, in line with a commercially con-
      ducted estimate seen by the authors. By mid-2021, the UN expects the number
      to be 211 million.

14   a citizenry that will double and outnumber Americans by 2050: "World Pop-
      ulation Prospects," UN Population Division, a unit of the Department of Eco-
      nomic Development and Social Affairs, 2019, https://population.un.org/wpp
      /Publications/Files/WPP2019_Highlights.pdf, p. 42.

15   what Nigerian scholars would later call subcolonialism: See Moses E. Ochonu, *Co-
      lonialism by Proxy: Hausa Imperial Agents and Middle Belt Consciousness in Nige-
      ria* (Bloomington: Indiana University Press, 2014). A good lay primer on Nigerian
      history, including the colonial era, can be found in Toyin Falola and Matthew M.
      Heaton, *A History of Nigeria* (Cambridge, UK: Cambridge University Press, 1989).

15   It had been founded in the 1700s by refugees fleeing bands of Muslim slave
      raiders: "A Brief Walk into the Homeland of One of Africa's Bravest People,"
      Pulse Nigeria (Lagos), November 16, 2017.

15   the last in the country to submit to the British Empire: Ibid.

15   For centuries, the residents had kept their ancestral religions: Gerald A. Neher
      with Lois Neher, *Life among the Chibok of Nigeria* (Elgin, IL: Brethren Press,
      2011).

16   Boxed in by a low wall with a single metal gate: Photos of the school published
      by news agencies after the attack.

16   the Chibok Government Secondary School was the first school to educate girls:
      Interview with Yakubu Nkeki, president of the Chibok Association, July 2019.

16   a book, a pencil, and a chili pepper: Photos of uniforms taken by Glenna Gor-
      don, "Abducted Nigerian School Girls," https://www.lensculture.com/articles
      /glenna-gordon-abducted-nigerian-school-girls.

16   a Ford pickup on an April evening in 1941: The *Gospel Messenger*, an early-twentieth-century periodical of letters from American missionaries stationed around the world, includes numerous letters from Ira Petre describing his journey to Chibok and his work in the town.

16   a local man they called a wash boy: Letter from Ira Petre, printed in the *Gospel Messenger.*

16   raised his children to speak fluent Kibaku: Interview in July 2019 with Ira's son Rufus Petre, who grew up in Chibok, speaking its language, but now lives in western Pennsylvania.

16   after a local early convert and chorister, Mai Sule: Ibid.

17   On April 13 . . . the community held an interfaith wedding: Interviews with K.A. and N.A., who also showed the authors photos of the event.

17   draping the decorative apostrophe: N.A. liked to sign her name Na'omi, as seen by the authors in her diaries.

17   It was just about noon: The WAEC government exam was meant to begin at 10:00 a.m., but for unknown reasons started late, so the students, already at a disadvantage with better school districts, were taking it hungry, near lunch hour, and in midday heat that reached 105 degrees that day, per historical weather reports from CustomWeather.com.

17   "Instructions to Candidates": Authors reviewed a copy of the test, stored in the WAEC headquarters in Abuja.

18   All around Naomi sat hundreds of students: Of note, a group of boys who took day classes at the school also took the test that day.

18   students in the nearby science lab were conducting a practical exam: Interviews with N.A., M.A.

18   Idle ceiling fans hung motionless over the desks: Photos of the burned-out exam hall provided by the Chibok Association, July 2019.

18   a teacher, probably Mr. Bulama, had ordered a student to wipe clean: Mr. Bulama served as proctor that day, per several students interviewed.

18   The nickname they called her, Maman Mu, Our Mother: Interviews with N.A, K.A., H.I., M.A., P.M.

19   Sadiya just wanted to be done with exams: Interviews with M.A. Authors also spoke with her father, Ali Maiyanga, in Yola, October 2018.

19   Government protects the lives and property: Authors reviewed a copy of the exam paper.

19   Malam Abba's men . . . had forgotten to bring enough water: In interviews, N.A., H.I., and P.M. relayed conversations the men had on their return journey.

19   Dozens of his Boko Haram fighters were squeezed onto the back of pickup trucks: Interviews with N.A., M.A., and H.I. Multiple witnesses in Chibok also described the men arriving to other media, including the above mentioned Amnesty report, "Nigerian Authorities Failed to Act on Warnings about Boko Haram Raid on School," https://www.amnesty.org/en/latest/news/2014/05/nigerian-authorities-failed-act-warnings-about-boko-haram-raid-school.

20   the fifth year that Malam Abba had been waging jihad against the Nigerian state: In an interview, M.A. described extensive biographical details of Malam Abba based on speeches and comments he made to other BH captives and members during her time in captivity.

20   Just one month after escaping from the country's most notorious jail: Some Nigeria specialists might dispute this designation, saying that the Kirikiri Maximum Security Prison in Lagos earns the title, but there is ample evidence to suggest that the conditions were worse at Giwa Barracks, whose very name instilled fear into residents throughout the northeast. The death toll at Giwa was staggering, even by the standards of Nigeria's troubled prison system. (See below.) Lastly, the dissident bandleader Fela Kuti, a former Kirikiri inmate

and an expert on the conditions in Nigeria's various jails, described his time in Maiduguri's main prison as tied with Benin's as the worst. Malam Abba, M.A., and others recalled him describing his escape from Giwa as part of the March 2014 attack on the facility.

20   the seat of a motorbike that trailed a safe distance: Interview with M.A.

20   Short and stocky with front teeth that jutted over his lower lip: Interviews with M.A. and N.A.

20   a red cap he called No Sulhu, or No Reconciliation: Ibid.

20   He was born Abba Mustapha: Aminu Abubakar, "Boko Haram's Leader Abubakar Shekau Has Been Injured in an Airstrike," Agence France-Presse, May 2017.

20   In three years and fourteen days: Ibid.

20   "Allah says the best martyrs are those who fight on the front": The authors reviewed dozens of propaganda videos made by Boko Haram.

21   Malam Abba would gird his fighters: Ibid.

21   Years in an overcrowded Nigerian prison, where inmates died of cholera: Morgue workers described this assessment repeatedly to international press and organizations, including Amnesty International, in the May 2015 report *Stars on their Shoulders. Blood on their Hands: War Crimes Committed by the Nigerian Military*, https://www.amnesty.org/download/Documents/AFR4416572015 ENGLISH.PDF, pp. 63–64.

21   a mission to steal, not kidnap: Interviews with N.A., M.A., P.M., H.I. The revelation was also contained in N.A.'s diaries. All of the Chibok students the authors interviewed described the men saying they had come to steal a brickmaker and debating what to do with the girls they hadn't expected to find.

21   Boko Haram had a sympathizer on the campus: Interviews with N.A., M.A., H.I.

21   They had a flatbed Mercedes 911 truck waiting: Sesay, *Beneath the Tamarind Tree*, 66.

22   almost the last of Borno State's 1,947 schools left open: Nigeria's Ministry of Education said in a statement on March 5, 2014, that it would close all schools "located within the high security risk areas of the northeast geo-political zone."

22   torching at least 50 schools: Emily Gustafsson-Wright and Katie Smith, "Abducted Schoolgirls in Nigeria: Improving Education and Preventing Future Boko Haram Attacks" (Washington, DC: *Brookings Institution*, April 2014).

22   scores of young women were reported missing as well: Joe Brock, "Boko Haram, Taking to Hills, Seize Slave 'Brides,'" Reuters, November 17, 2013.

22   parents had debated whether to keep their children in school: Interviews with K.A., R.A., and Chibok parent Mkeki Ntakai.

22   The school's administration had decided the classrooms should remain open: N.A., K.A., R.A.; Chibok principal Ababe Kwambura also spoke about the deliberation.

22   The exam season would be over on May 15: Listed on the WAEC 2014 Timetable.

22   started to beat the twitching rhythm of "A Girgiza Kwankwaso": Interviews with N.A., S.A., H.I, P.M.

23   Saratu was Naomi's dearest friend: Ibid.

23   girls had run from their dorms in a panicked stampede: The rising anxiety on campus was described by Sarah Topol in "If We Run and They Kill Us, So Be It," *Medium Matter*, October 2014.

24   "I pray for success on my final exams": Interviews with N.A., S.A., H.I.

25   "Where is your brickmaking machine?": Interviews with N.A., H.I., M.A., P.M, and M.D.; and Naomi's diaries.

25   Malam Abba raised his voice: Ibid.

25   sacks of grain and rice and containers of cooking oil: Ibid.; and Topol, "If We Run and They Kill Us, So Be It."

25   "What is in your hostel?": Mentioned in Naomi's diaries.

25   They loaded it onto a truck: Interviews with N.A., M.A.

26   the men returned to their trucks to retrieve their gasoline: Ibid.

26   The fire ripped through the principal's office, the teachers' quarters: Diaries.

26   "Shift a little! . . .": The story of Boko Haram burning the school buildings is
      described in Naomi's diaries and was corroborated by M.A., P.M., H.I., S.A.

Chapter 3: The Perfume Seller

28   Nigeria's military had issued statements confirming that they had killed him:
      "Boko Haram: Five Times Wen Abubakar Shekau Don Die," BBC Pidgin, July
      30, 2019.

28   Sitting in front of a golden curtain: Boko Haram video published to YouTube,
      since removed but stored on author's hard drive.

28   "In the name of the almighty Allah, killing is my mission now!": Ibid.

28   prisoners running free struggled to restraighten their legs and had to be car-
      ried: This video, also posted on YouTube, then removed, can be found in Jacob
      Zenn's digital archive, UnmaskingBokoHaram.com, under "Claim of Giwa Bar-
      racks Raid," March 2014.

29   Intelligence officials argued over whether he was alive, or dead: In an August
      2013 statement, the Nigerian military said, "The recent video released on 13
      August, 20013 [sic] by the purported sect leader was dramatized by an imposter
      to hoodwink the sect members to continue with the terrorism and to deceive the
      undiscerning minds."

29   The US government had offered a $7 million reward: "Abubakar Shekau: Up to
      $7 Million Reward," Rewards for Justice, June 2012, https://rewardsforjustice.
      net/english/abubakar_shekau.html.

30   while gnawing on his miswak chewing stick: Shekau is rarely seen without
      a miswak because of its strong associations with the prophet Mohammad,
      who recommended using it in several Hadiths. According to Zacharias Pieri and
      Jacob Zenn, "Under the Black Flag in Borno: Experiences of Foot Soldiers and
      Civilians in Boko Haram's 'Caliphate,'" Journal of Modern African Studies 56,
      no. 4 (December 2018): 645–75, at 655, Boko Haram fighters use miswaks to add
      to their credibility as self-proclaimed Salafis.

30   "It was said that I was killed, but here I am": An August 2013 video from Boko
      Haram that is archived at UnmaskingBokoHaram.com.

30   Shekau had demanded that all of Nigeria . . . implement Sharia law as a pre-
      condition: Interviews with M.Z.; see also Joe Parkinson and Drew Hinshaw,
      "Freedom for the World's Most Famous Hostages Came at a Heavy Price," Wall
      Street Journal, December 24, 2017.

30   to find a woman who claimed to be his mother: Interviews with two senior Ni-
      gerian security officials, who spoke on condition of anonymity, in Abuja in April
      2013. Separately, reporters also visited a woman claiming to be Falmata Abuba-
      kar in the village of Shekau in July 2019. Her neighbors and an official at the
      local government insisted that she is Shekau's biological mother. She was also
      identified as such by Chikah Oduah, in a June 14, 2018, documentary for the
      Voice of America, "Mother of Boko Haram Leader Speaks for First Time." More-
      over, the details she gave of Shekau's life match information on the man and
      his upbringing that is not well known, such as the year he was born, 1973. But
      there is a dispute over whether this woman is an imposter. Nigeria's State Se-
      curity Service (SSS) file claims Shekau's real mother died in 2013. In his video
      sermons, Shekau himself has been characteristically cryptic on the subject. The
      evidence suggests she is who she claims to be.

30   "Even his own mother said he was insane": Drew Hinshaw, "Nigerian Mili-
      tant Makes a Name for Himself Through Terror," *Wall Street Journal*, May 13,
      2014.

31   It was 1980 when Abubakar Shekau's parents sent him to study: Fatima Abu-
      bakar, also see Oduah, "Mother of Boko Haram Leader Speaks for First Time,"
      VOA.

31   He was seven years old: Emeka Okereke, "From Obscurity to Global Visibility:
      Periscoping Abubakar Shekau," *Counter Terrorist Trends and Analysis* 6, no. 10
      (November 2014): 17–22. The authors interviewed Okereke in January 2020;
      he described seeing the SSS file on Shekau, which states he was born in March
      1973. This matches the purported mother's recollections.

31   Home had been a thatched-roof cottage: Reporter visit to the village of Shekau,
      July 2019.

31   His mother, Falmata, sold peanuts: Ibid.

31   The new parents had been amazed at how much Shekau could recite back: In-
      terview with Fatima Abubakar, July 2019. By all accounts, Shekau had a prodi-
      gious gift for language (see below).

31   Yet he was also troubled: Interview with Fatima Abubakar, corroborated by her
      neighbors. Authors also interviewed Fatima Akilu, former director of behavioral
      analysis in Nigeria's Office of the National Security Adviser, in May 2014 and
      October 2017, who described Shekau as a brilliant but troubled young man
      based on the agency's own information on his youth: "His childhood friends say
      he was a very unstable, erratic figure." For more, see Colin Freeman, "Meet the
      Former NHS Psychologist Trying to Get Inside the Mind of Boko Haram," *The
      Telegraph* (London), June 2, 2015.

31 . Nigeria declared public elementary school free. Marg Csapo, "Universal Pri-
      mary Education in Nigeria: Its Problems and Implications," *African Studies
      Review*, March 1983: 91–106.

31   hoped that even the poorest families . . . would enroll their boys and girls:
      UPE (universal public education), as the free grammar schooling was called,
      was introduced by the first government of military leader Olesegun Obasanjo
      (1976–1979). However, it was largely implemented under the presidency of
      Shehu Shagari (1979–1983). During that time, Nigeria recruited educational
      consultants from elsewhere in the British Commonwealth, including England,
      Pakistan, and India. For this book, the authors spoke with former inspectors
      who described witnessing an explosion of students into a national school system
      that was simply unprepared. We were interested by the notion that an unam-
      biguously good idea (universal education) could also lead to such unfortunate
      outcomes (overcrowded schools whose failures later lent power to Boko Haram's
      message).

31   the number of students in school went up tenfold in two years: Csapo, "Univer-
      sal Primary Education in Nigeria," 91.

32   The people of Maiduguri liked to say: The authors conducted extensive interviews
      with Maiduguri residents during more than a dozen trips between 2014 and
      2020, including the contributions of our own Maiduguri-born reporter, Gbenga
      Akingbule.

32   home to just eighty-eight thousand people in 1960: *World Urbanization Pros-
      pects* (New York: United Nations Department of Economic and Social Affairs,
      2018).

32   its population had more than tripled: Ibid.

32   its commercial boulevard, Babban Layi—Broadway—buzzed: Jimeh Saleh,
      "How Boko Haram Attacks Have Changed the Maiduguri Where I Grew Up,"
      BBC News, May 1, 2012.

32 drifted into the jail cell of the imprisoned Nigerian band leader . . . Fela Kuti: The twenty-two-minute single "Akunakuna, Senior Brother of Perambulator," released in 1983, is based on a drumbeat Fela heard through his cell window while imprisoned in Maiduguri. Sola Olorunyomi, *Afrobeat: Fela and the Imagined Continent* (Lawrenceville, NJ: Africa World Press, 2002).

33 the preacher was shot by the police, who reportedly kept his ashes in a bottle: Lawan Danjuma Adamu, "Maitatsine: 30 Years After Kano's Most Deadly Violence," *Sunday Trust*, December 2010.

33 There are seventy-seven thousand words in the Quran: Quran Statistics, http://quranstatistics.weebly.com.

33 Shekau, a gifted student, memorized them quickly: Interviews with Fatima Abubakar, and in July 2019 with Kellumi Guni, a student in Shekau's Al-Majiri. Also see BBC News, "Nigeria's Boko Haram Leader Abubakar Shekau in Profile," May 9, 2014.

33 enunciating Quranic verses for alms or rice: Interview with Kellumi Guni. In interviews, M.Z. also spoke of Shekau's childhood as an Al-Majiri student. See also Oduah, "Mother of Boko Haram Leader Speaks."

33 He earned spare change sweeping the streets of the city's Government Reserved Area: Adam Nossiter, "A Jihadist's Face Taunts Nigeria From the Shadows," *New York Times*, May 18, 2014.

33 learning to speak Arabic, Hausa, Fulani, and English: Interview with Kellumi Guni. Also Okereke, "From Obscurity to Global Visibility," 18.

33 But he was also bursting with anger, quick to pick fights: Nossiter, "A Jihadist's Face."

34 his teacher deemed him his most troublesome student: Ibid.

34 In 1991, when he was eighteen, Shekau was expelled: Author interview with Okereke, January 2020; Nossiter, "A Jihadist's Face." Shekau may have moved in 1990, per International Crisis Group, "Curbing Violence in Nigeria (II): The Boko Haram Insurgency," April 2014, p. 19.

34 Nigeria sits on trillions of dollars' worth of oil and natural gas: Nigeria is well known to be Africa's largest crude-oil exporter, but its natural gas reserves are almost unfathomably large. The Nigerian Department of Petroleum Resources estimates that two hundred trillion cubic feet of natural gas lie below the surface.

34 "Nigeria's problem is not money, but how to spend it": Collins Olayinka, "At 80 Gowon Explains—'Nigeria's Problem Is Not Money, but How to Spend It,'" *Guardian Nigeria*, October 2014.

34 The central bank had been so thoroughly looted that the currency, the naira, once stronger than the dollar, was now worth a penny: Howard French, "Democracy to Despotism: Nigeria, in Free Fall, Seethes Under General," *New York Times*, April 4, 1998.

34 Prices for food were doubling every three months: The food inflation rate reached 38 percent in 1996, per Nigeria's National Bureau of Statistics, https://www.nigerianstat.gov.ng.

34 Blackouts lasted for weeks: Tim Sullivan, "African Politics: Nigeria Reborn?" Associated Press, October 19, 1998.

34 the middle class fled through airports where armed robbers swarmed planes on the runway: "Criminal Acts Against Civil Aviation," US Department of Transportation, 1999, p. 38; and "Criminal Acts against Civil Aviation," US Department of Transportation, 1993, p. 49.

34 had two decades later become the thirteenth poorest: French, "Nigeria in Free Fall."

34 By day, the expelled Shekau would walk downtown: Interviews with M.Z.

34 Diesel engines growled at downtown's main commercial plaza, Monday Market: The authors visited Monday Market numerous times and interviewed shopkeepers and shoppers.

34 shopkeepers nicknamed Shekau Kauwa Dima: M.Z.

34 Years later, he would dispatch several bombs into Monday Market: Gbenga Akingbule and Drew Hinshaw, "In Nigeria, a Market Braves Militant Attacks," *Wall Street Journal*, November 20, 2015.

35 Dressed only in white to signify moral purity: Nossiter, "Jihadist's Face Taunts Nigeria," *New York Times*. Videos from this time, archived on Unmasking BokoHaram.com, also show Shekau in white.

35 the Quran, a book he sometimes balanced on the handlebars of his motor scooter: Drew Hinshaw, "Nigerian Militant Makes a Name for Himself Through Terror," *Wall Street Journal*, May 13, 2014.

35 He would often stop by a nearby technical college: Interviews with Okereke and Z.M. International Crisis Group, "Curbing Violence in Nigeria (II): The Boko Haram Insurgency," April 2014, p. 19.

35 Lifting a Quran from the shelves, he would loudly recite passages: Interview with Z.M.

35 At night, Shekau would take his fundamentalist message to Maiduguri's poorest corner: Nossiter, "Jihadist's Face Taunts Nigeria." Descriptions of his preaching were also described by Z.M.

35 Railway Quarters, a shantytown built along abandoned train tracks, trapping the unwanted in a chockablock grid of tin roofs: Our colleague Gbenga Akingbule, grew up in Maiduguri and is familiar with the neighborhood. The authors also visited the area and spoke to other residents to confirm the description.

35 Maiduguri's population had doubled in the two decades since Shekau had moved to the city: "World Urbanization Prospects," United Nations Department of Economic and Social Affairs, 2018.

35 In its unlit streets, organized gangs ran prostitution rackets: Abdulkareem Haruna, "Nigeria: Why We're Demolishing 47 'Hotels' in Maiduguri—Attorney General," *Premium Times* (Abuja), June 1, 2018.

35 By the early 2000s, Nigeria had become a democracy, but the voting was rigged: *Nigeria's 2003 Elections: The Unacknowledged Violence*, Human Rights Watch, June 2004, https://www.hrw.org/report/2004/06/01/nigerias-2003-elections/unacknowledged-violence.

35 Typhoid and measles festered: Paul Francis and James Adewuyi Akinwumi, *State, Community, and Local Development in Nigeria*, World Bank Publications, 1996, p. 41.

35 a cholera epidemic . . . killed more than a thousand people: "Cholera Epidemiology and Response Factsheet Nigeria," UNICEF, 2014, https://www.unicef.org/cholera/files/UNICEF-Factsheet-Nigeria-EN-FINAL.pdf.

35 Mohammad Yusuf was a bright and charismatic village-born theology student: Abeeb Olufemi Salaam, *Boko Haram, Anatomy of a Crisis*, E-International Relations, 2013, p. 47. Alexander Thurston notes in *Boko Haram: The History of an African Jihadist Movement* (Princeton, NJ: Princeton University Press, 2018), p. 34, that it is "grimly ironic" that Yusuf was born at exactly the time when Nigeria's civil war was ending and an oil boom was approaching.

36 Public schools were the root of the country's ills, he argued: For more details on Yusuf's beliefs, see James J. Hentz and Hussein Solomon, *Understanding Boko Haram* (London: Routledge, 2017). Also see: Mike Smith, *Boko Haram: Inside Nigeria's Unholy War* (London: I.B. Tauris, 2015).

36 Maiduguri's poor and rich alike had begun to visit Yusuf: Shehu Sani, "Boko Haram: History, Ideas and Revolt," *Vanguard Nigeria* (Lagos).

36   political elite parked expensive SUVs next to the destitute cupping begging
      bowls: Ibid. The authors also interviewed Shehu Sani, January 2019.
36   hundreds would squeeze onto plastic mats: The authors reviewed videos from
      this time, some of which can still be found on YouTube: https://www.youtube
      .com/watch?v=xthVNq9OKD0&t=1750s and https://www.youtube.com/watch?v
      =f89PvcpWSRg.
36   Many were fresh arrivals to the city, villagers who converted: Christian Sei-
      gnobos, "Boko Haram and Lake Chad: An Extension or a Sanctuary?" *Afrique
      Contemporaine* 255, no. 3 (2015): 96.
36   disgust with the gambling, cocaine, and liquor: Andrew Walker, *"Eat the Heart
      of the Infidel": The Harrowing of Nigeria and the Rise of Boko Haram* (London:
      Hurst, 2018), 142–43. The authors also spoke with several graduates of the
      school, whose parties were indeed famous throughout Nigeria.
36   The congregation would sit enraptured by Yusuf: The authors reviewed videos
      from this time.
36   rattling off his worldview through a cheap sound system: Ibid.
36   The Earth was flat. . . . Evaporation was a lie: Joe Boyle, "Nigeria's 'Taliban'
      Enigma," BBC News, July 31, 2009.
36   It was blasphemy to sing the national anthem, a praise song to a secular state:
      Temitope B. Oriola, "Ideational Dimensions of the Boko Haram Phenomenon,"
      Studies in Conflict & Terrorism 41, no. 8 (2017): 606.
36   Shekau would sit next to him, face contorted in concentration. At times he
      would chime in: The authors reviewed videos from this time.
37   "Why would you love those people? Because they're good at football?": Élodie
      Apard, "The Words of Boko Haram: Understanding Speeches by Mohammed
      Yusuf and Abubakar Shekau," *De Boeck Supérieur* (Paris) 255, no. 4 (2015).
37   "When are we going to war?": Interviews with Z.M. and a Maiduguri-based as-
      sociate of the sect who spoke on condition of anonymity. The differences and
      continuity between the two are also addressed at greater length by Thurston in
      Boko Haram: The History of an African Jihadist Movement.
37   a month when the global financial crisis was fueling double-digit inflation and
      spiking food prices: Food prices rose 13 percent that month, and between May
      and October of that year, food prices had doubled. Nigeria's CPI: Food and Non
      Alcoholic Beverage Change, Nigeria's National Bureau of Statistics, https://
      www.nigerianstat.gov.ng.
37   police stopped a convoy of Boko Haram members on their way to a funeral: Mu-
      stapha Isah Kwaru and Ahmad Salkida, "Funeral Procession Tragedy: Police
      Shoot 17 in Maiduguri," *Daily Trust* (Abuja), June 12, 2009.
37   Boko Haram's most radical members had begun to stockpile weapons: Alexan-
      der Thurston, "Five Myths about Boko Haram," *Lawfare*, January 2018, https://
      www.lawfareblog.com/five-myths-about-boko-haram.
37   sandbagged commune outside Maiduguri that the local press dubbed "Afghan-
      istan": Bulama Bukarti, *The Origins of Boko Haram—And Why It Matters*,
      Hudson Institute, January 2020, https://www.hudson.org/research/15608-the
      -origins-of-boko-haram-and-why-it-matters. This commune has been described
      by other scholars as an al Qaeda–financed training camp, but Bukarti's work
      questions that thesis.
37   Another group had tried to board a flight to Afghanistan to fight alongside the
      Taliban: Authors interviewed imprisoned members of Boko Haram in Kuje Pris-
      on outside Abuja in August 2015.
37   In makeshift home laboratories, Yusuf's followers had been concocting fire-
      bombs, experiments that seemed to dare the authorities to respond: David Smith,
      "Nigerian Forces Storm Militant Islamist Mosque," *Guardian*, July 30, 2009.

37   Operation Flush II, "to ensure the security and lives of the people in the hinter-
      lands": Thurston, *Boko Haram*, 132–33.

37   Shouts escalated into a gunfight and when the shooting stopped, seventeen men
      were injured: Salkida, "Funeral Procession Tragedy."

37   Days later, hundreds of militants swarmed police stations, firing Kalashnikovs:
      Njadvara Musa, *Guardian Nigeria*, cited in Steve Coll's "Boko Haram," *New
      Yorker*, August 2, 2009.

37   Officers watched comrades die and retaliated in blind anger: Ibid. See also Mu-
      hammad Wahab, "Nigeria: Rights Worker Says Civilians Being Killed," Asso-
      ciated Press, July 30, 2009. Also see: *Spiraling Violence: Boko Haram Attacks
      and Security Force Abuses in Nigeria*, Human Rights Watch, October 2012, pp.
      58–71, https://www.hrw.org/report/2012/10/11/spiraling-violence/boko-haram-
      attacks-and-security-force-abuses-nigeria.

37   Troops fired mortar shells into the clinic and living quarters at Boko Haram's
      headquarters with mortars: David Smith, "Nigerian Forces Storm Militant
      Islamist Mosque," *Guardian*, July 30, 2009.

38   the local Red Cross alone removed nearly eight hundred bodies: Ibrahim Mshe-
      lizza, "More than 700 Killed in Nigeria Clashes: Red Cross," Reuters, August 2,
      2009.

38   police . . . paraded him shirtless before a crowd of jeering spectators: The in-
      terrogation video can still be accessed on YouTube: "Boko Haram leader Mo-
      hammed Yusuf Interrogation before his execution by Nigerian security agents,"
      SaharaTV, August 2009, https://www.youtube.com/watch?v=ePpUvfTXY7w.

38   Shot in the thigh during the Maiduguri street battles, Shekau recuperated in a
      safehouse: Interviews with Z.M. The authors also spoke with one of the inhabi-
      tants of this safehouse, who spoke on condition of anonymity, in May 2020.

38   Yusuf's most die-hard supporters and an army of new recruits: Hentz and Solo-
      mon, *Understanding Boko Haram*. The prevailing motivations for Boko Haram
      membership are complex, have changed over time, and are beyond the scope
      of this book. Commonly cited drivers include: financial reward, coercion and
      threats, frustration with Nigeria's status quo, religious fervor, and a desire for
      protection from the military. For more, see *Motivations and Empty Promises:
      Voices of Former Boko Haram Combatants and Nigerian Youth*, Mercy Corps,
      April 2016, https://www.mercycorps.org/sites/default/files/2019–11/Motivations
      %20and%20Empty%20Promises_Mercy%20Corps_Full%20Report_0.pdf.

38   Shekau became Nigeria's most wanted man: Yvonne Ndege, "Boko Haram 'Most
      Wanted' List," *Al Jazeera*, November 24, 2012.

38   hideouts in the Sambisa Forest: The authors reviewed aerial photographs of the
      Sambisa Forest from photojournalists aboard military flights over the region.
      Some of the US drone footage of the area was also described by officials who re-
      viewed it, speaking on condition of anonymity. See also: Azeez Olaniyan, "Once
      Upon a Game Reserve: Sambisa and the Tragedy of a Forested Landscape,"
      Arcadia no. 2 (Spring 2018), Environment & Society, Rachel Carson Center for
      Environment and Society, doi.org/10.5282/rcc/8176.

38   Parts of Sambisa had once been a game reserve during colonial times: Ibid.

39   Boko Haram sent car bombs into churches during Christmas services: Jon
      Gambrell, "Christmas Attacks in Nigeria by Muslim Sect Kill 39," Associated
      Press, December 25, 2011.

39   suicide bombers into crowded markets during morning traffic: Akingbule and
      Hinshaw, "Market Braves Militant Attacks."

39   Shekau's followers threw grenades into beer gardens at happy hour: Ioannis
      Mantzikos, "Boko Haram Attacks in Nigeria and Neighbouring Countries: A
      Chronology of Attacks," *Terrorism Research Initiative* 8, no. 6 (December 2014).

39   brutalized civilians for acts of collaboration as slight as selling a car to an off-
        duty policeman: A patient at the Federal Neuropsychiatric Hospital in Maidu-
        guri described this incident in an interview with the authors in May 2014.

39   systematically attack schools, the institution they most blamed for their coun-
        try's sinfulness: "Nigeria: Boko Haram Targeting Schools," Human Rights
        Watch, March 7, 2012, https://www.hrw.org/news/2012/03/07/nigeria-boko-haram
        -targeting-schools.

39   Gunmen ran through campuses, executing teachers and setting classrooms on
        fire: Ibid. See also, "Nigeria: How Boko Haram Killed 600 Teachers, Displaced
        19,000," *Vanguard Nigeria* (Lagos), October 5, 2015.

39   "Teachers who teach Western education? We will kill them!": Monica Mark,
        "Boko Haram Leader Calls for More Schools Attacks after Dorm Killings,"
        Guardian, July 14, 2013.

39   relied on civilian vigilantes who lashed out in rage and paranoia: *Stars on their
        Shoulders. Blood on their Hands: War Crimes Committed by the Nigerian Mil-
        itary*, Amnesty International, https://www.amnesty.org/download/Documents
        /AFR4416572015ENGLISH.PDF.

39   Soldiers barged into mosques and homes and rounded up thousands: "Nigeria:
        Senior Members of Military Must Be Investigated for War Crimes," Amnesty
        International, June 3, 2015, https://www.amnesty.org/en/latest/news/2015/06
        /nigeria-senior-members-of-military-must-be-investigated-for-war-crimes.

39   Giwa Barracks, a place that soldiers called the die house: Ibid. See also, Adam
        Nossiter, "Bodies Pour In as Nigeria Hunts for Islamists," *New York Times*, May
        7, 2013. Nigeria's military at the time denied that more than a handful of in-
        mates had died in their jails. Authors secured morgue records in Maiduguri in
        2014, which showed hundreds of bodies delivered from Giwa daily. On June 24,
        2013, soldiers carted out 115 bodies, according to these records; on June 29, 263
        bodies.

39   Prisoners fought over plastic bags of water: Drew Hinshaw, "Hundreds Killed in
        Jails Swelling with Islamist Suspects," *Wall Street Journal*, October 15, 2013.
        See also Nossiter, "Bodies Pour In"; and "They Didn't Know if I Was Alive or
        Dead," Human Rights Watch, September 2019, p. 1.

39   Desperate family members from neighborhoods throughout the city came to the
        jail: Ibid.

39   had charged fewer than six people for terrorism: Author interview with Fatima
        Akilu, then director of behavioral analysis at the Office of the National Secu-
        rity Adviser, in May 2014. See also: *Nigeria: Trapped in the Cycle of Violence*,
        Amnesty International, November 2012, https://www.amnesty.org/download
        /Documents/16000/afr440432012en.pdf, p. 34: "Of those detainees accused of
        being members of Boko Haram who have been charged with a criminal offence
        and brought to court since 2009, very few have had their cases heard."

39   police . . . arrested one of Shekau's wives: Jacob Zenn and Elizabeth Pearson,
        "Women, Gender, and the Evolving Tactics of Boko Haram," *Journal of Terror-
        ism Research* 5, no. 1 (2014): 48.

40   reports of kidnappings and forced marriages: Joe Brock, "Boko Haram, Taking
        to Hills, Seize Slave 'Brides,'" Reuters, November 17, 2013.

40   Girls as young as twelve began to disappear: Ibid. See also: "Boko Haram Ab-
        ducts Women, Recruits Children," Human Rights Watch, November 29, 2013,
        https://www.hrw.org/news/2013/11/29/nigeria-boko-haram-abducts-women
        -recruits-children.

40   "just wait and see what will happen to your own women": Mia Bloom and Hilary
        Matfess, "Women as Symbols and Swords in Boko Haram's Terror," *PRISM* 6,
        no. 1 (March 1, 2016): 113.

Chapter 4: Kolo and Naomi

41   dozens of gunmen: The journey into the forest was described in interviews with N.A., S.A., H.I., M.A., P.M., M.D., and Hannatu Stephens (hereafter referred to as H.S.) and is documented in the diaries.

42   Some were tearing strips off their school uniform: Interviews with Chibok students and parents, including discussions in May 2014 with some of the fathers and uncles who assembled their own convoy of motorbikes to follow Boko Haram into the forest in order to rescue their daughters. They turned back, unsuccessful, after being told by villagers they would be killed if they went any farther.

42   One by one, they jumped: Some fifty-three students escaped during the course of the kidnapping, according to the Chibok Association and official government records. Many of them jumped from the trucks, while others escaped after asking to be allowed to take a toilet break as the convoy paused en route. We spoke with one of these students at length, Kauna Bitrus, who escaped and was ultimately sponsored to study in the US. In 2020 she graduated from a community college in Maine.

43   Fire had gutted the pastel green exam hall: Interviews with K.A. and Rifkatu Ayuba (hereafter referred to as R.A.) and photos of the school after the attack.

44   another day when the temperature would top 105 degrees: Historical weather reports from Chibok on April 15, 2014, from CustomWeather.com.

45   From the day Naomi was born: Interviews with K.A., N.A., and R.A.

45   Nearly a quarter of children die: Gladys O. Aigbe and Ajibola E. Zannu, "Differentials in Infant and Child Mortality Rates in Nigeria: Evidence from the Six Geopolitical Zones," *International Journal of Humanities and Social Science* 2, no. 16 (Special Issue, August 2012), http://www.ijhssnet.com/journals /Vol_2_No_16_Special_Issue_August_2012/22.pdf. Data from the 1999 and 2008 Nigerian Demographic Health Survey showed that child mortality had risen almost 30 percent in northeastern Nigeria in a decade, and although more recent data don't exist, experts say the numbers after the outbreak of the insurgency are likely to be considerably higher.

45   in a sun-dried mudbrick home: Pictures of the Adamu house sent to the authors.

45   listening to the British Broadcasting Corporation's Hausa service: The service broadcasts to Hausa-speaking communities throughout West Africa. It was the corporation's first African-language service, launched in 1957, three years before Nigeria declared independence.

46   Sunday worship was a six-hour: Interviews with K.A., R.A., N.A., and pastor Enoch Mark, pastor of EYN (Ekklesiyar Yan'uwa a Nigeria, Church of the Brethren). Pictures of services taken by Glenna Gordon in *LensCulture*. The authors spoke with Church of the Brethren officials in Jos, Nigeria, and in Washington, DC, in March 2018.

47   Nigeria's religious divide had been turning more bitter: In 1999, the governor of Nigeria's Zamfara State, Ahmed Sani Yerima, introduced Sharia law. Eleven other states in northern Nigeria with majority Muslim populations, including Borno, followed suit, sparking riots between Muslim and Christian communities in the cities of Kano and Kaduna.

47   Newspapers were full of stories: "Where Teenagers Die by Rat Poison," *Nation*, June 23, 2018.

48   The men passed around packets of biscuits and cans of Malta Guinness: Interviews with N.A. and H.I.

48   Then the vehicles came to a halt in a village: Interviews with N.A. and H.I. Many of the hostages were in different vehicles, so not all witnessed the killing of the elderly villager. Throughout 2014 Boko Haram torched churches, village homes, and most infamously, the Buni Yadi school dormitory, where fifty-nine boys were burned alive.

Chapter 5: The Caretaker of an Orphan

50   the birthplace of Boko Haram: For more on Maiduguri's role as the birthplace of Boko Haram, see Abdulbasit Kassim and Michael Nwankpa, eds., *The Boko Haram Reader: From Nigerian Preachers to the Islamic State* (New York: Oxford University Press, 2018); Jacob Zenn, *Unmasking Boko Haram: Exploring Global Jihad in Nigeria* (Boulder, CO: Lynne Reinner, 2020); Brandon Kendhammer and Carmen McCain, *Boko Haram* (Athens, OH: Ohio University Press, 2018); and Alexander Thurston, *Boko Haram: The History of an African Jihadist Movement* (Princeton, NJ: Princeton University Press, 2018).

50   the hometown he called Yerwa, Blessed Land, a center of Islamic scholarship: Babagana Abubakar, "Origin and Meaning of Maiduguri," Academia, 2017, https://www.academia.edu/34560246/Origin_and_Meaning_of_Maiduguri.

50   one of the few people trying to talk to the Boko Haram leader: Interviews with Z.M.

51   Soldiers in bulletproof vests and helmets rested their rifles on sandbags: During reporting trips to Maiduguri between 2014 and 2020, the authors saw the extensive garrisoning of the airport.

51   damage the maintenance staff called "alligator cracks": These are interconnecting cracks and potholes on the runway that look like alligator skin and constitute a serious safety hazard for flights into the airport. The Federal Airports Authority of Nigeria concluded that Maiduguri airport's cracks had become so bad that the airport needed to be closed for resurfacing in 2005 and again in 2011.

51   Locals claimed that the landing strip had been lengthened: Walker, *"Eat the Heart of the Infidel,"* 142.

51   the only way for civilians to fly in was aboard an aid-agency propeller plane: The airport was shut to commercial traffic from early December 2013 until June 2015 after Boko Haram attacked the adjacent Nigerian Air Force base and destroyed three helicopters. See Ndahi Marama, "FG Re-opens Maiduguri Airport 18 Months after Closure," *Vanguard Nigeria* (Lagos), June 2015.

51   Shortly before takeoff, the lawyer's phone had received a message: Interviews with Z.M.

52   Former Nigerian president Olusegun Obasanjo called him "my son": Ibid.

52   representing the family of Boko Haram members who had been killed: Ibid.

52   The orphans there called him Barrister: Interviews with students and Future Prowess's longtime headmaster, Suleiman Aliyu, during an author visit in April 2019.

52   "Myself and the caretaker of an orphan will be in Paradise like this": The quote is taken from the Hadith collection Sahih al-Bukhari, which can be found online at https://sunnah.com/adab/7.

52   His students came from *both* sides of conflict: Interviews with Z.M.

53   But to the Boko Haram sympathizers on campus, he called it Islamiyah school: Ibid.

53   The guest was a Swiss diplomat with a pouf of red hair: Interviews with Z.M., who also provided photographs of the visit.

53   Behind the iron gates of a barely marked outdoor grill in Abuja: The authors visited this restaurant many times for interviews (and sustenance) between 2014 and 2020.

53   On first sight, the Swiss diplomat looked unremarkable: To build a fuller picture of the Swiss diplomat, the authors spoke to his colleagues in Switzerland, Western diplomats and aid workers based in Abuja, terrorism analysts, and Nigerian security officials whom he worked with regularly.

54   they had started calling him "the White Nigerian": Ibid.

54   His employers back in Bern ran a global program: Interviews with senior Swiss
     officials, including former president Micheline Calmy-Rey in November 2017
     and January 2020, and a former head of the Human Security Division, Thomas
     Greminger, in November 2017.
54   a tier-one target: Interviews with Swiss officials, who spoke on condition of an-
     onymity.
54   the two men began trading texts: Interviews with Z.M.
55   It didn't seem like a big moment: Ibid., and interviews with Swiss officials.

Chapter 6: @Oby

56   the thirtieth in a matter of hours: Twitter history sourced from the archive of
     @ObyEzeks and interviews with Oby Ezekwesili in Abuja in May 2014, Novem-
     ber 2017, and September 2019.
56   She had been walking across the lobby of her air-conditioned office: Ibid.
57   she had built up one of the country's most impressive CVs: Ibid. An abridged
     biography can be found on the World Bank Group website at https://blogs.world
     bank.org/team/obiageli-ezekwesili.
57   Madame Due Process: She earned the moniker during her role as special assis-
     tant to the Nigerian president on budget monitoring and in the Price Intelli-
     gence Unit from 2003 to 2005, where she instituted reforms to procurement and
     contracting. See: Folake Soetan, "'Madam Due Process': Oby Ezekwesili, World
     Bank, Former VP for Africa," *Ventures Africa* blog, September 22, 2012.
57   the European Union called it "the worst they had ever seen anywhere in the
     world": Human Rights Watch, *Criminal Politics: Violence, "Godfathers," and
     Corruption in Nigeria*, October 2007, https://www.hrw.org/reports/2007/nigeria
     1007/nigeria1007web.pdf.
57   a government town, consciously modeled to be West Africa's answer to Brasília
     or Washington, DC: An irony of capital cities is that they are often designed
     by foreigners. Just as France-born Pierre Charles L'Enfant designed Wash-
     ington, DC, Nigeria's Federal Capital Development Authority commissioned
     an American firm, WRMT, known for its work on the DC metro line, to help
     create the master plan for Abuja. See: Nnamdi Elleh, *Abuja: The Single Most
     Ambitious Urban Design Project of the 20th Century* (Weimar: VDG [Verlag und
     Datenbank für Geisteswissenschaften], 2001); and "The Founding of Abuja,
     Nigeria," *Building the World* blog, University of Massachusetts–Boston, http://
     blogs.umb.edu/buildingtheworld/founding-of-new-cities/the-founding-of-abuja
     -nigeria.
57   located almost exactly in the middle of the country to discourage a civil war or
     political unrest: The Aguda Panel tasked in the 1970s with finding a location
     for the new capital held security to be a chief requirement. The location chosen
     should not be "easily destroyed by a foreign enemy or in a civil war," including
     "local political disturbances and riots . . . and threats to Nigerian unity." See:
     Wale Adebanwi, "Abuja," in Simon Bekker and Göran Therborn, eds., *Capital
     Cities in Africa: Power and Powerlessness* (Cape Town, South Africa: HSRC [Hu-
     man Sciences Research Council] Press, 2012).
57   SUPPORT THE PRESIDENT, read pro-government billboards all over town: The au-
     thors saw the billboards on reporting trips to Nigeria in 2014.
57   The replies to her musings could be livid: See Oby Ezekwesili, Twitter archive
     @ObyEzeks.
58   "Validated by God and my Dad who taught me to never dignify nonsense": Ibid.
58   "She is literally always on Twitter": Drew Hinshaw, "Moved by Girls' Plight,
     Nigeria's Former Education Minister Helped Spark Call to Free Them, Her Son
     Recalls," *Wall Street Journal*, May 9, 2014.

58   If law enforcement doesn't find the missing within seventy-two hours: Julia Jacobo, "Why the First 72 Hours in a Missing-Persons Investigation Are the Most Critical, According to Criminology Experts," ABC News, October 8, 2018; and Hannah Moore, "Why the First 72 Hours Are Vital When Finding a Missing Person," BBC News, April 4, 2016.

58   She laced her posts with hashtags to see if one would catch fire: Oby Ezekwesili, @ObyEzeks, Twitter, April 15, 2014.

59   her son knocked at the door: Interviews with Oby and Chine Ezekwesili.

59   Nigeria's security chiefs filed into a wood-paneled meeting hall: Interviews with senior Nigerian officials who spoke on condition of anonymity. Also see Goodluck Jonathan, *My Transition Hours* (Kingwood, TX: Ezekiel Books, 2018), 28–36.

59   The State Department had called: Ibid.; and interviews with senior US State Department officials who spoke on condition of anonymity.

60   eventually earning a doctorate: Questions about the legitimacy of Jonathan's PhD became a campaign topic during the 2015 election after opponents accused him of forging his qualifications. The University of Port Harcourt later confirmed in a statement that Jonathan had completed his PhD in 1995.

60   with a 215-page thesis contrasting six different species of amphibious crustaceans: Goodluck Jonathan, "Ecology of *Parhyale hawaiensis* Dana 1853 and Other Intertidal Talitroids (Crustacean Amphipoda) in the Bonny and New Calabar Rivers Estuary" (PhD dissertation, University of Port Harcourt, Nigeria).

60   For most of his life, military dictators had governed his country: For a history of Nigerian military rule, see: Max Siollun, *Soldiers of Fortune: A History of Nigeria (1983–1993)* (Abuja, Nigeria: Cassava Republic Press, 2014); and Max Siollun, *Nigeria's Soldiers of Fortune: The Abacha and Obasanjo Years* (London: Hurst, 2019).

60   "I have come to preach love, not hate": Part of a speech Jonathan gave at Abuja's Eagle Square, September 18, 2010.

60   He was skeptical that an army of poor, unemployed youth: Author interview with President Jonathan, February 2015.

60   The president and his advisers had begun to coin a dark conspiracy theory: Ibid. and interviews with senior Nigerian officials who spoke on condition of anonymity.

61   dethroned South Africa as the continent's largest economy: Drew Hinshaw and Patrick McGroarty, "Nigeria's Economy Surpasses South Africa's in Size," *Wall Street Journal*, April 6, 2014.

61   Helicopters were circling: Author reporting trip to Abuja, May 2014.

61   One by one they assured him: The authors spoke in November 2017 to senior Nigerian officials present at the meeting in Abuja.

61   The kidnapping was a hoax: Ibid.

61   His office replied to Secretary Kerry with an assertion: Ibid., corroborated by interviews with senior US State Department officials.

61   "Anyone of us that belong to the Elite Class in Nigeria": Twitter archive @ObyEzeks.

62   The topic of her session was "Bring Back the Books": Interviews with Oby Ezekwesili.

62   But when Oby took the microphone, she was in no mood to celebrate: See part of the speech on YouTube, "World Book Capital 2014: Ezekwesili Advocates a Shift from Dependence on Oil," Channels Television, April 23, 2014.

62   Ibrahim Abdullahi, a lawyer with a penchant for books and bow ties: Interview with Ibrahim Abdullahi, January 2020.

63   he thumbed out a tweet: Ibrahim Abdullahi, @Abu_Aaid, Twitter, April 23, 2014.

63   "Use the hashtag #BringBackOurGirls to keep the momentum": Oby Eze-
      kwesili, @ObyEzeks, Twitter, April 23, 2014.
63   Nigerians began to retweet it in the thousands: Twitter analytics from Storyful,
      a social media intelligence company.
64   the small crowd included a few of the missing students' parents: The authors
      spoke to many attendees of the protest, including Oby Ezekwesili, several Chi-
      bok mothers, and representatives of the Chibok Association. We also reviewed
      television news footage of the march.
64   There was no water in the fountain: The authors, visiting Abuja since 2010,
      have never seen water in the fountain.
64   the Abuja Hilton hotel, whose $1,000-a-night tenth floor: Rooms at the Hilton
      ranged from $300 to more than $3,000 in 2015, according to the hotel's website.
      The tenth floor's King Executive Ambassador Rooms were priced at $1,230 at
      the time of writing.
64   it was the *New York Times* correspondent's hotel of choice: The authors enjoyed
      occasional drinks and buffet dinners at the Hilton's Bukka restaurant with the
      Times's West Africa Bureau chief, Adam Nossiter.
64   Enoch Mark, whose daughter Monica was among the missing: Ariel Zirulnick,
      "Nigerians Lambast Government Inaction as Hundreds of School Girls Remain
      Missing," *Christian Science Monitor*, May 1, 2014.
64   "If this happened anywhere else in the world": "Hundreds March over Nigeria
      Schoolgirl Kidnappings," Agence France-Presse, April 30, 2014.
64   a yellow placard with four pieces of A4 paper affixed: The placards are visible in
      photos and television footage of the march.
65   One of the fathers fell at Oby's feet: Interviews with Oby Ezekwesili.
65   "Hundreds of protesters marched in the streets": Adam Nossiter, "Nigerians
      Hold Protests over Mass Abductions," *New York Times*, April 30, 2014.

Chapter 7: "Timbuktu"

66   A tall, muscular figure: Aliyu's address to the hostages was described in inter-
      views with N.A. and more than a dozen of the other Chibok hostages, as well as
      in extracts of the diaries. The authors also reviewed footage of "Aliyu," or "Ali-
      yu al-Gombawee," archived by Boko Haram analyst Jacob Zenn at Unmasking
      BokoHaram.com, in which he rails against the Nigerian state using similar
      language.
67   He had fought with al Qaeda in the Sahara: Interviews with US intelligence
      officials briefed on Aliyu's background, who spoke on condition of anonymity.
68   The Chibok girls had been driven along dirt tracks: The story of the journey to
      "Timbuktu" was relayed by N.A., H.I., P.M., and M.A. and referenced in extracts
      of the diaries. Boko Haram had split the hostages onto different trucks, mean-
      ing some of the girls' recollections were slightly different.
68   Dawn light revealed the full scale of the camp: Hostages' descriptions of the
      camp were cross-referenced with images from the first proof-of-life video, shot
      very close to this location, which was never made public but was shared with
      the authors by Ahmad Salkida in July 2020.
68   corralled their victims under a vast tamarind tree: The hostages' time impris-
      oned under the Tamarind branches is also described in Isha Sesay's *Beneath the
      Tamarind Tree* (New York: Dey Street, 2019), 170–201.
69   the vast sprawl of wilderness to the north of Chibok: Azeez Olaniyan, "Once
      Upon a Game Reserve."
69   across the border with Cameroon: The authors visited Maroua on a reporting
      trip to northern Cameroon, July 2016.

69   Gunmen riding motorcycles and pickups could launch deadly raids: Zenn, *Unmasking Boko Haram.*

69   kept their members, their weapons, and their hostages hidden from: Ibid. Moki Edwin Kindzeka, "Sambisa Forest: An Ideal Hiding Place for Boko Haram," VOA, May 24, 2016.

70   The camps housed makeshift mosques and Quranic schools: Interviews with N.A. and other Chibok hostages. Their portrait was consistent with author interviews from 2014 to 2020 of dozens of other male and female hostages who later escaped similar Boko Haram camps.

70   Diesel electric generators powered MacBook laptops: Ibid.; and the four-part series written by Salem Solomon, "Boko Haram: Terror Unmasked," Voice of America, News, February 2017, https://projects.voanews.com/boko-haram-terror -unmasked.

70   a training ground they called Timbuktu: Ibid.; Twitter archive @ContactSalkida, April 14, 2018; and Drew Hinshaw, "Timbuktu Training Site Shows Terrorists' Reach," *Wall Street Journal*, February 1, 2013.

70   For the first few days, the insurgency seemed unsure of what to do with its captives: Interviews with N.A., H.I., S.A., M.A., M.D., and P.M.

70   The only explanation for their kidnapping came from the stocky commander: Interviews with N.A. and M.A.

71   Many of the young women sobbed for hours each day: Interviews with N.A., H.I., S.A., M.A., M.D., and P.M. and extracts from diaries.

71   she removed the Nokia from her wrapper and turned it on: Interviews with N.A., who brought the battered Nokia, still tied together with rubber bands, to show the authors. Other Chibok survivors, including P.M. and H.I., confirmed that N.A. kept her cell phone and volunteered to hide others for classmates afraid they would be caught.

Chapter 8: #BringBackOurGirls

72   Russell Simmons was finishing his morning yoga routine: Interview with Russell Simmons, April 2020. Russell Simmons, @UncleRush: "Be flexible. Practice smiling and breathing in difficult positions," Instagram, April 30, 2014.

72   in a cramped Manhattan dorm room: See Stacy Gueraseva, *Def Jam, Inc: Russell Simmons, Rick Rubin, and the Extraordinary Story of the World's Most Influential Hip-Hop Label* (New York: One World/Ballantine, 2005).

72   advocated silence as the place "from which all creativity springs": Russell Simmons with Chris Morrow, *Success Through Stillness: Meditation Made Simple* (New York: Penguin, 2014).

72   updated them constantly from TweetDeck: Many of Simmons's 2014 tweets note that they were posted by TweetDeck. He also used SocialFlow, another third-party tool that claims to "optimize the delivery of messages on social networks."

73   that would falter years later after a dozen women accused him of sexual assault: Elias Leight, "Russell Simmons Sexual Assault Allegations: A Timeline," *Rolling Stone*, February 2018. Simmons vehemently denies the claims.

73   Simmons's tweets hardly ever mentioned foreign affairs: Twitter archive @UncleRush.

73   *Good intentions lead to good outcomes*, he thought: Interview with Russell Simmons.

73   as @UncleRush popped out an eighty-one-character tweet: Twitter archive @UncleRush, April 30, 2014.

73   A few minutes later, he was tweeting again to plug Mission: G-Rok: Ibid.

74   Within minutes of tweeting #BringBackOurGirls, thousands of likes and retweets began cascading: Though Simmons was far from the first or only West-

erner to tweet the hashtag, analytics produced by Storyful concluded that his tweet was pivotal in the hashtag's evolution and in attracting engagement via "influencers."

74   Common, weeks short of signing a record deal with Def Jam: Sean Thompson, "Common Inks New Deal with Def Jam," *Vibe*, June 4, 2014.

74   Young Jeezy joined in: Twitter archives @Common, @Sno, April 30, 2014.

74   "WHY DO WE HAVE TO BEG THE MEDIA": Kim Moore, @SoulRevision, Twitter, May 1, 2014.

74   "There'd already be a Lifetime movie in the works": SugarCane Black @ellejkaye, Twitter, May 2, 2014.

74   At 1:27 p.m. from her estate in Saddle River: Mary J. Blige, @maryjblige, Twitter, April 30, 2014.

74   The teenager in the photo was not even from Nigeria: James Estrin, "The Real Story about the Wrong Photos in #BringBackOurGirls," *New York Times*, May 8, 2014.

75   By the time the day closed in New York, sixty-two thousand people: Analytics from Storyful.

75   The founder of Twitter had seen: Interview with Jack Dorsey, June 2020.

75   an argument Jack and his business partner, Evan Williams, had had: Nick Bilton, *Hatching Twitter: A True Story of Money, Power, Friendship, and Betrayal* (New York: Portfolio, 2014).

75   "If there's a fire on the corner of the street": Ibid.

76   a stock market valuation of $25 billion: Telis Demos, Chris Dieterich, and Yoree Koh, "Twitter IPO: Relief, Riches and a $25 Billion Finish," *Wall Street Journal*, November 7, 2013.

76   "It's as if they drove a clown car into a gold mine and fell in": Bilton, *Hatching Twitter*.

76   millions had tweeted a YouTube video tagged "Stop Kony": "KONY 2012: Invisible Children," YouTube, https://www.youtube.com/watch?v=Y4MnpzG5Sqc.

76   YouTube's first to top a hundred million views: Todd Wasserman, "'KONY 2012' Tops 100 Million Views, Becomes the Most Viral Video in History," *Mashable*, March 12, 2012.

76   Up to 40 percent of the entire world's population had watched: Will Richards, "Bob Geldof Says the Internet 'Breaking Down the World into Individualism' Means Another Live Aid Couldn't Happen Today," *NME* (*New Musical Express*), March 12, 2020.

76   charity release that became the fastest-selling pop single in history: Robert A. Bennett, "WHOEVER DREAMED THAT UP?" *New York Times*, December 29, 1985.

76   John Lennon, Joan Baez, and Jimi Hendrix all raised voices: The Scene discotheque on New York's West Forty-Sixth Street ran a series of benefit concerts for Biafra in 1968. John Lennon and Yoko Ono published a letter to Queen Elizabeth II returning Lennon's MBE because of the "imperialist wars" in Vietnam and Biafra. See: Eunice Rojas and Lindsay Michie, *Sounds of Resistance: The Role of Music in Multicultural Activism* (Westport, CT: Praeger, 2013), 353.

76   After two million deaths, Biafra surrendered: See Dan Jacobs, *The Brutality of Nations* (New York: Knopf, 1987). The death toll is still disputed: Nigerian journalist Chika Oduah wrote in "Nigeria Needs to Start Talking about the Horrors of the Biafra War, Fifty Years On," *Quartz*, June 2017: "Nigeria's civil war is such a sensitive topic in the country, that the government has never given an official death toll. Quoted figures range from one million to as high as six million." For further (though Biafra-sympathetic) reading, see Chinua Achebe, *There was a Country* (New York: Penguin, 2012).

77   Singers Chris Brown and Wyclef Jean, actresses Alyssa Milano and Mia Farrow: Twitter archives @chrisbrown, @wyclef, @alyssa_milano, and @MiaFarrow.

77   "Celebrities, World Leaders Unite to #BringBackOurGirls": *Huffington Post*, May 9, 2014.

77   Abuja Hilton, the gigantic 667-room modernist hotel complex: Details on the hotel taken from Hilton.com plus dozens of reporter trips to the hotel for interviews by the pool or in the Piano Bar, including with oil executives, diplomats, military commanders, and intelligence officers. The authors also spent a night on the tenth floor engaged in a drinking game with a fuel smuggler, an arms trader, a BP executive, and an associate of Equatorial Guinea's Teodorin Obiang, the president's flamboyant son who famously bought one of Michael Jackson's crystal gloves for $80,000, while a police officer counted a large stack of money nearby.

78   Oby Ezekwesili's protests had swelled: Chinenye Ugonna and Amina Mohammed, "Chibok Schoolgirls: Nigerian Women Continue Protest Despite World Economic Forum," *Premium Times* (Abuja), May 7, 2014.

78   Journalists from major networks: The authors were present for these protests.

78   President Jonathan was increasingly convinced: Interviews with senior Nigerian officials who spoke on condition of anonymity; and Goodluck Jonathan, *My Transition Hours* (Kingwood, TX: Ezekiel Books, 2018).

78   Quietly, she was deploying other, less public strategies: Interviews with Oby Ezekwesili and e-mail exchanges with Clinton adviser Cheryl Mills, October 2019.

79   @HillaryClinton posted a tweet: Twitter archive @HillaryClinton, May 4, 2014.

79   Nancy Pelosi held a moment of silence: "Representatives Boehner, Pelosi, and Cantor on Nigerian Kidnappings," C-Span, May 9, 2014, https://www.c-span.org/video/?319260-4/house-leaders-nigerian-abductions&event=319260&play Event.

79   Inside the chamber, Miami congresswoman Frederica Wilson: Helen Andrews-Dyer, "Why the Congresswoman Wears Red," *Washington Post*, May 19, 2015.

79   "I abducted your girls," said Abubakar Shekau: Adam Nossiter, "Nigerian Islamist Leader Threatens to Sell Kidnapped Girls," *New York Times*, May 5, 2014. The full fifty-six-minute video was removed from YouTube for violating terms of service but is still available at Zenn's archive, UnmaskingBokoHaram.com.

79   Within minutes of Shekau's image beaming onto Twitter: Analytics from Storyful.

80   In London, hundreds of demonstrators: Another crowd swarmed the United Nations. Emma Howard, "Global Protests over Abduction of Nigerian Schoolgirls," *Guardian*, May 9, 2014.

80   In a letter to the White House: Peter Sullivan, "Women Senators Urge Further US Action on Nigerian Kidnappings," *The Hill*, May 6, 2014.

80   Conservative talk-radio host Rush Limbaugh: Lucy McCalmont, "Limbaugh Hits #BringBackOurGirls," *Politico*, May 8, 2014.

80   Senator Dianne Feinstein was demanding: Press release, "Feinstein Statement on Kidnapped Nigerian Schoolgirls," Feinstein.Senate.gov, May 7, 2014, https://www.feinstein.senate.gov/public/index.cfm/press-releases?ID=83EF1187-52F5-405A-B337-9DFE9AC44138). Other senators were openly calling for action, with the most hawkish comments coming from John McCain, who told the *Daily Beast*, "In a New York minute I would (send in troops), without permission of the host country. I wouldn't be waiting for some kind of permission from some guy named Goodluck Jonathan." See Josh Rogin, "McCain: Send U.S. Special Forces to Rescue Nigerian Girls," *Daily Beast*, May 14, 2014.

80   "Even if it hits a billion tweets": Interview with a senior US official present, who spoke on condition of anonymity, in Washington, September 2019.

80   held hourlong Chibok sessions in the underground complex: Ibid., along with two other senior US officials present for these conversations, who spoke on condition of anonymity.

81   the president's morning intelligence summary: The workings and structure of "the brief" were described in interviews with three members of the NSC in Washington, as well as two senior State Department officials and other senior officials, between 2017 and 2019.

81   Each department needed to reach into its capabilities and present assets it could offer: Ibid.

81   The Justice Department could send FBI agents with experience defusing hostage crises: Details of the ultimate deployment came from interviews with eight officials on the team or diplomats working with it, all of whom spoke on condition of anonymity. The authors also spoke with officials at the NSC and State Department present for the discussions.

81   Department of Defense, whose $614 billion discretionary budget for the preceding year: "Fiscal Year 2013 Budget of the United States Government," Office of Management and Budget, Budget.gov, https://www.govinfo.gov/content/pkg/BUDGET-2013-BUD/pdf/BUDGET-2013-BUD.pdf.

81   The US had satellites and propeller planes that could scan the Sambisa area: Interviews with NSC and IDAT (Interdisciplinary Assistance Team) officials, who spoke on condition of anonymity, in 2018 and 2019.

81   they could use the "workhorse," an RQ-4 Global Hawk: Ibid.

81   gunmetal gray with a bulbous head like a dolphin: "RQ-4 Global Hawk Fact Sheet," US Air Force, October 2014; and Mike Ball, "Workhorse: US Air Force Global Hawk Reaches 500-Mission Milestone," *Unmanned Systems News*, November 2015.

82   the US might be able to rotate in another drone, named the Predator: Jeff Seldin, "US Drone Flying from Chad in Search for Missing Nigerian Girls," Voice of America, Washington, DC, May 22, 2014.

82   debating the deployment of more than a quarter of a billion dollars of sophisticated matériel: This is a conservative estimate. The RQ-4 unit price alone is $223 million. US Government Accountability Office, "Assessments of Selected Weapon Programs," March 2013, pp. 113–14.

82   "Why are we looking for some schoolgirls as opposed to looking for al Qaeda?": Described by a US security official present, who spoke on condition of anonymity.

82   "What's your threshold? . . . Is it five? Is it fifty?": Ibid.

82   To service drones, it would base some eighty troops, mostly air force, in the dictatorship of Chad: Seldin, "US Drone Flying from Chad."

82   The British would send a spy plane: David Smith and Harriet Sherwood, "Military Operation Launched to Locate Kidnapped Nigerian Girls," *Guardian*, May 14, 2014.

82   intelligence officers from France would scour the former French colonies: Interviews with Western diplomats in Abuja, London, and Paris from 2014 to 2018; see also: John Irish, "France Sends Agents to Nigeria, Boosts Ties to Fight Boko Haram," Reuters, May 7, 2014.

82   China pledged to send satellite coverage: Patrick Boehler, "China Pledges Help to Nigeria's Hunt for Boko Haram Militants," *South China Morning Post*, May 8, 2014.

82   Canada would send special forces: Geoffrey York and Kim Mackrael, "Canada to Aid Nigeria in Search for Abducted Schoolgirls," *Toronto Globe and Mail*, May 7, 2014.

82　　Israel would offer counterterrorism specialists: Obafemi Oredein, "Israel Offers to Send Counterterrorism Experts to Nigeria," *Wall Street Journal*, May 11, 2014.

82　　John Kerry would sell the plan to Nigeria's beleaguered president: Interviews with officials on Secretary Kerry's staff and President Jonathan's, who spoke on condition of anonymity. Katie Zezima, "U.S. Team to help Nigeria Find Nearly 300 Abducted Girls," *Washington Post*, May 6, 2014.

82　　On May 6, Goodluck Jonathan agreed: Ibid.

82　　following the news from Chibok on an encrypted iPad in the Oval Office: Interview with Denis McDonough, Barack Obama's chief of staff, August 2019.

82　　Susan Rice brought a stack of reading material to the president: Interviews with three NSC and two State Department officials, who spoke on condition of anonymity, between November 2017 and May 2019. One said, "It literally had a box: Check here to approve. Check here for more information."

83　　On top was a single piece of paper, with two checkboxes: Ibid.

83　　"Do everything we can," he said: Ibid. President Obama also repeated this comment to Al Roker during a *Today Show* interview the next day. See also: "U.S. to Help Nigeria in the Search for Kidnapped Girls," Obama White House Archives, May 2014.

83　　Within hours, the first American personnel were on planes: Interviews with diplomats and officials working on and with the IDAT team, who spoke on condition of anonymity.

83　　The NSC began scheduling regular video conferences to demand results: Ibid.

83　　"#BringBackOurGirls: Addressing the Threat of Boko Haram": Committee on Foreign Relations, United States Senate, May 2014, https://www.govinfo.gov/content/pkg/CHRG-113shrg94293/html/CHRG-113shrg94293.htm.

83　　drafted plans to rent the Jim Henson studio: Interview in September 2017 with Brett Bruen, former director of global engagement at the White House.

Chapter 9: The Blogger and the Barrister

87　　Ahmad Salkida was working at a supermarket as a security guard: Interviews with Ahmad Salkida from 2017 to 2020 in Abuja, plus his articles, blogs, and Twitter archive at @ContactSalkida. His story was corroborated by senior Nigerian security officials and members of the Swiss-backed Dialogue Facilitation Team.

87　　broken some of the best scoops on Boko Haram: Ahmad Salkida, "Genesis and Consequences of Boko Haram Crisis," Salkida.com, April 1, 2010; and Eromo Egbejule, "This Insider Gets the Best Scoops on Boko Haram," Ozy.com, April 6, 2020.

87　　on the ungoverned space of Twitter, he was one of the most important voices: Twitter archive @ContactSalkida.

87　　had asked him to run its newspaper: Salkida, "Genesis and Consequences."

88　　His business card said INDEPENDENT JOURNALIST: The authors have his card

88　　Anonymous callers would ring his phone: Committee to Protect Journalists, "Nigerian Reporter Threatened over Boko Haram Coverage," March 19, 2012, https://www.refworld.org/docid/4f702506c.html.

88　　kept an exact tally: Interviews with Salkida.

88　　could no longer visit his family village: Ibid.

88　　Sharjah, a coastal city made up almost entirely of immigrants: The authors visited Sharjah in January 2020.

88　　The government official on the other end of the call: Interviews with Salkida, corroborated by Nigerian officials, who spoke on condition of anonymity.

88　　Many of the world's most powerful governments, spearheaded by the United States: Jonathan Powell, *Talking to Terrorists: How to End Armed Conflicts*

(New York: Vintage, 2015); and Julie Browne and Eric S. Dickson, "'We Don't Talk to Terrorists': On the Rhetoric and Practice of Secret Negotiations," *Journal of Conflict Resolution*, March 10, 2010.

89   "We don't negotiate with evil; we defeat it": Warren Strobel, "Vice President's Objections Blocked Planned North Korean Nuclear Talks," Knight Ridder, Washington Bureau, December 20, 2003.

89   many nations choose whether to negotiate based on public opinion: Powell, *Talking to Terrorists*.

89   American intelligence officers downloaded imagery: Interviews with US officials, including the IDAT team.

89   The government of Goodluck Jonathan: Ibid.; and Joe Parkinson and Drew Hinshaw, "Freedom for the World's Most Famous Hostages Came at a Heavy Price," *Wall Street Journal*, December 24, 2017.

89   His blog, *News beyond the Surface*: Salkida.com.

89   five-fingered toe shoes: In meetings in Abuja over three years, Salkida wore only five-fingered toe shoes. His preferred brand is Vibram.

90   "Everything he does has to be in the public domain": Interview with a former Nigerian cabinet minister, August 2017 in Abuja.

90   He didn't want a first-class plane ticket: Interviews with Salkida and Nigerian officials.

90   felt the modest rumble of a Swiss train: Interviews with Zannah and Swiss officials. The authors also made the train trip in question.

90   thinking about the vast differences in temperature: Zannah Bukar Mustapha, Facebook, April 2014.

90   invitation-only courses: Interviews in Geneva in November 2017 with officials of SwissPeace, a government-backed NGO that helps stage the courses.

90   the Park Oberhofen, high up on an Alpine hillside: The authors visited the Park Oberhofen in November 2017, where we interviewed the receptionist and took in the spectacular views of Lake Thun.

91   Zannah was greeted by a host: Interviews with Zannah and with Swiss officials, who spoke on condition of anonymity.

91   Political Affairs Division IV: Interviews with Swiss officials from the division, who spoke on condition of anonymity. The authors also spoke to Simon Mason of SwissPeace and Thania Paffenholz of the Center on Conflict, Peacebuilding and Development in Geneva, November 2017.

91   In the hotel's restaurant, where a 12:30 lunch: Peace Mediation Course Schedule, Swiss Confederation, Directorate of Political Affairs.

91   The longest staircase on Earth: Ken Jennings, "The World's Longest Staircase Is in Switzerland," *Condé Nast Traveler*, June 4, 2018.

91   a briefing paper titled "Pawns of Peace": Peace Mediation Course Schedule.

91   Julian Hottinger had mediated conflicts from Sudan to Sri Lanka: Interviews with Swiss officials, who spoke on condition of anonymity; and "For Peace, Human Rights and Security: Switzerland's Commitment to the World," Swiss Confederation, Federal Department of Foreign Affairs, 2011.

91   "We are constantly surrounded by people yet always alone": Kathrin Ammann, "The Life of a Mediator," SWI, Swissinfo.ch, April 20, 2018.

91   Hottinger would be teaching the opening class: Peace Mediation Course Schedule.

92   Since the end of the Second World War, 148 ceasefires: "Mediating a Ceasefire, Interview with Julian Hottinger," *OSCE Magazine*, Organization for Security and Co-operation in Europe, July 23, 2019.

92   In reality, the effort was infinitely more complex: Julian Hottinger, "Engaging Armed Groups: The Relationship Between Track One And Track Two Diplomacy," Accord Issue 16, Conciliation Resources, London, 2005.

92   "There are no absolutely good or absolutely bad people": "Mediating a Cease-fire."

92   a role-play called the fence game: Interviews with Z.M. and Swiss officials present at the training, who spoke on condition of anonymity.

92   "You don't rush for the solution": Interviews with Simon Mason, SwissPeace.

93   "Never trust a breakthrough!": Interviews with Z.M. and Swiss officials.

93   flashed his newly issued green access badge: Interviews with Salkida and Nigerian government officials present, who spoke on condition of anonymity.

93   to the lobby of a dimly lit midprice hotel: The Rockview Hotel was a favored hangout for Abuja's political class and intelligence officers. The authors have conducted many interviews in it from 2011 to 2020.

93   Salkida walked into a carpeted lounge in Aso Rock's residential quarters: Interviews with Salkida and Nigerian government officials present.

93   Some in government suspected Salkida: "BOKO HARAM: Growing Threat to the U.S. Homeland," *U.S. House of Representatives Committee on Homeland Security*, September 13, 2013, pg. 135.

94   "I believe in you," the president said: Ibid. The authors also confirmed these quotes and details in a phone call with former president Goodluck Jonathan in November 2019.

94   married just two days before the kidnapping: Adenike Orenuga, "Goodluck Jonathan Gives Out Customised iPhones to Guests at Daughter's Wedding," *Daily Post* (Lagos), April 12, 2014.

94   if he brought a phone, they'd kill him: Interviews with Salkida.

Chapter 10: Thumb Drive Deal

95   the first images from a Global Hawk drone: Catherine Chomiak and Hasani Gittens, "U.S. 'Global Hawk' Drone Joins Search for Kidnapped Nigerian Schoolgirls," NBC News, May 13, 2014.

95   behind a barbed-wire fence running through a beech-and-pine forest: Details on the base are from Jeremy Scahill, "Germany Is the Tell-Tale Heart of America's Drone War," *Intercept*, April 17, 2015. The authors also spoke with officials and contractors familiar with the base and had several conversations in 2019 with Cedric Leighton, a widely cited surveillance specialist and consultant for Cedric Leighton International Strategies, LLC.

95   expat kids from the Ramstein American High School team were in spring training: Ramstein Royals Schedule and Record, Spring 2014, hosted by a Dick's Sporting Goods website called Gamechanger, https://gc.com/t/spring-2014/ramstein-royals-531a268fadd60ecf4c000001. The teenagers had lost their previous game during a difficult eighth inning but went 21–2 in a season full of lopsided victories.

95   Ramstein had been born as a Luftwaffe airstrip during the Second World War: Silvano A. Wueschner, "The History of Ramstein Air Base," *Kaiserslautern American*, March 16, 2012.

95   but was now America's vital perch in a global aerial surveillance program: Scahill, "Germany Is the Tell-Tale Heart."

95   Noncommissioned officers in battle uniforms . . . two thousand Dell monitors: "603rd Opens Doors to New AOC [Air and Space Operations Center]," Ramstein Air Base, 86th Airlife Wing Public Affairs, October 13, 2011. Videos of the base, created by the air force and reviewed by the authors, confirmed the setup.

95   streaming live footage from the coast of Greenland: Ibid.

96   officers watching the screens worked in pairs for eight-hour shifts: Interviews with Leighton.

96   analysts seated in US facilities elsewhere on the planet watched simultaneously:

Ramstein is the first point of arrival for US Air Force drone data, but the Chibok feed was more closely monitored in the embassy, among other stations, according to officials tasked with reviewing it.

96 chatting in small text boxes: Interviews between 2018 and 2020 with former US Air Force officials, who spoke on condition of anonymity, as well as Leighton.

96 They could zoom in or out in a Nigerian search zone the size of West Virginia: Officials working with the IDAT team in Abuja, who spoke on condition of anonymity, 2014 to 2019.

96 PTSD was not uncommon: For more, see Eyal Press, "The Wounds of the Drone Warrior," *New York Times*, June 13, 2018.

96 Many of its analysts had been monitoring individual terrorists for months or years: Former air force officials, and Leighton.

96 pickup trucks driving in a chain through wilderness: Officials working on or with the IDAT team in Abuja.

96 conflicting accounts of the number missing and discrepancies in the spellings of their names: Ibid.

97 mudflats sliced by the tire treads of motor scooters: Photographs from military aircraft flying over the search zone. US drone footage of the area was also described by air force and IDAT officials who reviewed it, speaking on condition of anonymity, 2014–2019.

97 Ahmad Salkida had made it into the heart of Boko Haram territory: Interviews with Salkida. Members of the Swiss-backed team were also briefed on his movements.

97 a small delegation authorized to speak on behalf of the Imam: Salkida said he spoke to a representative from the media, the Shura council, and one of the girls' guardians.

97 There, under the shade of a tamarind tree, were more than two hundred female faces staring back at him: The authors saw a screenshot from an unpublished video of this scene that is in the possession of Salkida.

98 Salkida asked for two tapes: Interviews with Salkida.

98 "Are you being fed?": This scene, relayed by Salkida, was also described in the diaries, as well as by N.A. and H.I. The authors also confirmed it in a call with President Goodluck Jonathan and spoke to an individual tasked with translating it for the president, who spoke on condition of anonymity.

98 "Every life that is lost pains me": The authors interviewed an official present in the room, then confirmed this comment in a call with President Jonathan.

99 President Jonathan arrived at the Élysée Palace in Paris: Martin Williams, "African Leaders Pledge 'Total War' on Boko Haram after Nigeria Kidnap," *Guardian*, May 17, 2014.

99 painted with angels said to represent guardians of the peace: Details on the extraordinary resplendent chamber in which heads of state declared war against a forest-encamped sect squatting in mudbrick huts can be found at "Visite du Palais de l'Élysée," https://www.olivierberni-interieurs.com/en/node/73.

99 Chad's president, Idriss Déby, . . . boasted: Interviews in 2017 and 2018 with Western and West African officials present, all of whom spoke on condition of anonymity.

99 had approved a ransom of just over €3 million: Tim Cocks, "Nigerian Islamists Got $3.15 Million to Free French Hostages: Document," Reuters, April 26, 2013.

100 Ahmad Salkida was sitting in a top-floor office: Interviews with Salkida and another official present.

100 At first, Shekau's negotiators had demanded that Nigeria cede territory: Ibid.

100 A swap was set for Sunday, May 18: Ibid. The authors confirmed details in a call with President Jonathan.

102 "Fuck it, I'm leaving": Interviews with Salkida.

102   Lying across the bed of his hotel suite: Interviews with Z.M. and author tour of
      the Hotel Oberhofen.
102   couldn't proceed until those "boulders in the road" were removed: This concept
      is discussed in the "Mediation Process Matrix," a document produced by Swiss-
      Peace in 2012. The Matrix is a tool for mediators to navigate complex peace
      processes. See: https://reliefweb.int/sites/reliefweb.int/files/resources/Mediation
      _Process_Matrix.pdf.
102   "I am trying to pick the lock": Interviews with Z.M.
103   instructor Julian Hottinger compared Boko Haram to Palestine: Ibid.
103   instructors told the group to layer up: Ibid. The authors also spoke with course
      instructors in Geneva in November 2017.

Chapter 11: Malam Ahmed's Rules

104   Even the thick, leafy branches of the tamarind: Interviews with N.A., H.I., H.S.,
      M.A., P.M., and Ladi Simon (whose name has been changed on her request,
      hereafter referred to as L.S.), as well as extracts from diaries.
105   "We should just pretend": Interviews with N.A.
105   breezy reassurances that they would be going home: One of the guards said
      "*wit, wit*" ("soon, soon" in Hausa) so often that the captives nicknamed him Soon
      Soon.
105   He had introduced himself as Malam Ahmed: The picture of Malam Ahmed was
      built in interviews with almost twenty of the Chibok hostages. He also features
      briefly in the diaries.
106   a limp that stemmed from his time in the Giwa Barracks: Interview with M.A.,
      who got to know Malam Ahmed better after she "married" into the sect. Torture
      with hot iron rods was also described repeatedly in an Amnesty International
      report, *"Welcome to Hell Fire": Torture and Other Ill-Treatment in Nigeria*,
      London, September 2014, https://www.amnesty.org/download/Documents/4000
      /afr440112014en.pdf.
106   Islamiyah school, a Quranic class: Interviews with several other female hos-
      tages who escaped Boko Haram between 2014 and 2020 confirmed that Boko
      Haram required women to take Quranic recitation classes. The authors spoke to
      Binta Umma and Maimuna Musa, both teenage captives who were forced into
      marriage then deployed as suicide bombers but survived. See: Joe Parkinson
      and Drew Hinshaw, "'Please, Save My Life': A Bomb Specialist Defuses Explo-
      sives Strapped to Children," *Wall Street Journal*, July 26, 2019.
106   The first lessons had covered the creation story of Adamu and Hauwa: The dia-
      ries contain notes in Hausa and Quranic passages from early lessons.
106   "The world is flat, and it is Allah, not evaporation, that causes rain!": Ibid.
106   they would have to obey strict rules: Interviews with N.A., H.I., H.S., L.S., M.A.,
      and P.M., as well as extracts from diaries.
107   Commanders were *qaids*. A *jasus* was a spy: Ibid.; and Omar S. Mahmood and
      Ndubuisi Christian Ani, *Factional Dynamics within Boko Haram*, Institute for
      Security Studies, Pretoria, July 2018, https://issafrica.s3.amazonaws.com/site
      /uploads/2018–07–06-research-report-2.pdf, p. 16.
107   A girl who had embraced the struggle could become an *emira*: Chika Oduah,
      "The Women Who Love Boko Haram," *Al Jazeera*, September 22, 2016.
107   Some of them were frightening, but others were childlike: Interviews with N.A.
      and other Chibok hostages. Also see the trove of videos leaked to Voice of Amer-
      ica in 2017, which show young fighters relaxing and joking around, as well as
      public executions and other atrocities committed by the group. VOA authenti-
      cated the videos by analyzing time stamps, checking references made by fight-
      ers in the footage, and correlating events described in news broadcasts heard in

the background. See: Salem Solomon, "Boko Haram: Terror Unmasked," Voice of America, News, February 2017, https://projects.voanews.com/boko-haram-terror-unmasked.

108 These guards were responsible for monitoring: Interviews with N.A. and other Chibok hostages.

108 the students had grown up preparing kuka soup: Interviews with N.A. and K.A.

109 the sky was filling up with aircraft sent to find them: Harriet Sherwood, "US Planes Begin Air Search for Kidnapped Nigerian Schoolgirls," *Guardian*, May 13, 2014; and Gabriel Gatehouse, "Kidnapped Nigeria Girls: British Military Planes Deployed," BBC News, May 18, 2014.

109 The sound was nothing they had heard before: Interviews with N.A., M.A., L.S., and H.S.

110 told the captives to remove any item with a bright color: Ibid.

110 One day, a plane dumped leaflets that floated to the ground: Ibid. The authors also spoke to Dr. Manaseh Allen, spokesman of the Kibaku Association, in May 2020 in Maiduguri, who was allowed to fly in a Nigerian Air Force jet over Sambisa in early 2015, which dropped what the military called "sensitization leaflets."

Chapter 12: The Mothers' March

112 Kolo Adamu looked in the mirror: Interviews with K.A., Philomena Adamu, and R.A.

112 The removal of hair was a mourning ritual: "Plight of Nigerian Widows," *Vanguard Nigeria* (Lagos), May 23, 2019.

112 Naomi's sandals lay tangled in the pile of shoes: Interviews with K.A., photos of the Adamus' home.

113 It was one of the wallet-size photos she had taken: Ibid. The photo of Naomi can be seen in Joe Parkinson and Drew Hinshaw, "Freedom for the World's Most Famous Hostages Came at a Heavy Price," *Wall Street Journal*, December 24, 2017.

113 she heard a car driving up, blasting a joyful and cathartic song: Journalist and author Adaobi Tricia Nwaubani, who has interviewed more than one hundred of the Chibok hostages' parents, said in discussions with the authors that another mother, Yana Galang, told a similar story about her daughter Rifkatu, raising the question of whether their shared grief has made some memories collective.

113 The next, she would hear speculation: Interviews with K.A. and R.A.

113 The idea to stage a march came from Saratu's mother, Rifkatu Ayuba: Ibid.

114 a prediction of the preacher of a Nigerian megachurch, T. B. Joshua: Adenike Orenuga, "'They Have to Be Released': Prophet T.B. Joshua Sees Vision about Abducted Chibok School Girls," *Daily Post* (Lagos), April 28, 2014.

114 Oby Ezekwesili was still leading daily marches: Twitter archive @ObyEzeks; and "Row Brews as 'Bring Back Our Girls' Rallies Banned in Nigerian Capital," *Daily Nation* (Nairobi), June 3, 2014.

114 replaced as a symbol of celebrity-backed global solidarity by the #IceBucketChallenge: Timothy Stenovec, "The Reasons the Ice Bucket Challenge Went Viral," *Huffington Post*, December 6, 2017.

114 Together they had recruited Mary Ishaya: Interviews with K.A., R.A., and Mary Ishaya.

115 Not a single journalist, foreign or local, covered the march: Ibid.

115 where they faced the torched shells of the hostels: At this point the burned-out remains of the classrooms and dormitory buildings had not been cleared or removed. For the mothers, it was like a monument to their daughters' kidnapping.

116 Local families would meet in church halls: Interviews with Yakubu Nkeki, head of the Chibok Association, between 2014 and 2020.

116 concluded that they should take a request to the Nigerian Air Force: Recalled by Victor Ibrahim Garba, a protest leader who represents the girls' families, interviewed by the authors in February 2015. See: Drew Hinshaw and Neanda Salvaterra, "Missing Nigeria Girls Fade from Campaign," *Wall Street Journal*, February 12, 2015.

116 Yana Galang, had started washing and folding: Tara Sutton, "'We Cry and Cry': Pain Endures for Mothers of Missing Chibok Schoolgirls," *Guardian*, October 3, 2019. See also the award-winning film by Joel Kachi Benson, *Daughters of Chibok*, VR360 Stories, 2019.

116 when Lydia John's eighteenth arrived, her parents held a small and somber gathering: Interviews with Lydia's parents, Rebecca and Samuel, Abuja, November 2017 and November 2019.

117 After the kidnapping, one mother stopped eating: Interviews with K.A., R.A., and Yakubu Nkeki.

117 eleven of the parents would never see their children again: "Missing Nigerian Girls: 11 Parents of Abducted Girls Die from Stress and Attacks," Agence France Presse, July 23, 2014.

Chapter 13: "You Must Write One Too"

118 with an image of Disney's Cinderella: The "Cinderella" diary is one of three books N.A. smuggled out of captivity, which constitute the diaries. Another is a similar forty-page exercise book with a blue cover with a cartoon image of a boy playing golf. A third diary, in a notebook given to the hostages later, was a dull sepia color, caked in dust; Naomi had to sew it together with string to prevent it from splitting apart.

118 Naomi had found herself sitting next to a girl: Interviews with N.A.

119 Lydia John had been such a diligent student: Interviews with Rebecca and Samuel John.

119 fleeing her hometown of Banki: Ibid.

119 The new girl spoke the best English: Ibid.; and interviews with N.A. and H.I.

119 wrote in bold, loopy cursive: When we discussed the diaries with Rebecca John in November 2017, she confirmed that they contained her daughter's distinctive handwriting.

119 Lydia sang in the school choir, the Glorious Singers: Interviews with Rebecca and Samuel John, N.A., H.I., P.M., L.S., and M.A.

119 When she was a child, a bout of measles had left her with an infected left eye: Interviews with Rebecca and Samuel John and N.A.

119 Visiting the local photo studio: The authors were shown the photo of Lydia, wearing white, by her parents in September 2017.

120 They braided their hair using a pin they'd carved with a razor blade from a piece of wood: A picture of the pins can be found in Adaobi Tricia Nwaubani's October 23, 2017, story for the BBC, "Chibok Diaries: Chronicling a Boko Haram Kidnapping."

120 Lydia had been spending a lot of time alone: Interviews with N.A., H.I., P.M., L.S., and M.A.

120 she had been a prayer warrior at her church: Interviews with Rebecca and Samuel John, and N.A.

120 It began: "First of all on 14th April 2014 on Monday": Extracts from the diaries.

122 As weeks dragged on, however, other material went into the pages: Ibid.

122 They copied passages from a small Bible: Ibid.

122   Nelson Mandela wrote his autobiography at night on scraps of paper: Karen Allen, "How a Secret Manuscript Became a Global Bestseller," BBC News, November 2, 2011.

123   Holocaust survivors kept similar diaries: Robin Shulman, "How an Astonishing Holocaust Diary Resurfaced in America," *Smithsonian Magazine*, November 2018.

123   the teenagers were keeping a record of injustice: To contextualize the psychological importance of keeping a diary in captivity, we spoke to Professor Anne Speckhard from Georgetown University in December 2017, who has done extensive research talking to former hostages of jihadist groups.

123   "Shake," a pulsating Nigerian dance-floor hit: "Shake" is the fifth track on Flavour N'abania's 2012 album *Blessed*, released by 2Nite Entertainment. The music video, which was shot in Johannesburg, features Flavour watching several women "shake" on top of a skyscraper.

123   The Malam's daily head count had come up short: The escape is mentioned in the diaries and was described by N.A., H.I., P.M., M.D., H.S., and Hauwa Ntkai (hereafter referred to as H.N.).

125   "Boko Haram is everywhere": There were several other unsuccessful escape attempts not described in this book, when hostages snuck away at night or during wash breaks. Some lasted several days.

Chapter 14: Tree of Life

126   a photo landed on a bank of monitors in the CIA station: Three senior US officials described this photo, as well as a British diplomat involved in the search. All spoke on condition of anonymity.

126   a group of sealed drop-ceiling rooms: Interviews with a US intelligence official and two members of the interdisciplinary assistance team.

126   The CIA had kept two agents on payroll in the region: Interviews with a senior US military official and a British diplomat.

126   The British had trained an informant close to Boko Haram: Interviews with two senior British officials.

127   "We're assuming it's not a rock band of hippies out there camping": Drew Hinshaw and Dion Nissenbaum, "U.S. Planes Searching for Boko Haram Abductees Spot Girls in Nigeria," *Wall Street Journal*, August 5, 2014.

127   all the way to President Barack Obama's morning briefing: Interviews with a senior intelligence official and a senior US State Department official.

127   In time, it would gain a nickname: the Tree of Life: The British coined the name for the picture, as described by two British diplomats, but it soon became common currency among the American team and other nations' representatives.

127   US drones had been flying missions over a vast area of the Sambisa Forest: Hinshaw and Nissenbaum, "U.S. Planes . . . Spot Girls in Nigeria."

127   More than fourteen thousand people were known to be dead: Nigeria Security Tracker, Council on Foreign Relations, https://www.cfr.org/nigeria/nigeria -security-tracker/p29483.

127   half a million were on the move: "Nigeria, Cameroon: Upsurge of Violence," Relief-Web, UN Office for the Coordination of Humanitarian Affairs, August 28, 2014, https://reliefweb.int/map/nigeria/28-august-2014-nigeria-cameroon-upsurge -violence.

127   so routinely ambushed in the countryside: Senator Iroegbu, "Military Keeps Mum After 315 Nigerian Troops Flee to Niger," This Day, *AllAfrica*, November 10, 2014.

127   the militants were encircling the city of Gwoza: "Nigerian Troops Flee Gwoza as Boko Haram Captures Tank and Driver, Commander Whereabouts Unknown," *Sahara Reporters*, New York, August 8, 2014.

127   fighters swept up hundreds more hostages: Testimonies of thirty women kidnapped and held by Boko Haram can be found in "Those Terrible Weeks in their Camp," Human Rights Watch, New York, October 27, 2013, https://features.hrw .org/features/HRW_2014_report/Those_Terrible_Weeks_in_Their_Camp/index .html.

128   the militants had seized six villages: Drew Hinshaw and Joe Parkinson, "The 10,000 Kidnapped Boys of Boko Haram," *Wall Street Journal*, August 12, 2016.

128   raided a village only a few hours from the school: International Federation for Human Rights, "Nigeria: Women Continue to Be Targeted by Boko Haram," July 2014, RefWorld, UN High Commissioner for Human Rights, https://www .refworld.org/docid/53c7c68710.html.

128   after the sect had grabbed three hundred more boys from an elementary school: Aminu Akubakar, "Boko Haram Seized 300 Children in 2nd 2014 School Attack,"Agence France-Presse, March 30, 2016.

128   The thirty-eight officials of the Assistance Team: Interviews from 2017 to 2019 with eight officials on the team or diplomats working with it. Though the book refers to the Interdisciplinary Assistance Team, US government officials knew it by its acronym, IDAT.

128   The crew had arrived in Abuja with their own office-support clerk: Ibid.

128   lodgings would be in the 667-room Hilton: The hotel's listing on Expedia states: "All 667 rooms feature comforts like Egyptian cotton sheets and down comforters, while conveniences include refrigerators and coffee makers."

128   Malala Yousafzai spent her seventeenth birthday in the hotel: "Malala Visits Nigeria on Birthday to Help Free Abducted Girls," NBC News, July 13, 2014.

128   Washington lobbyists stayed in a tenth-floor executive suite: The authors interviewed an employee of the Levick crisis-management firm over a plate of suya and several Heineken beers by the Hilton pool in July 2014.

128   a $1.2 million contract to "change the international and local media narrative": Megan R. Wilson, "Nigeria Hires PR for Boko Haram Fallout," *The Hill*, June 26, 2014.

128   four men who claimed to be senior Boko Haram members were holding hostage talks that went nowhere: Interviews with Z.M., a UN official, and a senior Nigerian official who were present at the talks.

128   an Anglican priest . . . bringing screenshots from Google Earth: The authors interviewed Stephen Davis in September 2017, "Mama Boko Haram" in Abuja in January 2019, and members of the Office of the National Security Adviser, as well as Jonathan spokesman Reuben Bati in Abuja, in July 2014 and September 2017.

129   unlimited dollar funds and the use of a military helicopter: Ibid.

129   who called herself Mama Boko Haram: Ibid.

129   boasted of having once cooked . . . for Abubakar Shekau: Egusi is a southeastern Nigerian dish. We leave it to readers to judge the credibility of Mama Boko Haram's account.

129   throwing money at the problem: The money included the US lobbyist contract, the money given to Davis, and other efforts not mentioned in the book. Additional funds were provided to Eeben Barlow's unit (mentioned below). "It became a money-making opportunity," said a senior official in the Jonathan administration, in November 2017.

129   "Bring back our girls! Now and alive!" Authors repeatedly attended these protests between 2014 and 2019.

129   to meet at the US ambassador's gated hilltop residence for tea: Interviews with team members and Western diplomats, May 2014 to September 2019.

129   warning that the air force "would simply bomb the baobab tree": Ibid.

130   cities as big as New York and as small as Bradford, England: Ewan Watt, "Worldwide Vigils on July 23 to Mark 100 Days Since Chibok Girls Were Kidnapped," *Theirworld* (children's rights blog), July 22, 2014, https://theirworld.org /news/worldwide-vigils-on-july-23-to-mark-100-days-since-chibok-girls-were -kidnapped.

130   congresspeople dialed the State Department demanding updates: In interviews in Washington in May, August, and September 2019, several senior State Department officials described congresspeople being very forceful. "They would call me very angry: 'You need to do this, do that,'" one recalled.

130   Michelle Obama's chief of staff had continued to ask about the Chibok girls: Interviews with two members of her staff in September and October 2017, as well as NSC members and others present at morning security meetings with Barack Obama.

130   McCain . . . had been privately lobbying the Pentagon: Interviews with two members of Secretary Kerry's staff, April and May 2019. White House chief of staff Denis McDonough told the authors in an August 2019 interview: "That would not be inconsistent with Senator McCain's view of American power."

130   The Nigerians . . . knew that the US wasn't trusting them with the full intelligence picture: Interviews with two Nigerian officials at Abuja's Office of the National Security Adviser, who spoke on condition of anonymity.

131   a process that took up to seventy-two hours: Interviews with three US defense and intelligence officials, as well as four Western diplomats and a senior official at the Nigerian National Security Advisers Office, between July 2014 and September 2019.

131   The photos the Americans brought in were lo-fi, grainy printouts: Ibid.

131   The Americans were withholding information: Ibid.

131   listening to intercepted calls in Nigerian languages they couldn't translate: Interviews with a Western diplomat, a US Department of Defense official, and an American intelligence official, November 2017 to September 2019.

131   hacked into e-mails, texts, and phone calls from Nigeria's military: Ibid.

131   The spying went both ways: Interviews with two members of the team in November 2017 and October 2019.

131   "You're not in Washington now!": A senior official at Nigeria's NSA Office who was present, interviewed in October 2017.

131   The British Sentinel R1 spy plane . . . broke down: Ben Farmer, "RAF Spy Plane Breaks Down on Way to Find Nigerian Schoolgirls," *The Telegraph* (London), May 19, 2014.

131   French intelligence struggled to confirm London's suspicion: Interviews with several Western diplomats.

132   No Nigerian officials ever recalled seeing the satellite images China promised to send: Since 2014, we have posed this question in repeated interviews with various Nigerian officials. None recalled receiving any satellite information from China.

132   the FBI hostage specialists flew home: Interviews with team members, an intelligence official, and Western diplomats.

132   The Nigerians were losing territory: "Boko Haram Crisis: Nigerian Soldiers 'Mutiny over Weapons,'" BBC, August 19, 2014.

132   Goodluck Jonathan himself asked John Kerry to send combat troops: Neanda Salvaterra and Drew Hinshaw, "Nigerian President Goodluck Jonathan Wants U.S. Troops to Fight Boko Haram," *Wall Street Journal*, February 13, 2015. The president said, "Are they not fighting ISIS? Why can't they come to Nigeria? Look, they are our friends. If Nigeria has a problem, then I expect the U.S. to come and assist us."

132   discussions about a possible special-forces rescue never gained traction: Interviews with multiple US officials who joined the video calls in which the subject was discussed, as well as with British officials who read and in some cases prepared their memos on the subject. Some British officials—including ex-premier David Cameron in his memoir *On the Record* (New York: HarperCollins, 2019)—say the UK was ready to rescue the girls, but the Nigerian government denied them access. This seems to be out of keeping with the reality that defense planners and diplomats saw on the ground, and Nigerian officials said they were more than open to the idea of a rescue, but doubted one could be successfully staged.

132   In August, officers in the CIA's Abuja station noticed that the women under the Tree of Life had been moved: Hinshaw and Nissenbaum, "U.S. Planes Searching."

132   drones spotted clusters of young women in the search zone about five times: Interviews with Western diplomats and US defense and intelligence officials.

133   drones circling Lake Chad were redeployed over Iraq and Syria: Ibid.

Chapter 15: The Senator's House

134   an enormous white mansion: The account of the hostages' time at Senator Ali Ndume's palatial home came from N.A., more than a dozen other Chibok students, and excerpts from the diaries. In an April 2015 story, "Nigeria's Chibok Girls 'Seen with Boko Haram in Gwoza,'" BBC News quoted a witness saying she had seen the Chibok hostages being held in a "large house" in Gwoza.

134   a toaster and an electric water kettle: The hostages' arrival at the senator's mansion is also described in Isha Sesay's *Beneath the Tamarind Tree* (New York: Dey Street, 2019), 213–18.

135   one of the young women who had joined Naomi's secret club of diarists: M.D. said in interviews that she didn't manage to smuggle her diaries out of the forest, but she penned another one after her release, and was kind enough to let us review it.

135   a central boulevard of what appeared to be a sprawling city: Interviews with several Gwoza residents who fled the city for camps, in Yola in October 2017 and in Maiduguri in July 2019.

135   it was a "ghost house": Interviews with M.A.

136   "This is Senator Ndume's house": Ali Ndume had been appointed to a presidential committee tasked with exploring talks with Boko Haram in 2011, but he was later charged with being a "sponsor" of the sect. He spent six years fighting the charges before being acquitted in 2017.

136   Boko Haram's new capital: "Boko Haram Declares 'Islamic State' in Northern Nigeria," BBC News, August 25, 2014. For more on the declaration of the caliphate, see Thurston, *Boko Haram*, 224–30. Gruesome footage of the sacking of Gwoza can be seen in Shekau's video "Declaring Islamic State in Gwoza Video," August 2014, archived at UnmaskingBokoHaram.com.

136   training academy for Nigeria's riot police: "Boko Haram Crisis: Nigeria Militants 'Seize Police Academy,'" BBC News, August 21, 2014.

136   the terrace-farmed mountains known as the Gwoza Hills: Joe Brock, "Boko Haram, Taking to Hills, Seize Slave 'Brides,'" Reuters, November 17, 2013.

136   Boko Haram had stormed into the city two weeks earlier: *"Our Job Is to Shoot, Slaughter and Kill": Boko Haram's Reign of Terror in North East Nigeria*, Amnesty International, April 14, 2015, https://www.amnesty.org/download /Documents/AFR4413602015ENGLISH.PDF. Also see: Defence Headquarters, "Update on Counter Terrorism Campaign: Military Takes Gwoza the Terrorists Phantom Caliphate Headquarters," March 27, 2015, https://twitter.com /defenceinfong/status/581436668827328512.

137 The emir's brother, trying to escape: Ibid.

137 residents, scampering up the mountainside to hide: Interviews with Gwoza residents who fled.

137 Fighters occupied the emir's palace: Isaac Abrak, "Killing and Preaching, Nigerian Militants Carve Out 'Caliphate,'" Reuters, September 10, 2014.

137 Insurgents fanned out into the hills: Amnesty International, "Our Job Is to Shoot."

137 The *hisba*, the religious police, punished sinners with public amputations: Interviews with Gwoza residents and the Chibok hostages. Public punishments can be seen in Salem Solomon, "Boko Haram: Terror Unmasked," Voice of America, News, February 2017, https://projects.voanews.com/boko-haram-terror -unmasked.

138 taken to the city's abattoir and slaughtered: Robyn Dixon, "Nigeria Survivors of Boko Haram Attack Recall 'Total Terror,'" *Los Angeles Times*, April 17, 2015.

138 had swept across Iraq and Syria: "Timeline: The Rise, Spread, and Fall of the Islamic State," Wilson Center, October 28, 2019, https://www.wilsoncenter.org /article/timeline-the-rise-spread-and-fall-the-islamic-state.

138 called on Muslims throughout the world to "pledge allegiance": Joas Wagemakers, "The Concept of Bay'a in the Islamic State's Ideology," *Perspectives on Terrorism* 9, no. 4 (August 2015): 98–106.

138 most wanted terrorist on Earth: Ruth Sherlock, "How a Talented Footballer Became the World's Most Wanted Man," *Daily Telegraph* (London), November 11, 2014.

138 Shekau dispatched an emissary: Interviews with Nigerian security officials, corroborated by Zenn's UnmaskingBokoHaram.com, which reviewed Kanuri and Hausa audio files from Boko Haram Shura Council meetings, including advice from a Syria-based Saudi Islamic State theologian, who told Nigerian jihadists that the world was not flat.

138 Islamic State fighters stormed into the Iraqi city of Sinjar: Loveday Morris, "Islamic State Seizes Town of Sinjar, Pushing Out Kurds and Sending Yazidis Fleeing," *Washington Post*, August 3, 2014.

138 the idea to mass-kidnap women came from Chibok: Islamic State's flagship *Dabiq* magazine featured an article called "The Revival of Slavery," referencing the Chibok kidnapping on page 15 of its November 2014 edition, titled "Remaining and Expanding."

138 it surrendered most of Borno State to the insurgency: "Boko Haram Militants Display Control of Captured Towns in Northeastern Nigeria," *Sahara Reporters*, New York, November 10, 2014; and "Terrorists Rename Another Captured Town," *Pulse Nigeria*, November 7, 2014.

138 a territory the size of Belgium: Drew Hinshaw, "Boko Haram Extends Its Grip in Nigeria," *Wall Street Journal*, January 5, 2015.

139 more than two thousand girls and ten thousand boys: "Boko Haram: 2,000 Women and Girls Abducted," Amnesty International, London, April 10, 2015, https://www.amnesty.org.uk/press-releases/boko-haram-2000-women-and-girls -abducted-many-forced-join-attacks-new-report; and Drew Hinshaw and Joe Parkinson, "The 10,000 Kidnapped Boys of Boko Haram," *Wall Street Journal*, August 12, 2016.

139 could be deployed as front-line fighters: Ibid.; and author interviews with dozens of former hostages—boys and girls—who escaped between 2014 and 2020 in Yola, Maiduguri, Mubi, and Madagali.

139 reports of an epidemic of sexual abuse began to filter out: "Nigeria: Boko Haram Abducts Women, Recruits Children," Human Rights Watch, New York, November 29, 2013, https://www.hrw.org/news/2013/11/29/nigeria-boko-haram -abducts-women-recruits-children.

139   a long speech about the evils of government schools: Interviews with N.A. and more than a dozen Chibok hostages, as well as excerpts from the diaries.

140   fragmented stories of mass rapes and conversions: Harriet Sherwood, "Boko Haram Abductees Tell of Forced Marriage, Rape, Torture and Abuse," *Guardian*, October 27, 2014. For more detail, see the October 2014 report by Human Rights Watch, *Those Terrible Weeks in Their Camp*.

140   They could be slaves, or they could enter what the sect called marriage: Ibid. See the documentary "Forced to Marry Boko Haram or Die," BBC News, June 29, 2015; also see the March 29, 2019, UN report of the UN secretary-general, *Conflict-Related Sexual Violence*, https://www.un.org/sexualviolenceinconflict /wp-content/uploads/2019/04/report/s-2019–280/Annual-report-2018.pdf.

140   They saw their arrangement as legitimate: Some of the best work on this topic has been done by Fatima Akilu, a psychologist who has helped deradicalize hundreds of young women forced to join the sect and marry fighters, and by Nigerian journalist and author Adaobi Tricia Nwaubani.

140   Rape was impossible within marriage, the group's theology held: Interviews with Chibok hostages, other women who escaped camps, and Boko Haram defectors.

140   women were locked into small spaces until they capitulated: A report by Mausi Segun for Human Rights Watch, "A Long Way Home: Life for the Women Rescued from Boko Haram," dated July 29, 2015, said that thirty women and girls were held inside a crashed airplane the fighters called Handak, https://www .hrw.org/news/2015/07/29/long-way-home-life-women-rescued-boko-haram.

140   short marriages that could be consecrated and ended quickly by a religious authority: Another method for religiously sanctifying rape, described by Boko Haram captives and defectors.

140   whose soldiers faced their own credible accusations of rape: *"They Betrayed Us": Women Who Survived Boko Haram Raped, Starved and Detained in Nigeria*, Amnesty International, May 2018, https://www.amnesty.org/download/Documents /AFR4484152018ENGLISH.PDF; and Dionne Searcey, "They Fled Boko Haram, Only to Be Raped by Nigeria's Security Forces," *New York Times*, December 8, 2017.

141   loudspeakers to proclaim that their harsh enforcement of fidelity: "Boko Haram: Terror Unmasked," four-part series, Voice of America, Washington, DC, February 2017. See also: Zacharias Pieri and Jacob Zenn, "Under the Black Flag in Borno: Experiences of Foot Soldiers and Civilians in Boko Haram's 'Caliphate,'" *Journal of Modern African Studies* 56, no. 4 (December 2018): 645–75, at 661.

141   often committed rape without even the pretense of a forced marriage: See: Christina Lamb, *Our Bodies, Their Battlefield: What War Does to Women* (New York: William Collins, 2020).

141   many joined willingly and believed in its cause: Hilary Matfess, *Women and the War on Boko Haram: Wives, Weapons, Witnesses* (London: Zed Books, 2017); and Chika Oduah, "The Women Who Love Boko Haram," *Al Jazeera*, September 22, 2016.

141   Some smuggled weapons, spied on the movements of military convoys: Interviews with former hostages; and Matfess, *Women and the War on Boko Haram*.

141   felt they had just as much freedom as at home: Adaobi Tricia Nwaubani, "The Women Rescued from Boko Haram Who Are Returning to Their Captors," *New Yorker*, December 20, 2018.

141   where most teenage girls marry before they turn eighteen in weddings arranged by their fathers: Annabel Erulkar and Mairo Bello, "The Experience of Married Adolescent Girls in Northern Nigeria," Population Council, New York, 2007, https://www.ohchr.org/Documents/Issues/Women/WRGS/ForcedMarriage/NGO

/PopulationCouncil24.pdf. In Nigeria's northeast, 57 percent of women twenty to forty-nine were married before they turned eighteen, per a 2017 factsheet from the Girls Not Brides NGO.

141 had once themselves been victims, kidnapped or coerced into joining: Hinshaw and Parkinson, "10,000 Kidnapped Boys"; and interviews with former captives.

141 waiting for a conversation with Naomi: This exchange was described by N.A. and was also witnessed by H.S., L.S., H.I., and several other hostages.

143 would leave a permanent scar: N.A. showed the authors the scar from the strike, visible on the top of her neck.

143 Lydia gathered some friends out of earshot: Described by N.A., H.I., P.M., and S.A.

143 "Dear my lovely Mum": An extract from the diary.

Chapter 16: Maryam and Sadiya

145 thatch and mudbrick *gidan tabo* home on the outskirts of Chibok: *Gidan tabo* is the Adamawa Hausa term for these mudbrick homes.

145 "like peas in a pod": Interview with M.A.'s father, Ali Maiyanga, October 2018.

145 their complementary natures making each other feel whole: Ibid.; and interviews with M.A.

146 eager to leave town and study at the college of her father's choice: Ibid.

146 Sadiya used those hours to talk to Baba: Interviews with M.A.

147 The students were locked inside the compound: Ibid.; and interviews with more than a dozen other Chibok hostages.

147 Malam Ahmed addressed the captives: Interviews with M.A.

147 Gwoza would be renamed Dar al-Hikma—House of Wisdom: Abdulbasit Kassim and Michael Nwankpa, eds., *The Boko Haram Reader: From Nigerian Preachers to the Islamic State* (New York: Oxford University Press, 2018), 321–26. Boko Haram also renamed the town of Mubi, just fifty miles from Chibok, calling it Madinatul Islam, or City of Islam.

147 Maryam and Sadiya stepped onto the road: Interviews with M.A

148 They were embracing one another to celebrate their victory: Interviews with M.A. and Rakiya Abubakar, hereafter referred to as R.Ab.

148 hostages had been *chosen*: Interviews with M.A. and R.Ab. On release, some of the Chibok hostages would claim that the first twelve girls to be married had chosen to do so and were not forced. The girls who married say they were too terrified to resist. Disentangling the idea of "choice" in the context of captivity to a terrorist group is extremely fraught.

148 They knew they had already been identified as more compliant: Interviews with M.A., R.Ab., N.A., H.I., and P.M.

148 even provide a means of escape: According to M.A., several of the girls who agreed to marry considered that it could be the fastest way to escape, because they were told that "wives" were given more freedom.

150 One of the twelve, Ummi, would be married to Malam Ahmed himself: M.A., N.A., R.Ab., H.S., and L.S.

150 The brides were not invited to their weddings: Interviews with M.A., R.Ab.

150 an emir, or senior commander, who was wanted by Nigerian and US authorities: M.A. said the marriages that night were conducted in the house of Mamman Nur, a veteran jihadist considered to be one of Boko Haram's most senior leaders before he was executed by hard-liners in 2018.

151 She wrote a quick note to Naomi Adamu: Interviews with M.A. and N.A.

151 Naomi felt little sympathy for Maryam: Interviews with N.A.

151 emiras, fighters' wives with a new status and privileges: Nwaubani, "Women Rescued from Boko Haram Who Are Returning to Their Captors"; see also: Matfess, *Women and the War*, for more on the status and privileges of fighters' wives.

151   he had dowries for each of them—15,000 naira, around $80: Interviews with
      M.A. and R.Ab.
152   arriving at the door: In interviews, the Chibok hostages said Boko Haram con-
      fiscated homes in Gwoza and offered them to members according to rank. This
      was corroborated by Salkida and members of the Dialogue Facilitation Team.

Chapter 17: "These Girls Are Global Citizens"

153   Oby Ezekwesili sat under a painting of George Washington: Interview with Eze-
      kwesili, September 2019, and with a senior White House official, September 2019.
153   How could it be that . . . there had been no news on the whereabouts of the
      Chibok girls?: Will Ross, "Chibok Abductions: Will Nigerian Schoolgirls Ever Be
      Freed?" BBC News, October 14, 2014.
153   America's rescue effort, announced with so much fanfare, had faded: Neither
      the international press nor social media seemed aware that the drones sent
      to find the Chibok girls had been repositioned to Iraq and Afghanistan or that
      most of the American personnel sent to help free them had returned home.
153   At night, she was still receiving calls from parents: Interviews with Oby Eze-
      kwesili.
154   R&B star Alicia Keys and her husband, the beatmaker Swizz Beatz: Mesfin
      Fekadu, "Alicia Keys Holds Protest for Lost Nigerian Girls," Associated Press,
      October 15, 2014; and Twitter archive @AliciaKeys, September 16, 2018.
154   another march, titled The Need For Speed: Interviews with Oby Ezekwesili,
      Twitter archive @ObyEzeks.
154   John Podesta, President Obama's senior counselor, was in his office: Interviews
      in September 2019 with a senior White House official, with Oby Ezekwesili, and
      with a member of Podesta's staff.
154   "These girls are global citizens": Ibid.
154   frozen into a kind of bureaucratic limbo: Ibid.; corroborated by authors' inter-
      views with numerous senior administration officials between 2017 and 2020.
154   John Kerry had plowed in, trying to convince officials around the table: The
      authors spoke to two senior State Department officials who served under Kerry,
      as well as an NSC official who witnessed this exchange. All of them spoke on
      condition of anonymity.

Chapter 18: Christmas

156   "Dear my lovely dad": Extract from diaries.
156   To celebrate Christmas: Ibid., expanded on by N.A. in interviews.
156   an attack across the border on Christian communities in Cameroon: A report
      by Felix Bate, "Boko Haram Militants Stage Attacks in Northern Cameroon,"
      Reuters, December 28, 2014, describes how the thousand fighters crossed the
      border and attacked five towns and a military camp before being repelled by the
      air force.
156   keeping a December calendar: Extract from diaries.
156   On Christmas Day back home in Chibok: Interviews with N.A., K.A., R.A., and
      members of the Chibok Association. The authors also spoke to Pastor Enoch
      Mark in Chibok in May 2016.
156   Kolo would prepare a pot of beef: Interviews with K.A.
156   Naomi was in tears as she dictated a letter to her father: Interviews with N.A.
157   They were writing in a new notebook given to them by the Malam: Ibid. The au-
      thors also reviewed this diary, which was longer than the others and contained doz-
      ens of scraps of paper, with notes, phone numbers, or Bible verses jotted on them.

157 "NEW: Ivory Complexion Care": Ivory Beauty Soap is a popular soap brand manufactured in Lagos.

158 tying them around her thigh in a makeshift pouch: In an interview, N.A. showed the authors the pouch she made from torn pink-and-white fabric that helped her secrete the diaries.

158 She learned to sit and sleep in a position: Interviews with N.A.

158 "Dad, I want to see you": Extracts from the diaries.

158 makeshift clinic that Boko Haram had established: Interviews with N.A., H.I., P.M., S.A., and H.S. The Boko Haram clinics were also described by other hostages and defectors in interviews.

159 Other students had been sent away to perform drudgery: Interviews with H.N., N.A, and M.A. Some other students were sent to the house of a commander in the town of Bama to work for Boko Haram wives, cooking, cleaning, and mending clothes.

159 Lydia John had been ordered to sweep: Interviews with N.A.

159 Ali Doctor, as the ten students called him: Ibid., and interviews with H.I., P.M., and S.A.

160 he had a family back in Maiduguri: Ibid.

161 a teacher from Chibok: The story of the teacher was recounted by several of the Chibok girls. The school's faculty said that none of the teachers were Boko Haram members or sympathizers.

Chapter 19: The Foreign Fighters

165 A strike force of South African mercenaries: The authors exchanged detailed e-mails with Eeben Barlow in April 2020 and consulted *Eeben Barlow's Military and Security Blog* (http://eebenbarlowsmilitaryandsecurityblog.blogspot.com) and two of his books.

165 blue fiberglass letters welcoming visitors to MAIDUGURI INTERNATIONAL AIRPORT: Author visits to the airport in 2014 and 2015.

165 had once fought as special forces for the apartheid government's Defense Force: Barlow's autobiographical memories of his combat years can be found in Eeben Barlow, *Executive Outcomes: Against All Odds* (Pinetown, South Africa: 30° South Publishers, 2018). The authors also spoke with Nigerian soldiers and officials who interacted with him.

165 arrived with sacks of dry food, utility belts, medical kits: Ibid., confirmed in author e-mails with Barlow.

166 had invited some of the world's most controversial foreign soldiers: "Nigeria Drafts in Foreign Mercenaries to Take on Boko Haram," Voice of America, Washington, DC, March 12, 2015.

166 two thousand troops from neighboring Chad: "Boko Haram Crisis: Chad's Troops Enter Nigeria," BBC News, February 3, 2015.

166 a column of Toyota pickups: The Chadian Army, the most feared in the region, had mastered the use of Toyota Hiluxes as an offensive weapon and counterinsurgency tool. In the 1980s, the Chadian Army, with just four hundred vehicles, inflicted a heavy defeat on the Libyan Army in the so-called Toyota War. In total, 7,500 Libyans were killed, and military equipment worth $1.5 billion was destroyed or captured. Chad lost only 1,000 men and very little equipment. For more information, see Ignacio Yárnoz, "Toyota Wars and the Next Generation in Counter Insurgency Strategies," Global Affairs, Strategic Studies, Universidad de Navarra, Pamplona, Spain.

166 joined by Ukrainian helicopter pilots: Adam Nossiter, "Mercenaries Join Nigeria's Military Campaign Against Boko Haram," *New York Times*, March 12, 2015.

166   sent three Tornado surveillance jets: "Royal Air Force Tornados Deployed to West Africa," DefenceWeb, September 8, 2014.

166   Vladimir Putin's Russia and Angela Merkel's Germany: Joe DeCapua, "Analysts Weigh Nigeria-Russia Arms Deal," Voice of America, Washington, DC, December 10, 2014; and "Merkel Supports the Fight Against Boko Haram," Bundesregierung [The Federal Government], January 19, 2015, https://www.bundesregierung.de/breg-en/news/merkel-supports-the-fight-against-boko-haram-421898.

166   Abubakar Shekau was on YouTube: Martin Ewi, "What Does the Boko Haram–ISIS Alliance Mean for Terrorism in Africa?" Institute for Security Studies, Pretoria, March 17, 2015, https://issafrica.org/iss-today/what-does-the-boko-haram-isis-alliance-mean-for-terrorism-in-africa.

166   "Oh Muslims, this is a new door opened by Allah": Statement from Islamic State spokesman, Abu Mohammed al-Adnani, according to the Search for International Terrorist Entities (SITE) Institute's monitoring group, which tracks jihadist statements; see: Nossiter, "Mercenaries Join Nigeria's Military Campaign."

166   few fighters managed to reach northeastern Nigeria: Interviews with Nigerian security officials, who spoke on condition of anonymity.

166   Istanbul, whose bustling airport, allowed thousands of recruits: Ayla Albayrak and Joe Parkinson, "Turkey Struggles to Halt Islamic State 'Jihadist Highway,'" Wall Street Journal, September 4, 2014.

167   pursued by an awkward coalition: The authors interviewed hunters involved in combat against Boko Haram in May 2014 and August 2015. Their efforts go underappreciated as a factor in the war; see: Drew Hinshaw, "Nigerian Hunters Join Effort to Find Missing Girls," Wall Street Journal, May 19, 2014.

167   corked in Black Is Beautiful face cream: Barlow, Executive Outcomes.

167   As white minority rule crumbled, Barlow had looked for other battlefields: Ibid.

168   the Jonathan administration had dramatically expanded his contract: Author correspondence with Barlow, April 2020.

168   a tactical concept called relentless pursuit: Barlow details his tactical and strategic concepts in his book, Composite Warfare: The Conduct of Successful Ground Force Operations in Africa (Pineland, South Africa: 30° South Publishers, 2016), appendix 6.

168   "The Western response was hashtag-save-the-girls": "Eeben Barlow: Inside the World of Private Military Contractors," Al Jazeera, January 5, 2020. Barlow reiterated this sentiment in his exchange with the authors.

168   In one nightmare, she saw a warplane: Interviews with N.A.

168   a new camp called Sabil-Huda, or Path of Guidance: Ibid., confirmed in interviews with more than a dozen hostages. Neither the Nigerian nor Western media mentioned Sabil Huda until 2018, when the army's Operation Deep Punch cleared parts of the camp. See: "Nigerian Troops Clear Boko Haram Sabil Huda Hideout, Free 19 Women, 27 Children," Sahara Reporters, New York, February 13, 2018.

169   screams of neighbors over the walls of their prison compound: Interviews with N.A., M.A., H.I., and P.M.

169   Sabil-Huda was far larger than "Timbuktu": Ibid.

169   cramped metal structures, just large enough for a few bodies to lie down inside: Ibid. Some images and video of Sabil-Huda showing small structures erected under trees, including houses and a small factory producing ammunition, would be published in 2018 when the Nigerian Army cleared the camp. See: "Update on Operation Deep Punch II," Defence Headquarters Nigeria, February 13, 2018, https://defenceinfo.mil.ng/update-on-operation-deep-punch-ii-troops-clear-boko-haram-out-of-sabil-huda-hideout-in-sambisa-forest.

170   "I think the war might start today": Interviews with N.A. and H.I.

171   some sort of rockets or possibly heavy gunfire: The attack and its aftermath was described by N.A., with parts corroborated by L.S., H.I., and P.M.

173   on a camp radio, interviews with some of the parents: Ibid.

173   Naomi knew for sure that Kolo would have joined: Interviews with N.A.

174   the melody of one of their favorites, "Dogorana," meaning "Trust": See Mama Agnes, "Dogarana," YouTube, September 2014, https://www.youtube.com/watch ?v=2p_lpO4LHVc. In the video Mama Agnes, flanked by a clutch of dancers in brightly colored dresses, sing about holding their faith.

Chapter 20: Agents of Peace

175   Under strip lighting in a conference room on the second floor: Interviews with Z.M., as well as Swiss officials and other members of the team, who spoke on condition of anonymity; corroborated by pictures from the Swiss embassy's on-line magazine, *Swiss Gazette* no. 31, September 2017, https://www.eda.admin .ch/dam/countries/countries-content/nigeria/en/Swiss%20Gazette_31_E.pdf.

175   a large color-coded organization chart of Boko Haram's leaders: The authors have reviewed the unpublished presentation, "Mapping the Boko Haram Nebulous," 2015.

175   Zannah scanned the names and recognized only a few: Interviews with Z.M. and with other people present, who spoke on condition of anonymity.

175   the Barrister had been gleaning clues to his partner's backstory: Ibid.

176   funded by a discreet Swiss government agency: Interviews with Swiss and Nigerian officials, all of whom spoke on condition of anonymity.

176   The Swiss diplomat had neither enormous resources nor special access: Ibid.

176   and a tiny spy agency: According to the Swiss Federal Intelligence Service's report *Switzerland's Security 2019*, https://www.newsd.admin.ch/newsd/message /attachments/57076.pdf, the service comprised 316 full-time positions.

176   dispatched exactly one person to deal with Boko Haram: Interviews with Swiss and Nigerian officials.

176   often eccentrics from the fringes of the conflict: Interviews with Z.M. and other members of the team.

177   a more idealistic term—Agents of Peace: Ibid.

177   had to assume that their phones would be tapped: One of the team members said in an interview in July 2019 that they were initially very disciplined in their communications, using code and avoiding certain words or names that could attract attention. But as the years went on, he said, "We started not to give a shit."

177   an article of faith: absolute discretion: Switzerland's Federal Department of Foreign Affairs said in the December 23, 2015, post "Focus on Mediation" on its website: "Mediation tends to take place behind the scenes; discretion and stamina are the order of the day." See: https://www.eda.admin.ch/countries/japan/en/home /news/news.html/content/eda/en/meta/news/2015/12/23/meditation_im_fokus.

177   "Everything is being done transparently but quietly": E-mail reviewed by the authors.

177   "The people who talk the most have the least to contribute": Interviews with Z.M. and other members of the team.

178   trading thoughts over encrypted phone apps: The authors reviewed WhatsApp chat groups in which the Chibok girls, or CBGs, were discussed among mediators, who spoke on condition of anonymity.

178   talk their way onto the governor's private jet: The authors flew with Borno State governor Kashim Shettima on his jet as part of press visits in 2014 and 2015, until the airport reopened.

178   an acceptance speech for the Nobel Peace Prize: Interviews with Z.M. and other members of the team.

178　a razor-sharp analyst of jihadist groups: Ibid.

179　At the other end of the line was Roger Federer: Ibid.

179　Neither man used a Twitter account: Z.M. opened an account, @zbmustapha1, in 2009 but didn't send his first tweet until April 4, 2017, https://twitter.com /zbmustapha1/status/849163640645312514.

179　they had come to accept that the fleeting storm: Interviews with Z.M. and other members of the team.

179　They were the boulder in the road: "Mediation Process Matrix," SwissPeace, Bern, 2012, https://www.swisspeace.ch/fileadmin/user_upload/pdf/Mediation/Mediation -Process-Matrix_2016.pdf.

179　Jonathan had given the Swiss embassy official clearance: Interviews with Swiss and Nigerian officials.

180　"Trust is the fuel of the mediator": "Mediation Process Matrix."

180　makeshift classroom for the study of Boko Haram: Interviews with Z.M., other team members, and Swiss officials.

180　They mapped its leadership structure: Ibid., and "Mapping the Boko Haram Nebulous."

180　brochures advertising courses: The embassy displays brochures for courses at Ecole hôtelière de Lausanne, where a one-year master's degree in Global Hospitality costs $36,000.

181　An American from West Point: Interviews with Z.M. and other members of the team.

181　the neighboring embassy of Denmark, whose staff were intrigued: The authors spoke to officials at the Danish embassy in August 2017.

181　assumed that every call was listened to: Interviews with Z.M.

181　"Don't write your real name": Interviews with Nigerian officials and members of the team.

182　an encrypted fax machine: The 007 was described by Swiss officials in Nigeria and Bern, including serving and former ambassadors, who spoke on condition of anonymity. In late 2019, after the US assassination of Iranian general Qassim Soleimani, American and Iranian officials were able to communicate to de-escalate the prospect of war through the Swiss ambassador, who relayed their messages via the 007 kept in a sealed upstairs room of the Tehran embassy. For more details, see Drew Hinshaw and Joe Parkinson, "Swiss Back Channel Helped Defuse U.S.-Iran Crisis," *Wall Street Journal*, January 11, 2020.

182　upper floor of a nondescript government office building: Ibid. The building is also opposite a bakery and an art-house cinema called Cinema Rex. Dozens of bicycles are parked outside, many of them unchained.

182　a staff of about seventy officials: Interviews with Human Security Division and Foreign Ministry officials in Bern in November 2017, January 2020, and March 2020. The authors also interviewed Thomas Greminger, a founding figure of the division, in November 2017.

182　From this cramped gray-paneled office: Ibid.

182　sent dozens out into the world to help solve intractable conflicts: Ibid.

182　Many Swiss people had no idea the division existed: Ibid.

182　a fleeting moment, "the end of history": See Andreas Graf and David Lanz, "Conclusions: Switzerland as a Paradigmatic Case of Small-State Peace Policy?" *Swiss Political Science Review* 19, no. 3, special issue, *Natural Born Peacemakers?: Ideas and Identities in Foreign Policies of Small States in Western Europe* (September 10, 2013): 410–23.

183　"You cannot stay anymore in this world hiding behind your mountains": Interviews with President Calmy-Rey in November 2017 and January 2020.

183　could issue visas to wanted men whom other states called terrorists: Interviews with Swiss Foreign Ministry officials. See: Simon Bradley, "Hamas Visit to Switzerland Sparks Outcry," SWI, SwissInfo.ch, January 19, 2012. For example, the PKK (Kurdistan Workers' Party), listed as a terrorist group by the US and EU, not only sends members to Geneva, but "continues to recruit and launch fundraising campaigns in Switzerland." See: Peter Siegenthaler, "PKK Creates Rift Between Swiss Kurds and Turks," SWI, SwissInfo.ch, November 3, 2015.

184 Colombian FARC insurgency sent Swiss presidents Christmas cards: Interviews with President Calmy-Rey, November 2017 and January 2020.

184 One government-backed NGO started inviting wanted militants: Officials of the NGO, which the authors visited in October 2018, asked that it not be named, claiming that a US Defense Department official joked about sending a Tomahawk missile into their next gathering.

184 backchannel contacts helped Swiss leaders schedule face time with US secretaries of state: Interviews with President Calmy-Rey, November 2017 and January 2020.

184 Greminger had his officers call the Islamist extremists: Ibid., as well as other Swiss ministry officials, who spoke on condition of anonymity.

184 their foreign ministry knew how to reach the militants who'd taken them: Ibid.

184 At one point, the government suggested opening talks with Osama bin Laden: Clare O'Dea, "Call for Dialogue Stirs Trouble for Calmy-Rey," SWI, Swiss Info.ch, August 27, 2008.

184 some 650 peace agreements: Marko Lehti, *The Era of Private Peacemakers: A New Dialogic Approach to Mediation* (London: Palgrave Macmillan, 2019), chapter 2.

184 more than a dozen accords from Sudan to Indonesia: See: Thomas Greminger, *Swiss Civilian Peace Promotion: Assessing Policy and Practice*, Center for Security Studies, Federal Institute of Technology, Zurich, 2011, https://css.ethz .ch/content/dam/ethz/special-interest/gess/cis/center-for-securities-studies/ pdfs/Swiss-Civilian-Peace-Promotion.pdf; and David Lanz et al., "Understanding Mediation Support Structures," SwissPeace, Bern, October 2017, https:// www.swisspeace.ch/fileadmin/user_upload/pdf/Mediation/swisspeace_MSS _study_18_Oct_Final.pdf.

185 an industry of war that was daunting to untangle: For more, see: David Keen, *Useful Enemies: When Waging Wars Is More Important Than Winning Them* (New Haven, CT: Yale University Press, 2012).

185 The division labeled Nigeria a tier-one target: Interviews with Swiss Foreign Ministry officials.

185 The individual they sent was not a career diplomat: The authors spoke to colleagues and associates of the Swiss diplomat, all of whom spoke on condition of anonymity.

Chapter 21: The General

186 talking live to Christiane Amanpour: Christianne Amanpour, "Nigeria Pres. Backs Talks for Girls If Boko Haram 'Can Deliver,'" CNN, July 21, 2015, https:// www.cnn.com/videos/world/2015/07/21/intv-nigeria-amanpour-muhammadu -buhari.cnn.

186 "What about the girls, the famous Chibok girls who were kidnapped so long ago?": Ibid.

186 four months earlier on a promise to vanquish Boko Haram: "Nigeria Elections: Winner Buhari Issues Boko Haram Vow," BBC News, March 2015.

186 Buhari had landed in Washington for his first visit as president: John Campbell, "Buhari Visit to Reset the Bilateral Relationship," Council on Foreign Relations blog, July 20, 2015, https://www.cfr.org/blog/buhari-visit-reset-bilateral -relationship.

187   "We can not claim to have defeated Boko Haram without rescuing the Chi-
      bok girls": "President Buhari's Inaugural Speech," *Guardian Nigeria*, May 29,
      2015, https://guardian.ng/features/president-muhammadu-buharis-inaugural
      -speech.

187   President Buhari had once been Major General Buhari, a soldier-dictator: Clif-
      ford. D. May, "Deposed Nigerian President Is Under Arrest," *New York Times*,
      January 4, 1984.

187   who used his twenty months in power to jail journalists: One of General Bu-
      hari's most controversial laws was Decree No. 4, a measure that permitted the
      government to jail journalists even for truthful stories if the articles criticized
      or embarrassed the government. See: Clifford D. May, "Nigeria's Discipline
      Campaign: Not Sparing the Rod," *New York Times*, August 10, 1984.

187   execute drug dealers, order the expulsion of nearly a million immigrants: "Mu-
      hammadu Buhari, Nigeria's President in Profile," BBC News, December 2014.

187   imprison dozens of politicians for corruption: Buhari's government even tried to
      abduct politicians who had fled abroad. In 1984, his government attempted to
      repatriate former cabinet minister Umaru Dikko, accused of looting as much
      as $1 billion, from London by drugging him, sticking him into a specially made
      crate, and putting him on a plane back to Nigeria—alive—but he was discov-
      ered in the crate, claimed as diplomatic baggage, before the plane took off.

187   "The trouble with Nigeria is simply and squarely a failure of leadership": Chi-
      nua Achebe, *The Trouble with Nigeria* (Portsmouth, NH: Heinemann, 1984), 1.

187   known as Nigeria's conscience: Victor Ehikhamenor, "Africa's Voice, Nigeria's
      Conscience," *New York Times*, March 25, 2013.

187   as a young cadet at an infantry college in Aldershot, England: According to
      biographer Lawal Jimi Adebisi in *Buhari: The Making of a President* (Sebas-
      topol, CA: Safari Books, 2017), Buhari attended Mons Officer Cadet College, a
      precursor to Sandhurst, whose alumni include two other Nigerian generals who
      became presidents, Olusegun Obasanjo and Sani Abacha.

187   Buhari launched a nationwide War Against Indiscipline: John Campbell, "Ni-
      geria's War Against Indiscipline," Council on Foreign Relations blog, October 4,
      2016. In 2016, as a civilian chief of state, president Buhari relaunched the effort
      and rebranded it (without frog jumps and whips) as "Change begins with me."
      It was similarly unsuccessful. See Isa Sanusi, BBC Hausa, "Can Buhari Get
      Nigerians to Queue Again?" September 15, 2016.

187   his close friends in the military seized power for themselves: "Army Officers
      Said to Overthrow Nigeria's Ruling Military Council," Associated Press, August
      28, 1985.

188   emerged as a democracy, laden with debt and broken infrastructure: Norimitsu
      Onishi, "Nigeria's Military Turns Over Power to Elected Leader," *New York
      Times*, May 1999.

188   More than a million people had been made homeless by war: "Historic Nigeria
      Election Won by Retired General Buhari," Deutsche Welle, April 1, 2015.

188   as he boarded a flight: Jonathan was following in Charles Darwin's footsteps to
      an island famed for its flightless blue-footed boobies, marine iguanas, and giant
      tortoises. See William Wallis, "What Is Next for Nigeria?" *Financial Times*, July
      24, 2015.

188   begun calling him Baba Go-Slow: Daniel Magnowski, "Buhari Goes from Nige-
      ria's Change Champion to 'Baba Go Slow,'" Bloomberg, July 1, 2015.

189   with a pair of crying mothers: "Buhari, Wife Meet Mothers of Chibok Girls,"
      Premium Times (Abuja), June 12, 2015.

189   President Buhari had gathered his most senior advisers: Interviews with senior
      Nigerian officials, including people present, all of whom spoke on condition of
      anonymity.

189 He wanted the Chibok girls returned within his first hundred days as president: Sani Tukur, "100 Days: Chibok Girls Still Missing as Buhari Tackles Boko Haram Insurgency," *Premium Times* (Abuja), September 6, 2015.

189 It was 94 degrees on the morning of July 20: Historical weather reports from CustomWeather.com.

189 had spent the morning with Joe Biden: Senior US officials present or briefed on the meeting, who spoke on condition of anonymity, described a gregarious and long-winded Vice President Biden rambling to a jet-lagged and taciturn President Buhari.

189 leaving with the regret that Africa hadn't gotten more bandwidth: Ibid.

189 Obama welcomed Nigeria's new leader as a "man of integrity": Video clips and transcript available at "Remarks by President Obama and President Buhari of Nigeria Before Bilateral Meeting," obamawhitehouse.archives.gov, July 20, 2015, https://obamawhitehouse.archives.gov/the-press-office/2015/07/20/remarks -president-obama-and-president-buhari-nigeria-bilateral-meeting.

189 When the cameras: Interviews with US and Nigerian officials present.

190 a dramatic increase in military assistance the US was offering: "Obama Pledges Support for Nigeria's Fight Against Militants," BBC News, July 20, 2015.

190 the Pentagon was pouring hundreds of troops: Nick Turse, "You Might Not Know Where Chad Is, but the US Military Has Big Plans for It," *Mother Jones*, November 2014.

190 were bringing new tactical vests and night-vision goggles: Interviews with US officials.

190 it created a new post of special coordinator for the Boko Haram conflict: The Obama administration appointed veteran ambassador and Africa hand Dan Mozena. Eric Schmitt, "African Training Exercise Turns Urgent as Threats Grow," *New York Times*, March 8, 2015.

190 much more American intelligence . . . supplied to the Nigerians: Interviews with senior US officials.

190 It appeared in a local newspaper, next to a posting for "Crew at Chipotle": Elizabeth Janney, "44 Job Openings in Columbia," Patch, October 6, 2015, https:// patch.com/maryland/columbia/44-job-openings-columbia-0.

190 pledged to "crush" it with a renewed ground advance: Daniel Magnowski, "Buhari Promises Campaign to Crush Boko Haram in Nigeria," Bloomberg, April 1, 2015.

191 Alpha Jets—with the weapons removed: The Alpha Jets were delivered in two batches by Air USA Inc., based in Quincy, Illinois, which specializes in air-combat training and ordnance according to "Nigerian Air Force Receiving Additional Alpha Jets," DefenceWeb, March 2015.

191 read a license plate from fifteen thousand feet: Major Emily Potter, "Comparing Apples to Elephants," US Army, December 10, 2013, https://www.army.mil /article/116700/comparing_apples_to_elephants. Also see the work of Sean D. Naylor, including "The Army's Killer Drones: How a Secretive Special Ops Unit Decimated ISIS," Yahoo! News, March 2019.

191 the Night Stalkers: Interviews with US officials. Also see: Joseph Trevithick, "Army's Elite Night Stalkers Quietly Stood Up a New Unit Ahead of Getting New Drones," *War Zone* newsletter, TheDrive.com, February 9, 2019; and Joshua Hammer, "Hunting Boko Haram."

191 trying to convince John Kerry: Neanda Salvaterra and Drew Hinshaw, "Nigerian President Goodluck Jonathan Wants U.S. Troops to Fight Boko Haram," *Wall Street Journal*, February 13, 2015.

191 Twitter had finally deleted the group's account: Omar S. Mahmood, "More than Propaganda: A Review of Boko Haram's Public Messages," West Africa Report 20, Institute for Security Studies, Pretoria, March 2017.

191   "But as my wife reminds *me*, you volunteered for it": Interviews with US offi-
      cials present.

Chapter 22: Shadrach, Meshach, and Abednego

192   When Lydia John was a senior, she auditioned: Interviews with Lydia's parents,
      Rebecca and Samuel, in Abuja in October 2017 and November 2019.

192   The Glorious Singers accepted only the best: Several of the Chibok hostages
      spoke about the Glorious Singers.

192   Singing was a collective release for the town's residents: Ibid., and interviews
      with K.A., R.A., and members of the Chibok Association.

192   a student would often pull out a drum: Interviews with Chibok students. The
      authors also visited the young women at the New Foundation School at the
      American University of Nigeria in Yola in November 2017 and February 2018,
      after their release, and saw them drumming, singing, and dancing to traditional
      songs.

193   Lydia wasn't a natural performer, but she did have an exceptional voice: Inter-
      views with Rebecca and Samuel John, and N.A.

193   Music was the preserve of the *arna* and made God quake with rage: Although
      the Boko Haram fighters sang the *nasheed*, all other forms of music were con-
      sidered haram and banned. Women, especially, were not allowed to sing. For
      more, see: Zacharias Pieri and Jacob Zenn, "Under the Black Flag in Borno:
      Experiences of Foot Soldiers and Civilians in Boko Haram's 'Caliphate,'" *Jour-
      nal of Modern African Studies* 56, no. 4 (December 2018): 645–75. The same
      interpretation was taken by the Taliban, who banned music when they ruled
      Afghanistan until 2001. See Sean Carberry, "From a Land Where Music Was
      Banned," National Public Radio, February 3, 2013.

193   she and Naomi would write down lyrics they didn't want to forget: Interviews
      with N.A.

193   They had been filling them up with hymns: Extracts from the diaries.

193   Ukuba, an abandoned village: The camp at Ukuba was described by more than
      a dozen of the Chibok hostages in interviews.

194   Naomi would choose which songs to perform: Interviews with N.A.

194   clusters of hostages would sing through cupped hands: Ibid., and interviews
      with H.I., P.M., S.A., and H.N.

195   The girls could hear the *rijal* share rumors that Shekau had been badly injured:
      In August 2015, Chad's president Idriss Déby announced that Shekau had been
      killed in an airstrike. He was alive and reappeared some days later in an audio
      recording. But Nigerian intelligence officials and members of the Dialogue Fa-
      cilitation Team would later claim that Shekau was badly injured and had trav-
      eled into Chadian territory, then to Libya for treatment to his leg. Those claims
      are impossible to verify, but several of the Chibok hostages said that when they
      met him, he had an obvious injury to his leg. See David Zounmenou and Segun
      Rotimi Adeyemon, "Dead or Alive: The Seven Lives of Boko Haram's Leader,"
      Institute for Security Studies, Pretoria, September 2015, https://issafrica.org
      /amp/iss-today/dead-or-alive-the-seven-lives-of-boko-harams-leader.

195   They were the *arnan daji*, "pagans in the forest": The term and its origin were
      described by N.A., P.M., H.I., and S.A.

195   a collective identity, speaking as "we" more than "I": Teachers and counselors
      to the freed Chibok hostages said that it took years before many of them re-
      learned to respond to questions with the word "I." The authors spoke to Somi-
      ari Fubara Demm, the NFS (New Foundation School) counselor, several times
      between 2018 and 2020.

195 *Bashin Gaba*, "Debt of Hatred": A popular Hausa-language movie made in "Kannywood," the northern city of Kano's film hub and answer to Lagos's much bigger national "Nollywood."

195 But the girls had already buried their diaries: The burying of the diaries was described by N.A., H.N., and L.S.

196 watched as a guard whipped every single one with an electrical wire: Ibid.

198 the one who most strenuously refused to get married: Interviews with N.A., H.S., and H.I.

198 the result of a secret abortion: Interviews with N.A.

198 five girls snuck off to the banks of a small stream: The authors spoke to three of the five girls who snuck away, N.A., H.I., and Yanke Shettima, hereafter referred to as Y.S.

198 see if it fit the camera phone: Ibid. The hostages said the phone was made by Tecno, a Chinese manufacturer known for cheap handsets that are popular in Nigeria.

198 waiting for a bar to appear, one of the girls had an idea: Ibid. Not mentioned in the book are the extreme lengths to which Naomi went to keep her phone charged, using sunlight to warm the battery, for example. There was also electricity in the senator's mansion, but no reception, as cell towers had been dismantled during the conflict.

Chapter 23: Sisters

200 Maryam and Sadiya were standing close to each other: The scene of the public executions was described by M.A. The authors also interviewed other former hostages—men and women—who described similar scenes, and reviewed gruesome video footage of Boko Haram punishments, including executions and amputations, from Salem Solomon, "Boko Haram: Terror Unmasked," Voice of America, News, February 2017, https://projects.voanews.com/boko-haram-terror-unmasked, and other sources.

200 An old man was sentenced to a hundred lashes for using drugs: For more on the choreography of these punishments, see: Pieri and Zenn, "Under the Black Flag in Borno."

201 "Whoever does not watch all the punishments will be beaten": Interviews with M.A.

201 relocated to new camps elsewhere in the northeast: Ibid., and interviews with other Chibok hostages.

201 full-length black niqab that covered everything but the eyes: Ibid.

201 women wearing precious stones on henna-dyed fingers: Journalist and author Adaobi Tricia Nwaubani has spoken to women who married senior fighters, who describe the luxurious gifts they received after looting raids. For more, see her December 20, 2018, piece in the *New Yorker*: "The Women Rescued from Boko Haram Who Are Returning to Their Captors."

202 The Boko Haram wives viewed them with suspicion: Interviews with M.A., R.Ab.

202 Abu Walad, a fighter who helped guard the Chibok girls: Abu Walad is identified and described by M.A. and several of the other Chibok hostages whom he helped to guard.

203 Sadiya was always accompanied by her co-wife: Senior fighters often took several wives. Adaobi Tricia Nwaubani describes the four wives of senior commander Mamman Nur in "The Women Rescued from Boko Haram Who Are Returning to Their Captors."

204 Maryam worried that Sadiya was becoming too sympathetic: Interviews with M.A.

205 The phone lit up with footage of thousands of people marching: Ibid.
207 But as she lifted the niqab, some rushed to greet her: Ibid., and interviews with N.A., H.I., and P.M.

Chapter 24: The Imam

208 Masked fighters were standing on a dirt track: The journey to see Shekau was described by N.A., H.I., P.M., and S.A.
208 The driver was Tasiu: Tasiu, aka Abu Zinnira, rose to become Boko Haram's head of communications and appeared, always masked, in the group's propaganda videos.
208 Gaba Imam, the House of the Imam, was Shekau's hideout: The Chibok hostages say that Gaba Imam was a fluid location, wherever Shekau, who moved frequently, was hiding.
209 in the other, two Qurans, which he stacked on his lap: Each of the hostages present described the holy books. Shekau has used them as props in several appearances, including a 2016 video wherein he is seated ostentatiously behind dozens of leather-bound holy books embossed in gold Arabic script. "Abubakar Shekau: Video Mentioning Chibok Girls and Denying Death," September 25, 2016, UnmaskingBokoHaram.com, https://unmaskingbokoharam.com/2019/04/08/abubakar-shekau-video-on-ceasefire-and-mentioning-chibok-girls-september-25-2016.
210 journey to the Imam had begun at midnight the previous night: Interviews with N.A., H.I., and P.M.
212 Rather, he wanted a simple trade: Shekau repeatedly said in videos that he wanted jailed Boko Haram fighters in return for the Chibok hostages, including in his first appearance after their abduction in April 2014, when he said, "By Allah, we will not release them until you release our brothers." Full speech translated in Abdulbasit Kassim and Michael Nwankpa, eds., *The Boko Haram Reader: From Nigerian Preachers to the Islamic State* (New York: Oxford University Press, 2018).
212 one of his own wives had been arrested: Jacob Zenn and Elizabeth Pearson, "Women, Gender, and the Evolving Tactics of Boko Haram," *Journal of Terrorism Research* 5, no. 1 (2014).
213 eat any of the food and cook with any of the utensils: The idea, later voiced explicitly to the hostages, was to help them regain weight and look healthy for the television cameras if they were released.

Chapter 25: The Escapade

216 Naomi, Hauwa, and Saratu were crawling through waist-high grass: The story of the covert supply line was recounted by members of the so-called *arnan daji* and by some of the hostages who were being starved into submission.
217 some locals would trap, box, and deep-fry them: In a 2017 NPR dispatch from Maiduguri, "So What Does a Deep-Fried Grasshopper Taste Like?" Ofeibea Quist-Arcton said the delicacy, spiced with powdered chili, tastes like "a crunchy prawn, eaten with its shell." Another young reporter decided to embed with a group of grasshopper catchers in 2007 for an article, "Dusk to Dawn Lives of Grasshopper Catchers in Maiduguri," published in Abuja's *Sunday Trust*. His name was Ahmad Salkida.
217 The guards had become lackadaisical: It may seem bizarre that such high-value hostages would be able to forage alone for food or firewood, but after two years in the most heavily guarded parts of Boko Haram territory, where the militants

had total control, there had been no successful escapes, and their guards had become more relaxed.

218 Malam Ahmed had been depriving them of enough food: Described by H.S., L.S., H.N., and other hostages deprived of food.

219 "Mix salt with hot water" to wash yourself when he falls asleep: Concoctions using hot water and salt, and also using a local brand of analgesic known as Alabukun, lime and potash, are popular in northeastern Nigeria as emergency contraception, even though scientific research confirms that they don't work. For more, see Anthony Idowu Ajayi, et al., "Use of Non-Emergency Contraceptive Pills and Concoctions as Emergency Contraception among Nigerian University Students: Results of a Qualitative Study," *BMC Public Health*, BioMed Central, London, October 4, 2016, https://bmcpublichealth.biomedcentral.com/articles/10.1186/s12889-016-3707-4.

219 Lydia John . . . announced that she too was giving in: Lydia's eventual capitulation to marriage was described by M.D., L.S., and H.N., and also reported by N.A.

220 her classmates feared she'd been possessed by spirits: Lydia was not the only hostage who appeared to suffer such an episode in captivity. M.A. described another Chibok captive with a high fever who began to scream uncontrollably. Boko Haram treated her with herbal potions and smoke from burning coals. She said the militants told her the use of these herbs was encouraged by the Prophet Mohammad, who wrote, "Utilize the black seed, for without a doubt, it is a cure for all," Hadith Al-Bukhari 7:591.

220 In Lydia's absence, she had stopped keeping a diary: The diaries are undated except for the Christmas dispatch but appear to have finished in the early months of 2015.

Chapter 26: The Deadline

224 standing against a cream-colored wall, her face: The authors have seen stills of the proof-of-life videos handed to Ahmad Salkida.

224 Nearby, Saratu, Hauwa, Palmata, and the other holdouts sat under a tree: The story of the abortive hostage exchange was relayed by N.A., H.I., and P.M.

224 sending the more pliable captives away on trucks: Ibid.

225 They were burial shrouds that Boko Haram used to wrap their dead: Islamic custom dictates that the dead are to be washed and wrapped in a shroud before being placed on their right side and the body covered with wood and earth. For more, see Phil Hazlewood, "Gravediggers of Maiduguri: Burying Boko Haram and the Past," Agence France-Presse, February 6, 2016.

226 a battleship-gray C-130 Hercules military transport plane: Interviews with Salkida and Nigerian government officials.

226 Inside the C-130 cabin sat three handcuffed Boko Haram members: Ibid.

226 jailbreak that freed more than seven hundred: Ahmad Salkida said the qaid to be exchanged was senior commander Ibrahim Agaji, allegedly responsible for the infamous Bauchi prison break in 2010. See: Scott Stearns, "Gunmen Stage Massive Prison Break in Northern Nigeria," Voice of America, September 7, 2010.

226 Another had been Shekau's "accountant": Identified by Salkida as Mohammed Shuibu, Boko Haram's money man and communications chief, whom he had known in Maiduguri before the war.

227 In the suburbs of Dubai, the exile had come to feel at home: Interviews with Salkida, who said his family had to unlearn some of the habits they'd picked up in the Gulf when they returned to Abuja. "In Nigeria, if you try to walk across a Zebra crossing, you'll be run over," he said.

227 skeptical . . . of the man behind this well-sourced Twitter handle: "This Is Ah-
     mad Salkida, the 'Wanted' Journalist Who Knows BH Inside Out," *The Cable*
     (Abuja), August 2015.
227 Are you a member of the insurgency?: Mansur Liman, "Girls 'Precious' to Boko
     Haram," Africa Live Round-Up, BBC Hausa, April 14, 2015.
227 Would he like to come in for a meeting?: Interviews with Salkida and Nigerian
     officials.
227 the president had authorized a new round of negotiations: One of the details
     revealed in a post published on Medium.com by Nigeria's government, May
     9, 2017, "Chibok Girls: A Timeline," https://medium.com/@TheAsoVilla/chibok
     -girls-a-timeline-571a36e7c173.
227 husky voice he knew as Abu Zinnira. Naomi knew him as Tasiu: Interviews with
     Salkida and Nigerian government officials.
228 Salkida on a Thuraya satellite phone to discuss whether a deal could be possi-
     ble: Salkida says he was in regular contact with Tasiu, who would in turn relay
     the progress of negotiations to Shekau. One of the conditions of the swap was
     that the Chibok girls would be brought close to the exchange point just outside
     Maiduguri so it could be done quickly, because each side was worried about
     being ambushed by the other.
228 A convoy carried them to a three-bedroom apartment: Interviews with Salkida.
228 wall surrounding the compound had moved farther north: The authors have
     driven around Maimalari Base several times but have never been granted per-
     mission to enter.
228 Operation Lafiya Dole, Peace by All Means: "Army Boss Launches 'Operation
     Lafiya Dole,'" *Pulse Nigeria*, July 20, 2015.
228 would sit alongside the US intelligence officers: Interviews with senior US and
     Nigerian officials.
228 a graveyard filling up with bodies marked by plastic placards: Joe Parkinson,
     "Nigeria Buries Soldiers at Night in Secret Cemetery," *Wall Street Journal*, July
     31, 2019.
229 the accommodation was almost unfathomably luxurious: Interviews with Salk-
     ida, who said that the days spent with these commanders were a wellspring of
     fresh insights into the group, its workings, and its key personalities.
229 the trucks halted at a small set of ramshackle houses: Interviews with N.A.,
     H.I., and P.M.
230 began to *ping* with voicemails: Interviews with Salkida.
230 arrived from Aliyu, the murderous commander: According to Salkida, Aliyu had
     been an understudy to Ibrahim Agaji before he was arrested.
230 a ransom of €500,000 to sweeten the offer: Interviews with people involved in
     the deal.
230 The cash was in Boko Haram's preferred currency: Salkida, a Nigerian cabinet
     minister, and several Boko Haram analysts confirmed that the insurgency pre-
     fers euros because they are easy to launder.
231 traded easily in the former French colonies surrounding Nigeria: Euros are
     widely accepted tender in the fourteen Francophone countries, including Chad,
     Niger, and Cameroon, that share the common CFA (Communauté Financière
     Africaine, or African Financial Community) currency. The CFA is pegged to the
     euro and printed by the French Central Bank.
231 furious debate emerged over the value of the hostages: Interviews with Salkida
     and Nigerian government officials familiar with the deal. One Nigerian intelli-
     gence officer described it thus: "The deal got jacked."
231 never agreed to or even discussed these demands: In the Nigerian government's
     2017 Medium.com post "Chibok Girls: A Timeline," it claimed that at the last

minute the captors issued a new set of demands, never previously discussed or negotiated.

231 He refused to fly back to the capital on the C-130: Interviews with Salkida.

231 he penned a cryptic tweet: Twitter archive @ContactSalkida.

232 Naomi and her classmates had begun cooking: Interviews with N.A., H.I., and P.M.

232 Tasiu showed up covered in dust: Ibid.

Chapter 27: The Windows Are Closed

234 phones of the Dialogue Facilitation Team *pinged* with messages: Interviews with members of the team.

234 the Swiss diplomat, could be summoned at any time: Ibid., and Nigerian government officials.

235 practically a resident of the Swiss embassy: Ibid.

235 "Meet the diaspora": "Mediating a Ceasefire," Organization for Security and Co-operation in Europe, July 23, 2019.

236 The Barrister had visited followers, sympathizers and former companions of Shekau: Interviews with Z.M. and members of the team.

236 once offered refuge to Osama bin Laden and Carlos the Jackal: Jonathan Schanzer, "Pariah State: Examining Sudan's Support for Terrorism, Foundation for Defence of Democracies, July 5, 2012, https://www.fdd.org/analysis/2012/07/05/pariah-state-examining-sudans-support-for-terrorism.

236 These men were out of the fight: For more on the Khartoum-based diaspora of Boko Haram, see: "Nigerian Jihadists in Sudan and the Sahel," in Jacob Zenn, *Unmasking Boko Haram: Exploring Global Jihad in Nigeria* (Boulder, CO: Lynne Reinner, 2020), 19–47.

236 a gardened oasis that visitors entered by walking under a floral archway: Pictures on Swiss government website under "Embassy of Switzerland in Sudan," https://www.eda.admin.ch/khartoum.

236 had talked for four full days over cups of sweet tea: Interviews with members of the team.

236 a philosophical concept they called "the windows": Ibid., and interviews with Swiss officials.

237 militants the Dialogue Team had painstakingly cultivated were dead: Ibid.

237 The army knew him as the Nyanya bomber: Interviews with Nigerian military officials.

237 Tahir was kneeling in the middle of prayer: Interviews with members of the team.

237 Tahir was beyond the reach of Swiss diplomacy: The team would later learn that Tahir was initially detained by the directorate of military intelligence, the DMI, and he would later be handed off to other security agencies and shuffled deeper into the system.

238 bitter turf war over strategy and access to the billions: The competition between security agencies was an open secret in Abuja and preceded Buhari, who tried to squelch it by placing a longtime friend and ally from his hometown, Lauwal Daura, to run the SSS (State Security Service). By the end of 2015, Daura was unquestionably in charge of the Chibok file, but the interagency rivalry persisted, officers from both agencies openly undermining each other and hurling mutual allegations of sabotage. The rivalry is tackled by Okechukwu Innocent Eme, "Inter-Security Agency Rivalry as an Impediment to National Counter Terrorism Strategy (NACTEST)," African Heritage Institution, Enugu, Nigeria, 2018, https://media.africaportal.org/documents/Inter-Security-Agency-Rivalry.pdf.

238   One of the agencies had launched an internal task force: Interviews with members of the team.

238   same impression, that Buhari had too little information: Interviews with Nigerian and Swiss officials and with a senior US official, who said Buhari responded with bromides that suggested he had little idea of the reality on the ground. One Western ambassador who met Buhari said, "I always wondered how much his staff told him."

238   "message management": Interviews with members of the team and Swiss officials.

239   began to call him the Glue: Ibid.

239   he would see a city utterly transformed: Interviews with Z.M.

239   Vast tent cities stretching for miles: By the middle of 2015, there were twenty-one camps for internally displaced people in Maiduguri alone. See: "Burden of Feeding Over 500,000 IDPs Getting Unbearable—Borno Govt," *Vanguard Nigeria* (Lagos), May 19, 2015.

239   Most of the new arrivals were children: "The Plight of Women and Children in Nigeria's IDP Camps," UN High Commissioner for Refugees, March 8, 2016, https://data2.unhcr.org/en/news/12180.

239   old flour mills: "Nigeria: Exploratory Mission Report," ReliefWeb, February 2016, https://reliefweb.int/report/nigeria/nigeria-exploratory-mission-report -february-2016.

239   hundreds of young children were near starvation, their bellies swollen: "Displaced by Boko Haram, 450 Nigerian Children Die of Malnutrition," *Premium Times* (Abuja), February 21, 2016.

239   the last psychiatrists left in the city: The authors visited Maiduguri's Federal Neuropsychiatric Hospital in May 2014, where they met the city's nine overworked psychiatric doctors and a few of their fifty-two thousand active patients. One was a six-year-old girl who had seen Boko Haram murder her parents and was now frequently interrupting her class by shouting: "They're here!" See Drew Hinshaw, "Boko Haram's Mounting Toll in Trauma," *Wall Street Journal*, May 30, 2014.

240   Billboards from the election held months earlier still hung eerily over the city: Author visits to Maiduguri in 2015.

240   They brought stories of forced marriage, children born of rape: Author visits to Maiduguri camps run by the government and the Red Cross. See: Mausi Segun, "A Long Way Home: Life for the Women Rescued from Boko Haram," Human Rights Watch, July 29, 2015, https://www.hrw.org/news/2015/07/29/long-way -home-life-women-rescued-boko-haram.

240   bombed so many times that its security guards had lost count: In 2015 alone, Maiduguri's Monday Market was hit four times by suicide bombers, leaving dozens dead. See Gbenga Akingbule, "In Nigeria, a Market Braves Militant Attacks," *Wall Street Journal*, November 20, 2015.

240   dispatching children wearing suicide vests, usually small girls: According to data from UNICEF, the deployment of child suicide bombers began in 2014 and rose rapidly to reach 236, of whom 191 were girls, in 2017. For more, see: Jason Warner and Hilary Matfess, "Exploding Stereotypes: The Unexpected Operational and Demographic Characteristics of Boko Haram's Suicide Bombers," Combating Terrorism Center at West Point, US Military Academy, August 2017, https://ctc.usma.edu/wp-content/uploads/2017/08/Exploding-Stereotypes -1.pdf. See also: Joe Parkinson, "Sent as Boko Haram Suicide Bombers, These Girls Broke Away," WSJ Video, *Wall Street Journal*, July 26, 2019.

240   Some of them were as young as ten years old: Joe Parkinson and Drew Hinshaw, "'Please, Save My Life': A Bomb Specialist Defuses Explosives Strapped to Children," *Wall Street Journal*, July 26, 2019.

240   scrutinize the latest videos of men riding stolen army vehicles: Interviews with Z.M.

241   a loud thud shook the buildings. It was an explosion: Ibid.

241   Ammar appeared, smiling, in his freshly laundered school uniform: Ammar, thirteen years old at the time of writing, is still in school. When he graduates, he wants to be a journalist.

Chapter 28: "I'm Not Going Anywhere"

245   Kolo Adamu had made the perilous journey: The trip to the presidential villa was described by K.A., R.A., Oby Ezekwesili, and Yakubu Nkeki, head of the Chibok Association.

245   more than six hundred days had passed: Abiodun Alade, "Chibok Girls, 600 Days After: Buhari, Mama Taraba, Oby, and the Lingering Trauma," *Vanguard Nigeria* (Lagos), January 17, 2016.

245   which cost more than their monthly earnings: A round-trip bus ticket from Chibok to Abuja cost around 30,000 naira, or $70, in 2016 and took around ten hours each way. The mothers would usually make around 5,000 naira, or $13, per month from selling vegetables, according to K.A.

245   the mothers had made it to Abuja: Ibid.

245   in the baking midmorning sun: Verified with historical weather reports from CustomWeather.com.

246   The scene was beginning to feel more like a wake: The authors visited the highway median at least once in each year between 2014 and 2019.

246   "Ma, can you tell me what happened to our girls?": Interviews with Oby Ezekwesili.

246   written repeatedly to Buhari's office: Ibid.

246   one of which read #CRYINGTOBERESCUED: Image published in Alade, "Chibok Girls, 600 Days After."

246   police officers refused to let the marchers through: Interviews with Oby Ezekwesili. Twitter archive @BBOG_Nigeria, January 13, 2016.

247   dressed in their finest *atampa* dresses: Images and videos of the protest can be seen in the Twitter archive of @BBOG_Nigeria.

247   a cascade of real-time dispatches: Twitter archives @ObyEzeks, @BBOG_Nigeria, January 13, 2016.

247   exhausted from the journey, fainted in the heat: Interviews with K.A., R.A., and Oby Ezekwesili.

247   sarcastic memes lambasting Buhari: Twitter archive #BringBackOurGirls.

247   marble-clad banquet hall: Images of the meeting were published in local news reports and are in the Twitter archive of @BBOG_Nigeria, January 13, 2016. Also see: "Photos: Tears Flow as President Muhammadu Buhari Meets Parents of Chibok Girls," *Guardian Nigeria*, January 14, 2016.

247   "I regret to inform you that the president will not be able to see you": Interviews with K.A., R.A., Oby Ezekwesili, and Yakubu Nkeki, head of the Chibok Association.

248   the women's affairs minister, Aisha al-Hassan: Chibok families and Oby Ezekwesili said Ms. al-Hassan's speech was offensive, as she upbraided the families for not giving the government enough notice of their planned petition.

248   "You understand how I feel?": Interviews with Esther Yakubu, Oby Ezekwesili, K.A., and R.A.

248   "You are the remote control of my life": The authors saw a copy of Dorcas's school notebook, which was photographed by American photographer Glenna Gordon in 2015, while she was working with Nigerian reporter Julius Emmanuel. It was not published.

248   "Ladies and gentlemen, please rise for the president": Shortly after President
      Buhari entered the room, all journalists were told to leave. Alade, "Chibok: 600
      Days After."
249   he put down the microphone and left: Video captured from the scene and posted
      on Twitter by @BBOG_Nigeria on January 13, 2016, shows the confused mur-
      murs of parents as the president hastily leaves the room with his entourage.
249   they felt it meant their daughters might be dead: Interviews with K.A. and R.A.
249   "Our President @MBuhari MISSED OPPORTUNITY": Twitter archive
      @ObyEzeks, January 14, 2016, https://twitter.com/obyezeks/status/6876992
      44849991680. Oby later asked, in an April 2016 op-ed, "Why Has the World
      Forgotten about Nigeria's Missing Girls?" in Toronto's *Globe and Mail*, in which
      she lamented, "Despite that flurry of chants and protests, the reality is that the
      arduous work of actually locating and rescuing the girls has failed."

Chapter 29: The Fire

250   coordinates scrawled on a paper: Interviews in 2019 and 2020 with Nigerian Air
      Force pilots based in Yola, who spoke on condition of anonymity.
250   the hideout was nearly invisible: The authors reviewed multiple Nigerian Air
      Force videos of strikes on targets in Sambisa from 2015 to 2017 showing a dense
      canopy camouflaging the forest floor.
250   surveillance planes were closing in on him: Interviews with Nigerian and US
      officials, who spoke on condition of anonymity.
250   spy plane they called the Beechcraft: Three modified Beechcraft King Air 350
      planes were delivered from the US to Nigeria in 2014, according to Guy Martin,
      "Recent Nigerian Military Acquisitions," DefenceWeb, March 30, 2015.
250   ask the US to take much higher-resolution photos: Interviews with Nigerian
      and US officials.
251   for the raid that killed Osama bin Laden: "The Most Elite U.S. Special Forces,"
      CBS News, https://www.cbsnews.com/pictures/the-most-elite-u-s-special-forces
      /2/. The Night Stalkers would later be involved in the 2019 raid that killed
      Islamic State leader Abu Bakr al-Baghdadi. See: Sebastien Roblin, "Revealed:
      Inside the Black 'Nightstalker' Special Ops Helicopters Used in the Raid That
      Killed Baghdadi," *National Interest*, November 2019.
251   fenced-off airstrip across Nigeria's eastern border: More about the base, cryp-
      tically named Contingency Location Garoua, can be found in Joshua Hammer,
      "Hunting Boko Haram: The U.S. Extends Its Drone War Deeper into Africa with
      Secretive Base," *Intercept*, February 25, 2016.
251   drones fed their footage to the CIA station in Abuja: Interviews with US and Ni-
      gerian officials, who confirmed that the station chief would usually meet Abba
      Kyarri, President Buhari's powerful chief of staff.
251   The pilot was flying an Alpha Jet: Interviews with US officials and sources in-
      side the Nigerian Air Force. See "Nigerian Air Force Arms Two Recently Ac-
      quired Alpha Jets," DefenceWeb, February 2016, https://www.defenceweb.co.za
      /aerospace/aerospace-aerospace/nigerian-air-force-arms-two-recently-acquired
      -alpha-jets/; and Tyler Rogoway, "This Man Owns the World's Most Advanced
      Private Air Force After Buying 46 F/A18 Hornets," April 2020, https://www
      .thedrive.com/the-war-zone/32869/this-man-owns-the-worlds-most-advanced
      -private-air-force-after-buying-46-f-a-18-hornets.
251   restricted American businesses from selling Nigeria arms: The Leahy Law and sub-
      sequent amendments, first introduced in 1997, prohibit the US defense sector from
      providing military assistance to countries committing human-rights violations.
251   paid a local car-parts manufacturer $13,000 to cheaply weld: The rockets were

welded onto the wings by Innoson Inc., a Nigerian manufacturer. For more detail, see: "Nigerian Air Force Receiving Additional Alpha Jets," DefenceWeb, March 26, 2015; Sebastien Roblin, "Nigeria's Tiny, Low-Tech Alpha Jets Have Flown in Brutal Wars Across Africa," WarIsBoring.com, July 28, 2016; and Kieron Monks, "Nigeria's First Car Maker Takes to the Skies in Fighter Jets," CNN, March 30, 2016.

251   only a half hour to reach his target: Interviews with NAF pilots based in Yola.

251   Boko Haram had hidden heavy Soviet Dushka antiaircraft guns: Described by the Chibok hostages and Boko Haram defectors.

251   shot down, captured, and beheaded with an ax: The grotesque video is available at Zenn's UnmaskingBokoHaram.com and features Aliyu jabbering angrily into the camera before executing the pilot.

251   Air force officers had become skittish: Interviews with NAF pilots.

252   Sunday mornings always felt cloudier: The airstrike was described by H.S., L.S., H.N., and several other Chibok hostages, and cross-referenced with images from a video released by Boko Haram shortly after.

252   Her boyfriend, Ibrahim, a Muslim from central Chibok: Interviews with H.S.

252   she had loved the required reading of "Chike and Chukwu": Ibid.

253   Ladi Simon called her the smiling type: Interviews with L.S.

253   Hannatu had been beaten on the hands: Interviews with H.S.

253   tin-roofed structures that could shelter more than a dozen: Interviews with H.S., L.S., H.N., and M.D.

253   They had nicknamed the turboprop Buhari: We can safely assume that Buhari was the Beechcraft.

254   "Before I knew it I just saw the fire": Interviews with H.S., L.S., H.N., and several other Chibok hostages, cross-referenced with video images of the aftermath of the strike at "Boko Haram: Message to the Families of the Chibok Girls and Federal Government," August 14, 2016, UnmaskingBokoHaram.com.

255   just two mass graves, shallow and unmarked: Ibid.

255   "they are just videoing us with their phones": Interviews with H.N.

255   a few seconds of an Alpha Jet quickly climbing: The plane can be seen flying away at 10:40 of the 11-minute video, "Boko Haram: Message to the Families," UnmaskingBokoHaram.com.

256   Somebody was shouting in Kibaku: Interviews with N.A., H.I., P.M., and S.A.

257   Medics hovered over Hannatu: Interviews with H.S.

257   the only instrument available was a pair of scissors: Ibid.

257   the metal would remain embedded in their flesh: several of the Chibok hostages still have shrapnel lodged in their bodies to this day.

258   "Yes," Hannatu told Abubakar Shekau. "You are the Imam": Interviews with H.S.

Chapter 30: "Never Trust a Breakthrough"

259   Zannah's phone buzzed with a number he didn't recognize: Interviews with Z.M.

260   intelligence foretelling Boko Haram's gradual rupture: Interviews with members of the team, who spoke on condition of anonymity.

260   Nigeria's military was advancing: James Butty, "Nigeria Army Chief: Boko Haram Capabilities 'Virtually Eliminated,'" Voice of America, September 16, 2016.

260   Shekau and his rivals had been circulating audio files: M.A. describes her husband sharing audio files, referred to as *elan*, wherein leaders, including Shekau and Mamman Nur, were openly criticizing one another. In retreat, the Imam saw enemies everywhere, was traveling with more bodyguards, and wore a bulletproof vest. He had come to suspect that followers plotting his assassination had planted a tracking device on him.

260   Islamic State's leaders were receiving missives: Jacob Zenn, "Boko Haram's
      Conquest for the Caliphate: How Al-Qaeda Helped Islamic State Acquire Terri-
      tory," *Studies in Conflict and Terrorism* 43, no. 2 (2018): 89–122.
261   Swiss team was analyzing tapes, encrypted texts: Interviews with members of
      the team.
261   "Just In: Link to Video on #Chibokgirls": Twitter archive @ContactSalkida, August
      14, 2016, https://twitter.com/A_Salkida/status/764723790102335489?lang=pt.
261   On the screen stood Tasiu, carrying a Kalashnikov: "Boko Haram: Message to
      the Families," UnmaskingBokoHaram.com. A full translation into English can
      be found in Abdulbasit Kassim and Michael Nwankpa, eds., *The Boko Haram
      Reader: From Nigerian Preachers to the Islamic State* (New York: Oxford Uni-
      versity Press, 2018).
262   Nigerian Air Force publicly denied that the airstrike ever happened: Ronald
      Mutum, "Air Force Denies Killing Chibok Girls in Air Strikes," *Daily Trust*
      (Abuja), August 17, 2016. The air force also released a statement claiming the
      video was a "ruse" and that the bodies had been staged to appear as if they had
      been killed in a strike.
262   although senior officers later conceded: Interviews with Nigerian officials.
262   The incident wasn't reported to senior Obama administration officials: Inter-
      views with senior US officials.
262   White House Council on Women and Girls . . . was not told: No member of Mi-
      chelle Obama's senior staff interviewed by the authors recalled being told about
      the strike.
262   Nigeria's army declared Ahmad Salkida wanted: "Nigerian Journalist Wanted
      over Boko Haram Video," *Al Jazeera*, August 15, 2016; also see: Twitter archive @
      HQNigerianArmy, August 14, 2016, https://twitter.com/HQNigerianArmy/status
      /764876630389063680.
262   Oby Ezekwesili had begun to fire off a series of pressing tweets: Twitter archive
      @ObyEzeks, August 14, 2016.
262   "How can any leader watch the recent video": The tweet received only seventy
      -two retweets and twenty-four likes, https://twitter.com/obyezeks/status/7648928
      57870217216.
262   contained two important revelations: Interviews with members of the Team.
263   "the ripeness of conflict": Concept described in SwissPeace's "Mediation Pro-
      cess Matrix," Bern, 2012, https://www.swisspeace.ch/fileadmin/user_upload/pdf
      /Mediation/Mediation-Process-Matrix_2016.pdf.
263   Zannah received an even clearer signal: The story of the audio recordings was
      relayed by Z.M. Some members of the team would later cast doubt on the impor-
      tance of this exchange to the mediation effort.
264   the more the Swiss team learned about Boko Haram: Interviews with members
      of the team.
264   President Buhari's health was deteriorating: Festus Owete, "President Buhari
      Is Ill," *Premium Times* (Abuja), June 4, 2016.
264   in his absence, messages went unanswered: Interviews with the team.

Chapter 31: Meeting in Manhattan

266   dressed in a starched black kaftan: US State Department, Remarks by Pres-
      ident Obama after Bilateral Meeting with President Muhammadu Buhari of
      Nigeria, September 20, 2016 (includes an uncredited photo).
266   under sticky, 80-degree weather: Verified with historical weather reports from
      CustomWeather.com.
266   "I wish you well in your retirement": Remarks Following a Meeting with President
      Muhammadu Buhari of Nigeria in New York City, American Presidency Proj-

ect, September 16, 2016, https://www.presidency.ucsb.edu/documents/remarks
-following-meeting-with-president-muhammadu-buhari-nigeria-new-york-city.

267 François Hollande, also attending his last UN meeting: "Buhari Seeks More
Cooperation with France in Agriculture, Mining," *Vanguard Nigeria* (Lagos),
September 21, 2016.

267 Johann Schneider-Ammann, a bespectacled former mechanical engineer: Peter
Siegenthaler, "Swiss President Looks Ahead to 2016," SWI, Swissinfo.ch, January 1, 2016.

267 stated reason for the meeting was not the Chibok issue: "Buhari Meets French,
Swiss Leaders, Seeks Support," *Vanguard Nigeria*, September 21, 2016.

267 $1 billion of Nigerian money had vanished into Swiss banks: The Swiss gov-
ernment recognizes that $723 million of government money illicitly gained by
General Abacha was held in Switzerland, as was an additional $321 million,
initially deposited in Luxembourg. "The Swiss Federal Council, and the International Development Association on the Return, Monitoring, and Management
of Illegally Acquired Assets Confiscated by Switzerland to Be Restituted to the
Federal Republic of Nigeria," March 2016, https://www.newsd.admin.ch/newsd
/message/attachments/50734.pdf.

267 died of a heart attack, reportedly overdosing on Viagra: Abacha had preexisting
health conditions and struggled to climb a flight of stairs, but according to Max
Siollun, *Nigeria's Soldiers of Fortune: The Abacha and Obasanjo Years* (London:
Hurst, 2019), chapter 13, the Viagra overdose became a standard backstory,
repeated in the local and foreign press.

267 the question of up to $5 billion: Abbas Jimoh, Itodo Daniel Sule, and Haruna
Ibrahim, "Investigation: N1.4trn [$4.6 Billion] Abacha Loot Returned in 18
Years," *Daily Trust* (Abuja), February 16, 2020.

267 agreed to return $321 million of the looted funds: "Switzerland and Nigeria
Conclude Agreement on the Return of USD321 Million of Abacha's Assets,"
World Bank, March 8, 2016, https://star.worldbank.org/corruption-cases/sites
/corruption-cases/files/Swiss%20Federal%20Council%20press%20release%20
-%20Letter%20of%20Intent.%20March2016.pdf.

267 the most important, off-the-record agenda item: Interviews with Swiss and Nigerian officials, who spoke on condition of anonymity.

268 "Do you want this?": Ibid.

268 first tranche of a €3 million ransom: The amount was confirmed by a senior
Swiss Foreign Ministry official, two Nigerians who took part in the negotiations,
and two former Nigerian cabinet ministers. Other members of the team refused
to confirm or deny it, saying it was the most secret part of the operation.

268 phase two, de-escalation: The authors interviewed a Red Cross operations official involved, who spoke on condition of anonymity, in November 2018, as well
as team members. Buhari's office declined to engage on the plan, which may
have had limited backing on both sides.

269 A call came in to the global headquarters of the Red Cross: Interviews with
three Red Cross officials in Switzerland and Nigeria, November 2017, December 2017, and November 2018.

269 conference room decorated with four Nobel Peace Prizes: The authors visited
this room in December 2017 and saw the Nobels, which include the first ever
awarded.

269 from a field officer in the Maiduguri Subdivision: The authors visited the Red
Cross Maiduguri Subdivision twice, in April 2018 and July 2019, and spoke to
officials.

269 what its fifteen thousand employees do: Annual Report, 2017, International
Committee of the Red Cross (ICRC), May 2017, https://www.icrc.org/en/docu-
ment/annual-report-2017.

270   The folders would stretch for three and a half miles: ICRC Archives, https://
      www.icrc.org/en/archives.
270   one government with which the Red Cross remains umbilically linked: For a
      history, often critical of the Red Cross's attachment to Switzerland and its neu-
      trality, see: Caroline Moorehead, *Dunant's Dream: War, Switzerland, and the
      History of the Red Cross* (New York: HarperCollins, 1999).
270   Naomi had been lying on the forest floor: Interviews with N.A. and H.I.
270   she killed an ant to eat the crumb of food it was carrying: Ibid.
270   Boko Haram was splitting apart and running out of food: Dionne Searcey, "Boko
      Haram Falls Victim to a Food Crisis It Created," *New York Times*, March 4, 2016.
      In a 2016 report, "Boko Haram Food Crisis Demands Cooperation and Account-
      ability," https://www.hrw.org/news/2016/12/01/boko-haram-food-crisis-demands-
      cooperation-and-accountability, Human Rights Watch stated: "Scenes like these
      haven't been seen here since the 1967–70 war with secessionist Biafra."
271   familiar voice was blasting through speakers strapped to the back of a pickup:
      Many of the hostages described this technique used by Boko Haram to broad-
      cast Shekau's instructions and teachings.
271   Nigerian military had cut off his supply lines and captured their food stocks:
      See: Human Rights Watch, "Boko Haram Food Crisis," which reported that gov-
      ernment officials refused to comment on the food supply for displaced people
      because it had become a "state secret."
271   the sinewy grass called *yakuwa*: Interviews with L.S. and H.S., who said, "No-
      body taught us to eat it, it was just desperation that forced us."
271   There was a Swiss man on the line, and he wanted to talk to them: This ex-
      change, described by team members, was confirmed with Rebecca Mallum in
      March 2020.

Chapter 32: Praise God

273   "You, what's your name?": Interviews with L.S., corroborated by H.S., N.A., H.I.,
      and P.M.
273   Ladi as a trendsetter who'd research fashion tips: Interviews with H.S.
273   Her face had been chiseled away by months of malnutrition: Shocking images of
      Ladi on her release, looking bewildered and gaunt in a pale green dress, show
      how close she was to death. H.S.; see also: "Reports Differ on Conditions for
      Release of Nigeria's Chibok Girls," Voice of America, October 14, 2016.
273   That, she assumed, was why her name had been called: Interviews with L.S.
      and other Chibok hostages that appear to be corroborated by members of the
      team, who said in interviews that they requested that the first batch of girls
      released should be the weakest, the most severely ill.
274   A militant commander addressed the group: Interviews with L.S.
274   "If I don't come back, you know I have made it out": Ibid.
274   she felt a growing anger: Interviews with N.A. and H.I.
275   The roll call added up to twenty-two names, and an argument ensued: Inter-
      views with L.S., corroborated by N.A. and other Chibok hostages who saw the
      hostage return.
275   a helicopter with its lights switched off: A no-fly zone was part of the cease-fire
      that allowed the hostages to be released, according to members of the team, plus
      Swiss and Nigerian officials.
275   Zannah and Pascal ran through the crucial stages: Interviews with members of
      the team.
275   crisply ironed kaftan he'd picked out, with an embroidered kufi cap: The authors
      have reviewed photos from that exchange, provided by Z.M.

275   he said nothing about it to his wife: Interviews with Z.M.
276   roads there were well known for improvised bombs: Just a few months earlier, the Nigerian Army had found factories for manufacturing IEDs in the same area Boko Haram selected for the handover. See Kingsley Omonobi, "Boko Haram: Military Discovers Bombs [*sic*] Factories in Borno's Kumshe," *Vanguard Nigeria* (Lagos), March 3, 2016.
276   he would send his most battle-hardened fighters: One member of the team said of the Boko Haram unit chosen to escort the girls, "These guys don't play."
277   nothing on him except for a cell phone, the list, and the jangling keys: Interviews with members of the team.
277   The first vehicle carried the money: Ibid., and other sources familiar with the deal.
277   One of the drivers was a man from Chibok: Hurray Peter's place in the team of drivers was confirmed by members of the team, by the Red Cross, and by L.S.
277   progress was being continuously tracked from Geneva in real time: Interviews with Red Cross officials.
277   gunshots: Interviews with team members.
278   he could see more fighters than he could attempt to count: Interviews with members of the team.
278   she could make out men clambering from four or five cars: Interviews with L.S.
279   "I thought, *Wow, maybe this is actually real?*": Ibid.
279   The deal had called for twenty hostages to be released, but there were twenty-one: Interviews with members of the team, corroborated by Chibok hostages.
280   shouting, "Delete the photos!": Some members of the team later said this slip by Z.M., which was against the arranged protocols, could have imperiled the whole effort.
280   meat pies and juice boxes of Five Alive to lift their blood sugar levels: Interviews with L.S., corroborated by other hostages and members of the team.
281   the passengers descended and saw a small group of men on the tarmac: Ibid.

Chapter 33: The One You've Been Looking For

282   Maryam was confronting the final, most heart-wrenching item: The story of Maryam's escape was recalled in several interviews by M.A. She is the only witness apart from her infant son, but where possible, we have tried to corroborate through other sources.
283   Maryam saw her sister's eyes: Ibid.
283   *Hasbunallahu wa ni'mal wakil*: Spoken by the Prophet Ibrahim (Abraham) when he was thrown into a fire by temple priests at Babylon (Quran, Surah Al 'Imran 3:173).
284   Maryam had wanted to name the child Khairat if it was a girl: Maryam said she had been hoping for a girl, partly because she knew that Abu Walad wanted a boy.
284   overwhelmed from the physical pain and the emotion of the birth: From the stress and trauma, Maryam was unable to feed, and became worried that Ali might die until she was offered some goat milk.
284   "I know how a normal naming ceremony would have looked": Hausa naming ceremonies usually take place seven days after the child's birth. The father offers kola nuts, and a ram or goat is slaughtered and prepared for guests.
285   the body of an infant boy, no older than four, dressed in polyester pants: Interviews with M.A.
285   they could be killed by Boko Haram lookouts or vigilante patrols: Ibid. M.A. said that Shekau's fighters were under orders to kill anyone suspected of fleeing his territory.

286   the small town of Pulka: Pulka is a small town near Gwoza on the front line of the insurgency. It was held by Boko Haram until the army recaptured it in 2015, building a large base that enabled the town to become a haven for thousands of internally displaced people.

286   They'd been spotted by soldiers: M.A. said the group couldn't approach the town from the road because they would certainly have been shot before she had a chance to say she was from Chibok.

287   "*Oga!*" she shouted. Sir! "It is me": Maryam's escape was covered by local media and some international outlets, including Chandrika Narayan, "Missing Chibok Schoolgirl Found with Baby, Nigerian Army Says," CNN, November 5, 2016. Photo showed a gaunt-looking Maryam, in a gray hijab, carrying Ali and holding a bottle of Coca-Cola before being escorted to a military helicopter. Maryam said that the soldiers at first did not believe she was a Chibok girl and began to test her with questions on the names of the other hostages. She described how five senior officers arrived with pictures of the captives provided by their families, and when they saw her image they began to laugh and embrace. "Then the soldiers came to take pictures with me. I finally started to feel that I was saved."

Chapter 34: "Now There Is Trust"

288   Pascal was scrutinizing a new video: Interviews with team members, who spoke on condition of anonymity.

288   "Kill the infidels, detonate bombs! Kill, slaughter and abduct!" Michelle Faul and Haruna Umar, "'Kill All the Infidels and Detonate Bombs Everywhere,'" Associated Press, December 31, 2016.

288   "Geneva people, all your plans according to Allah are not up to a cobweb!": Transcript published in "Boko Haram Leader Shekau Threatens President Buhari and Mocks #BringBackOursGirls Campaigners," SaharaReporters.com, September 25, 2016.

288   parents who were asking questions: Marama Ndahi, "Chibok Parents Ask Buhari Where 197 Missing Girls Are," *Vanguard Nigeria* (Lagos), April 2017.

289   more food was starting to reach the starving insurgency through Red Cross deliveries: Interviews with the members of the team and Swiss and Nigerian officials.

289   Shekau was executing some of his most senior commanders: Several of the Chibok hostages recounted that Aliyu fled but was captured by Shekau's fighters and beheaded. Shekau himself talks about the killing of Tasiu, and the audio is archived at Zenn's UnmaskingBokoHaram.com. See also: Ken Karuri, "Boko Haram Spokesperson Killed After Plot to Oust Leader," Agence France-Presse, February 25, 2016.

289   he'd tweeted for the first time, posting an article: Twitter archive @ZbMustapha1, April 4, 2017, https://twitter.com/zbmustapha1/status/849163640645312514.

290   Muhammadu Buhari, who looked sicker with each public appearance: "President Buhari: I've Never Been So Sick," BBC News, March 10, 2017.

290   Naomi was lying flat on a piece of tarpaulin: Interviews with N.A.

290   He had been killed in an airstrike: Ibid., plus H.S., L.S., H.I., and P.M.

290   hadn't brought enough money: Ibid. N.A. said one of the commanders, Ammar, had returned saying the government didn't bring enough money for twenty-two girls. In Shekau's 2016 recording confirming the killing of Tasiu, he accuses him of plotting a coup with Baba Ammar.

291   stacks of leather-bound books: Shekau appeared with stacks of leather-bound books around this time. See: "Abubakar Shekau: Video Mentioning Chibok Girls and Denying Death," September 25, 2016, UnmaskingBokoHaram.com, https://

unmaskingbokoharam.com/2019/04/08/abubakar-shekau-video-on-ceasefire
-and-mentioning-chibok-girls-september-25–2016.

292 It was Abubakar Shekau, descending from the driver's seat of a truck: Interviews with N.A. and H.S.

292 He did not even flinch as a screaming rocket hit a Toyota: Ibid.

293 "I felt I'd become a leader": Interviews with N.A.

293 second release was more practiced than the first: Interviews with members of the team and Nigerian officials.

293 A caravan of Red Cross Land Cruisers, longer this time: Photos of the convoy were leaked, then published as "Photos of 82 Chibok Girls Boarding a Helicopter from Banki to Abuja," SaharaReporters, May 7, 2017.

293 Number 62 was easy to pronounce: The list, with the title Rescued Chibok Girls scrawled across the top in blue Biro, was leaked and also published on the SaharaReporters Twitter account, @SaharaReporters, May 8, 2017, https://twitter .com/SaharaReporters/status/861484967125282817.

293 five Boko Haram prisoners fished from maximum-security jails: The exchanged prisoners, led by a militant called Shuaibu Moni, appeared in two videos after their release, threatening to bomb Abuja: "Two Videos of Shuaibu Moni after Chibok Exchange," May 12, 2017, and March 7, 2018, UnmaskingBoko Haram.com, https://unmaskingbokoharam.com/2019/04/08/boko-haram-video -of-shuaibu-moni-after-chibok-release-may-12–2017.

293 lined up in two rows on the other side of the road: Almost all of the eighty-two can be seen in photos taken by Z.M. at the exchange point and shared with the authors.

294 one of the men handed a black bag to a militant: Interviews with N.A. Several other hostages saw this exchange. In one of the pictures shared by Z.M., a militant is seen carrying a brand-new black backpack.

295 the five prisoners, dressed in dark-colored kaftans: Images from the exchange point, courtesy of Z.M.

295 Zannah knew they were much smaller fish who wouldn't threaten Shekau's authority: Interviews with the team and Nigerian officials.

295 "Today is a happy day!": Several of the freed women recall singing, and M.D. remembers singing this song.

Chapter 35: Naomi and Kolo

296 Kolo Adamu was sitting on her porch: Interviews with K.A. and R.A.

296 "They advised me not to lose hope": Ibid.

297 A picture taken by a curious bystander: An image leaked to SaharaReporters of the dozens of young women in Red Cross vests being airlifted from Banki airbase. See SaharaReporters, "Photos of 82 Chibok Girls Boarding a Helicopter."

297 "We will be like the Angels!" blasted Mama Agnes: Interviews with K.A. and R.A.

298 Naomi was settling into a gold-rimmed chair: Interviews with N.A. and other hostages. Naomi can be seen in images released by the presidency, wearing a gray polo shirt.

298 "Welcome, our dear girls, back to freedom": "82 Chibok Girls Freed: Buhari Meets The Released School Girls," Channels Television, YouTube," May 2017, https://www.youtube.com/watch?v=02lQPXYxQVI.

298 A television news crew captured the moment: In a video clip by Channels Television that can be seen on YouTube, "President's Chief of Staff Receives 82 Released Chibok Girls," May 7, 2017, https://www.youtube.com/watch?v=TKxDl _fgw2M, Naomi can be seen at 0:53–0:58, posing for the cameras in her school uniform.

298   A military helicopter had whisked Naomi: The helicopter, serial number NA558, was a Russian-made Mi-17.

299   under the thick green-and-white fabric of a Nigerian flag: Dozens of images were released by the presidency and the media invited to cover the event. See "Nigeria Chibok Girls: Freed 82 Meet President Buhari," BBC News, May 7, 2017.

299   a fixed expression that suggested more bewilderment than relief: Ibid.

299   Pascal had decided not to attend: Interviews with the team.

299   Zannah wouldn't have missed it for the world: Z.M. can be seen smiling in many of the photos, just three seats to President Buhari's right.

299   They reminded him of Shadrach, Meshach, and Abednego: Interviews with Z.M.

299   Kolo sat squished into the bucket seat of a small bus: Interviews with K.A. and R.A.

300   After what felt like an eternity, the moment had come for their reunion: Adaobi Tricia Nwaubani, "The Chibok Girls Are Still in Custody, and Their Parents Are Still Desperate," *The World*, Public Radio International, May 17, 2017.

300   three special-purpose buses: Interviews with K.A., R.A., and members of the Chibok Association.

300   wearing a yellow floral-print *atampa* dress: Interviews with N.A.

300   Is the Earth flat?: The authors spoke to a psychologist who interviewed the hostages when they were released into the custody of the DSS in August 2017. He spoke on condition of anonymity.

300   tests for signs of rape or sexual abuse continued for days: Psychologists and medical professionals who examined the hostages, and members of the team briefed on the official reports, all said there was no evidence that, contrary to reports in the press, any of the released girls had been raped.

300   Parents began to emerge from the folding doors to a chorus of screams: Video footage of the reunion can be seen on OakTV's YouTube channel, "Minute by Minute: How 82 Chibok Schoolgirls Captured by Boko Haram Reunited with Families," May 2017, https://www.youtube.com/watch?v=Vzv4jn3DZOM.

301   Her confusion quickly turned to fear: In the OakTV video, Naomi can be seen running in panic past other embracing families as she searches for Kolo.

301   she knocked an NPR reporter's smartphone from her hands: Ofeibea Quist-Arcton, "That Amazing Moment When 82 Chibok Girls and Families Reunited," National Public Radio, May 22, 2017.

301   "To err is human, to forgive is divine": Ibid.

301   Kolo felt Naomi's face with her hand: Interviews with K.A. and N.A.

Selected Bibliography

Achebe, Chinua. *The Trouble with Nigeria*, reissue edition. Enugu, Nigeria: Fourth Dimension Publishing Co., 2012.

Bilton, Nick. *Hatching Twitter: A True Story of Money, Power, Friendship, and Betrayal*. New York: Portfolio, 2014.

Brahimi, Lakhdar, and Salman Ahmed. *In Pursuit of Sustainable Peace: The Seven Deadly Sins of Mediation*. New York: Center on International Cooperation, 2008.

Calmy-Rey, Micheline. *La Suisse que je souhaite* [The Switzerland I want]. Lausanne, Switzerland: Éditions Favre, 2014.

Campbell, John. *Nigeria: Dancing on the Brink*. Lanham, MD: Rowman & Littlefield, 2010.

Chouliaraki, Lilie. *The Ironic Spectator: Solidarity in the Age of Post-Humanitarianism*. Cambridge, UK: Polity Press, 2013.

Cole, Teju. "Captivity." *New Yorker*, May 6, 2014.

Debos, Marielle. *Le métier des armes au Tchad: Le gouvernement de'entre-guerres*. Paris: Éditions Karthala, 2013.

Falola, Toyin, and Matthew M. Heaton. *A History of Nigeria*. Cambridge, UK: Cambridge University Press, 2008.

Fawehinmi, Feyi, and Fola Fagbule. *Formation: The Making of Nigeria from Jihad to Amalgamation*. Abuja, Nigeria: Cassava Republic Press, 2020.

French, Howard. *A Continent for the Taking: The Tragedy and Hope of Africa*. New York: Penguin Random House, 2005.

Habila, Helon. *The Chibok Girls: The Boko Haram Kidnappings and Islamic Militancy in Nigeria*. New York: Penguin, 2017.

Hentz, James J., and Hussein Solomon, eds. *Understanding Boko Haram: Terrorism and Insurgency in Africa*. Abingdon, UK: Routledge, 2017.

John, Elnathan. *Born on a Tuesday: A Novel*. New York: Black Cat, 2016.

Kane, Cheikh Hamidou. *L'aventure ambiguë*. Paris: Julliard, 1961.

Kassim, Abdulbasit, and Michael Nwankpa, eds. *The Boko Haram Reader: From Nigerian Preachers to the Islamic State*. New York: Oxford University Press, 2018.

Kendhammer, Brandon, and Carmen McCain. *Boko Haram*. Athens, OH: Ohio University Press, 2018.

Lamb, Christina. *Our Bodies, Their Battlefields: War through the Lives of Women*. New York: Scribner's, 2020.

Maier, Karl. *This House Has Fallen: Midnight in Nigeria*. New York: Public-Affairs, 2000.

Matfess, Hilary. *Women and the War on Boko Haram: Wives, Weapons, Witnesses*. London: Zed Books, 2017.

Nossiter, Adam. "A Jihadist's Face Taunts Nigeria from the Shadows." *New York Times*, May 18, 2014.

Nwaubani, Adaobi Tricia. *Buried Beneath the Baobab Tree*. New York, Katherine Tegen Books, 2018.

———. "The Women Rescued from Boko Haram Who Are Returning to Their Captors." *New Yorker*, December 20, 2018.

O'Brien, Edna, *Girl: A Novel*. New York: Farrar, Straus & Giroux, 2019.

Obadare, Ebenezer. *Pentecostal Republic: Religion and the Struggle for State Power in Nigeria*. Chicago University Press, 2018.

Ochonu, Moses E. *Colonialism by Proxy: Hausa Imperial Agents and Middle Belt Consciousness in Nigeria*. Bloomington: Indiana University Press, 2014.

Okeowo, Alexis. *A Moonless, Starless Sky: Ordinary Women and Men Fighting Extremism in Africa*. New York: Hachette, 2017.

Osaghae, Eghosa E. *Crippled Giant: Nigeria since Independence*. Bloomington: Indiana University Press, 1998.

Salkida, Ahmad. "Genesis and Consequences of Boko Haram Crisis." Salkida.com, April 2010; currently available at Google Groups, https://groups.google.com/g/usaafricadialogue/c/aj_Lg2kXoRo?pli=1.

Searcey, Dionne. *In Pursuit of Disobedient Women: A Memoir of Love, Rebellion, and Family, Far Away*. New York: Ballantine Books, 2020.

Sesay, Isha. *Beneath the Tamarind Tree: A Story of Courage, Family, and the Lost Schoolgirls of Boko Haram*. New York: HarperCollins, 2019.

Rice, Susan. *Tough Love: My Story of the Things Worth Fighting For*. New York: Simon & Schuster, 2019.

Simon, Joel. *We Want to Negotiate: The Secret World of Kidnapping, Hostages, and Ransom*. New York: Columbia Global Reports, 2018.

Siollun, Max. *Soldiers of Fortune: A History of Nigeria (1983–1993)*. Abuja, Nigeria: Cassava Republic Press, 2014.

———. *Nigeria's Soldiers of Fortune: The Abacha and Obasanjo Years*. London: Hurst, 2019.

Smith, Mike. *Boko Haram: Inside Nigeria's Unholy War*. London: I.B. Tauris, 2015.

Thompson, Derek. *Hit Makers: The Science of Popularity in an Age of Distraction*. New York, Penguin Press, 2017.

Thurston, Alexander. *Boko Haram: The History of an African Jihadist Movement*. Princeton, NJ: Princeton University Press, 2017.

Topol, Sarah. "If We Run and They Kill Us So Be It." *Medium*, October 14, 2014, https://medium.com/matter/if-we-run-and-they-kill-us-so-be-it-but-we-have-to-run-now-ad48d0b28994.

Walker, Andrew. *"Eat the Heart of the Infidel": The Harrowing of Nigeria and the Rise of Boko Haram*. London: Hurst, 2016.

Zenn, Jacob. *Unmasking Boko Haram: Exploring Global Jihad in Nigeria*. Boulder, CO: Lynne Reinner, 2020.

Zenn, Jacob, Abdulbasit Kassim, Elizabeth Pearson, et al., eds. *Boko Haram beyond the Headlines: Analyses of Nigeria's Enduring Insurgency*. West Point, NY: Combating Terrorism Center, US Military Academy, 2018.

Index

Chibok Government Secondary
  School for Girls (*continued*)
  demolition of, 305
  description of, 16–17
  dorms, as safe place, 23–24
  evening after finals, 22–24
  graduation requirements, 9–10
  isolation of, 12
  keeping open as calculated risk,
    22
  Naomi's enrollment in, 47
  as outpost for Western
    education spread, 22
  security guard for, 11
Chibok Local Government Area, 14
Chibok Parents Association,
  305–306
Chibok schoolgirls
  Adamu, Naomi. *See* Adamu,
    Naomi
  adjusting to new normal,
    105–106
  Ali, Maryam. *See* Ali, Maryam
  Ali, Sadiya. *See* Ali, Sadiya
  at American University,
    302–305, 311–312
  Ayuba, Saratu. *See* Ayuba,
    Saratu
  creating video for hostage
    negotiations, 97–98
  daring escape, 42
  food supply for, 290–291
  freeing of. *See* extraction
  indoctrinated into Boko
    Haram's beliefs, 106–107
  Ishaya, Hauwa. *See* Ishaya,
    Hauwa
  killed by military airstrike,
    253–255, 261–262
  mental/interpersonal transition
    of, 303–304
  missing or dead, 306
  mutiny of, 123–124
  night of kidnapping, 1, 9–12,
    25–27
  not wishing to be rescued, 306
  parents. *See* parents of missing
    girls
  Patience, 123, 124
  postcapture, 300–305, 311–312
  rescue of, 3–5
  reuniting with parents, 300–301
  routine punishments, 253
  running from kidnappers, 11

Salkida seeing captured, 97–98
  in Sambisa Forest. *See* Sambisa
    Forest
  as secret diarists, 121–123
  slavery versus marriage, 140
  starvation of, 270–271, 273–274
  Tabitha, 123, 124
  Ummi, 150
  US intervention in finding,
    80–83
China, 82
Church of the Brethren, 46–47, 64,
  115, 156, 173–174, 199, 252
CIA station, Abuja, 126–127, 251
Clinton, Hillary, 79
Colombian FARC insurgency,
  183
Common (rapper), 74
communal fasting. *See* fasting
confidence building, 268, 289
@ContactSalkida, 87–88, 102,
  226–227

Dapchi, Nigeria, 308
Dar al-Hikma, 147. *See also*
  Gwoza, Nigeria
Dauda, Mary, 11, 135
Dauwa, Wamdo, 17
Déby, Idriss, 99
de-escalation, 268
Def Jam Records, 72
Dialogue Facilitation Team. *See
  also* Holliger, Pascal; hostage
  negotiations; Mustapha,
  Zannah, "the Barrister"
  building trust on both sides,
    263–264
  classroom study of Boko Haram,
    180–181
  clearance for hostage talks,
    179–180
  decoding Boko Haram structure,
    175–176
  formation of, 176–178
  "message management" by, 238
  military air campaign and, 259
  negotiating release, 234–235.
    See also extraction
  "the windows" philosophy, 236
  working their way to Boko
    Haram core, 235–236
diaries
  Adamu, Naomi and, 5, 125,
    157–158

scouring help from anyone, 88
Swiss embassy and, 179
Jonathan administration. *See also*
    Jonathan, Goodluck; Nigerian
    government
expanding Barlow's contract,
    167–168
throwing money at the problem,
    129
Joshua, T. B., 114

Kardashian, Kim, 76
Kerry, John, 59, 82, 154–155
Keys, Alicia, 154
Khomeini, Ayatollah, 33
Kibaku Area Development
    Association, 63
Kony, Joseph, 76

lesbianism, 139
Lewis, John, 79
Limbaugh, Rush, 80

Madame Due Process. *See*
    Ezekwesili, Oby
Maharishi Mahesh Yogi, 73
Maiduguri, Nigeria, 32–33, 35, 229,
    239–241
Maiduguri International Airport,
    51, 165, 228, 239–240
malam, defined, 20
Malam Abba, 20–21, 25–27, 70–71
Malam Ahmed
    Ali, Maryam and, 148–149
    encouraging hostages to marry,
        139–140, 195, 218–220
    furious with girls' indiscipline,
        195–196
    as hostages' official guardian,
        105–106
    identifying insubordinate girls,
        210
    joining for meals, 108–109
    lecturing against singing, 193
    Naomi and, 141–143, 158,
        197–198, 290
    offering dowries to hostages, 151
    responding to insubordination,
        196–197
    restarting Quranic classes, 291
    at Ukuba settlement, 193
    Ummi and, 150
    warning girls to shelter from
        planes, 110–111

warning to runaways, 123–124,
    125
Mallum, Rebecca, 271–272,
    279–280
Mama Agnes, 18, 113, 120, 160,
    174, 194, 297
Mandela, Nelson, 122–123, 176
Mark, Enoch, 46, 64, 156
marriage. *See* forced marriages
Maurer, Peter, 270
May, Theresa, 266
McCain, John, 130
mediators. *See* Ferobe, Tijani;
    Holliger, Pascal; Mustapha,
    Zannah, "the Barrister";
    Nasrullah, Fulan; Salkida,
    Ahmad; Umar, Tahir
"message management," 239
Mills, Cheryl, 78
modern aid industry, 76
Monday Market, 34–35, 240
Musa, Palmata, 195, 211–212,
    215–217, 221–222, 232, 311
Mustapha, Abba. *See* Malam Abba
Mustapha, Zannah, "the Barrister."
    See also Dialogue Facilitation
    Team
    as attorney, 51–52
    background, 50–52, 175–176
    Boko Haram orphans and, 241
    decoding Boko Haram structure,
        176
    extraction agreement and, 275
    at extraction point, 279–281
    at freed hostages press
        conference, 299
    greatest accomplishment of,
        53
    Holliger and, 54–55, 175
    learning of military air
        campaign, 259–260, 262–263
    Maiduguri and, 50, 239–241
    meeting with Boko Haram
        exiles, 236
    negotiating release, 234
    negotiations training in
        Switzerland, 90–93, 102–103
    at orphanage, 52–53, 239, 241
    as public face of release success,
        310
    receiving Nansen Prize, 310
    at second extraction, 293
    Shekau and, 263
    under surveillance, 181

United Nations, 62, 266–267
United States
American Interdisciplinary
Assistance Team (IDAT),
128–129, 130–131
CIA station, Abuja, 126–127
Defense Department, 81
difficulties operating in Nigeria,
154–155
distrusting Nigerian
government, 131
global aerial surveillance
program, 95–96
intelligence officers'
deployment, 190
Justice Department, 81
National Security Council,
80–83
providing fighter jets to Buhari,
190–191
rescue effort, 127, 153
State Department, 81, 190
White House strategy, 80–82
Usman dan Fodio, 15

Walad, Abu, 124, 149, 152,
202–203, 284, 285–287
War Against Indiscipline, 187
War on Terrorism, 309
Western education, 20, 22, 35–36

Western public opinion, 130
White House Council on Women
and Girls, 262
Williams, Evan, 75
Wilson, Frederica, 79
"the windows," 236
Winfrey, Oprah, 76
women, joining Boko Haram
willingly, 141
World Book Capital, 62
World Economic Forum, 61, 77

Yakubu, Esther, 248
Yama, Margret, 11
Yar'Adua, Umaru, 60
Yousafzai, Malala, 2, 100, 128, 312
YouTube. *See also* Twitter
of aerial bombing of schoolgirls,
255
Aliyu and, 67, 261
captive schoolgirls using, 306
executions on, 38, 185, 251
Kony and, 76
Shekau's use of, 28, 30, 39,
147–148, 166
Yusuf, Mohammad, 35–38, 87–88

Zinnira, Abu. *See* Tasiu (Abu
Zinnira)
Zuckerberg, Mark, 76

About the Authors

JOE PARKINSON is the Africa Bureau Chief for the *Wall Street Journal* and a Pulitzer Prize finalist currently based in Johannesburg. Born in the UK, he is one of the *Journal*'s most seasoned foreign correspondents, has reported from more than forty countries, and has won numerous international awards.

DREW HINSHAW is a former Fulbright scholar to West Africa, where he spent a decade as a journalist, most of that time for the *Wall Street Journal*, which nominated his reporting for a Pulitzer four years consecutively. He has also written for the *New York Times Magazine*, *Time*, *Al Jazeera*, the *Atlantic*, and *Rolling Stone*. He is from Atlanta.